POLYMER GRAFTING AND CROSSLINKING

POLYMER GRAFTING AND CROSSLINKING

Edited by

AMIT BHATTACHARYA
JAMES W. RAWLINS
PARAMITA RAY

WILEY

A JOHN WILEY & SONS, INC. PUBLICATION

Published by John Wiley & Sons, Inc., Hoboken, New Jersey
Published simultaneously in Canada

Library of Congress Cataloging-in-Publication Data:

Polymer grafting and crosslinking / edited by Amit Bhattacharya, James W. Rawlins, and Paramita Ray.
p. cm.
Includes index.
ISBN 978-0-470-40465-2 (cloth)
1. Crosslinking (Polymerization) 2. Crosslinked polymers–Industrial applications.
I. Bhattacharya, Amit. II. Rawlins, James W. III. Ray, Paramita.
TP156.P6P585 2009
668.9—dc22

2008026774

10 9 8 7 6 5 4 3 2 1

CONTENTS

PREFACE

We are very fortunate to have the opportunity to edit a book in the grafting and crosslinking arena. While several excellent books on polymers are available, our experience in research over the years has revealed the need for a book which emphasizes the basics as well as applied orientation for beginners and those who want to increase their knowledge. An attempt has been made to present the subject lucidly by adopting a non-mathematical approach.

The book covers various mechanistic methods, kinetic factors, analytical evidence, and diverse applications. We extend our appreciation to the numerous publishers who have graciously accorded us permission to use figures and data from their publications.

This book is intended for beginners in this area, advanced students, as well as teachers. It is expected that they possess an adequate background in polymer technology. However, due to the book's wide coverage and simple style, researchers can also benefit from it. The book consists of twelve chapters. All topics have been selected with great care bearing in mind the needs of the students. The first five chapters provide a vivid introduction to the basic concepts. The remaining chapters are devoted to applications.

Needless to say, we owe a great debt of gratitude to our spouses for their encouragement, and to the authors of the various chapters. We are deeply indebted to Dr. P. K. Ghosh, Prof. S. N. Bhattacharyya, and Prof. B. N. Misra for their suggestions and assistance in completing this book. It would have been difficult for us to complete this book without the assistance of our students. Last but not the least, we express our heartiest thanks to readers of the book.

DR. A. BHATTACHARYA
PROF (DR.) JAMES W. RAWLINS
DR. P. RAY

CONTRIBUTORS

Amit Bhattacharya, Central Salt and Marine Chemicals Research Institute (Council of Scientific and Industrial Research), G.B. Marg, Bhavnagar, Gujarat-364002, India, bhattacharyaamitl@rediffmail.com

Christopher M. Fellows, School of Science and Technology, The University of New England, Armidale, NSW 2351, Australia, cfellows@une.edu.au

Eun Seok Gil, Department of Biomedical Engineering, Tufts University, Medford, MA 02155

Inderjeet Kaur, Chemistry Department, Himachalpradesh University, Himachalpradesh, India, ij_kaur@hotmail.com

Tao Lu Lowe, Department of Pharmaceutical Sciences, Thomas Jefferson University, 130 South 9th Street, Philadelphia, PA 19107

Sharathkumar K. Mendon, University of Southern Mississippi, Hattiesburg, MS 39406, sharathkumar.mendon@usm.edu

Gauri P. Misra, Department of Pharmaceutical Sciences, Thomas Jefferson University, 130 South 9th Street, Philadelphia, PA 19107

Mohamed Nasef, Chemical Engineering Department, Universiti Teknologi Malaysia, 81310 UTM, Skudai, Johor, Malyasia, mahmoudeithar@fkkksa.utm.my

Mahammad Safikur Rahman, Ahmedabad Textile Industry's Research Association, Ahmedabad 380015, India, chem@atira-rnd-tex.org or safikur_r@hotmail.com

James W. Rawlins, University of Southern Mississippi, Hattiesburg, MS 39406, james.rawlins@usm.edu

Paramita Ray, Central Salt and Marine Chemicals Research Institute (Council of Scientific and Industrial Research), G.B. Marg, Bhavnagar, Gujarat-364002, India, paramita022002@yahoo.co.in

Achintya Sen, Clariant Chemicals (India) Limited Div. Masterbatches, Kolshet, P.O. Sandoz Baug, Thane 400607, Maharashtra, India, aksen004@ yahoo.com or Achintya.Sen@clariant.com

Jeremy Swanson, University of Southern Mississippi, Hattiesburg, MS 39406, jeremy.swanson@usm.edu

James Whittermore IV, University of Southern Mississippi, Hattiesburg, MS 39406, iv.whittemore@usm.edu

1

INTRODUCTION

Amit Bhattacharya and Paramita Ray

We deal what we are. This saying serves to introduce the discussions on polymers. A quantum leap in this area brought about the Industrial Revolution in the 19th century. Recognition should be given to Herman Staudinger, who, in 1922, proposed the first explanation that polymers contain long chains of relatively simple repeating units. This date marks a turning point in the history of polymers, for it was then that the term "macromolecule" was first used. Thus the word becomes self-explanatory when broken down: The word "poly" means many, and "meros" means parts. The macromolecular concept was formulated by Staudinger, who received the Nobel Prize in 1953. Polyethylene, for example, is a polymer that contains large numbers of [$-CH_2-CH_2-$] units. In 1970, the first experiments with polyacetylene took place. Polyacetylene is a long string of molecules chained together, with one unit of acetylene repeated over and over again. In 1977, the trio of Mac Diarmid, Shirakawa, and Heeger joined to focus in this arena to revolutionize electronics and received the Nobel Prize in the 21st century.

Polymers play an essential role in the emergence of the modern world. Though started in the middle of the last century, today, the uses of polymer systems are legion. Technology from commodities to rockets is based on productive research on polymers. The polymer industry developed as population growth created increased demands for natural products that could not readily be met because of their limited supplies. "Osmosis" and "reverse osmosis" techniques that regulate life in every aspect are based on polymers.

Abbe Nolet had observed the "osmosis" phenomenon in the year 1748 in pig bladders, with natural polymers. Later on, synthetic ones came to replace natural polymers.

Polymer Grafting and Crosslinking, Edited by Amit Bhattacharya, James W. Rawlins and Paramita Ray

The history of the “reverse osmosis” membrane started at UCLA in 1959. Samuel Yuster and two of his students, Sidney Loeb and Srinivasa Sourirajan, produced a functional synthetic reverse osmosis membrane from cellulose acetate polymer. The ability to make high-quality biocompatible materials in the biomedical field is at the heart of this revolution.

Traditionally, people in the medical field view polymers as components of devices such as inhalers and catheters, inert bioprostheses, or transdermal patches. Polymers have also proactive role, i.e., they are used as integral parts of therapeutics. New drugs, as well as drug delivery systems based on polymers, have the potential to counter many diseases. The application of synthetic polymers for gene therapy has also been investigated. They may provide a safer way of gene delivery than use of viruses as vectors. Polymeric materials have also been used for biosensors, in testing devices, and for bioregulation.

Textile products made of polymers have always satisfied aesthetic requirements. Synthetic polymers replace natural polymers (e.g., wool and cotton) to help clothe growing populations. The success of aramid fibers has also spawned a variety of other polymer fibers based on nylon, terephthalates, and polyethylene, for example.

Polymers are also identified with insulation. This property has contributed to the enormous success of plastics in insulated shielding for wires and other safety functions. Polymers are also used in the conducting or semi-conducting fields for such things as plastic batteries, light emitting diodes, and sensors.

Polymers touch every aspect of our lives. The nylon toothbrush, the plastic bucket, or the polystyrene umbrella handle are all polymers. Knowingly or unknowingly, every individual today relies on polymers to meet his needs.

Though polymers are legion, sometimes they cannot fulfill the demand, depending on their properties. Improvements in polymers are tremendously important because they will widen the scope of application. There are two main approaches: construction of new molecules that are likely, from their molecular composition, to have the desired properties, and modification of properties of existing large-scale polymers.

Modification of polymers has received greater attention in light of the scarcity of starting materials required for the synthesis of new monomers to deliver better polymeric materials. In other words, modification is essential to meet various challenges, as it is very diffcult to get new polymers. The next generation awaits polymer modification as it opens up new possibilities. Surface and bulk properties can be improved easily by modifying conventional polymers. Sometimes, balancing of properties is needed, and this is possible only through modification of polymers.

Polymer modification is required to bring specific properties to the modified material, such as enhanced thermal stability, multiphase physical responses, compatibility, impact response, flexibility, and rigidity. Modifications make an insoluble polymer from a soluble one or vice versa. Thus polymer modification improves the processibility of the polymers. One of the recent directions regarding polymer modification is intended to avoid their nonbiodegradable nature.

We have already referred to giant molecules and indicated that they are very long molecular chains of carbon atoms, often literally "branched" like a tree or linked together crossways in various directions. These chains can be provided by nature, as in the case of cellulose, or are built up from single short molecules known as "monomers" by uniting them into a chain to give "polymers," which is known as polymerization. When monomers of different kinds are united, the product is known as a "copolymer," and the process of uniting different monomers is copolymerization. Depending on the alignment sequence of different monomers in a polymer chain, copolymers are diversified in their property profiles.

Copolymers with an alternate sequence of monomers are alternating copolymers, whereas random copolymers do not follow any definite sequence in alignment patterns of the constituent monomers. Rarely may we expect any property gain from such compositions. Block copolymers are composed of long sequences (blocks) of one monomer unit, covalently bound to sequences of unlike type. Depending upon the chemical nature of the monomers block copolymers may be di-, tri-, tetra-, or even multi-block.

Generally, blocks are spontaneously self-assembled into a diversity of mesophases with the size scale governed by the chain dimensions. Dissimilar blocks exist as microdomains and exert their properties in the system. Thus glassy microdomains serve to anchor rubbery segments permitting the material to be used as melt processable adhesives and rubbers, i.e., thermoplastic elastomers such as SBS or ABS. Block copolymerization is one of the keys to achieve polymeric products with the desired property profiles.

The prime techniques for polymer modifications are grafting, crosslinking, blending, and composite formation, which are all multicomponent polymer systems. Such materials have attracted considerable attention in the industrial field as they combine a variety of functional components in a single material. Micro- and macrophase separation directly affect the physical and chemical properties in such multicomponent polymers. Graft copolymers are branched macromolecules in which the branches are of a different type from the backbone. They have a variety of potential applications resulting from the wide range of properties available when different polymer chains are connected to form hybrid branched macromolecules.

Multidirectional chain extension of polymers leading to the formation of network structures may be defined as crosslinking and viewed as a stabilization process. Crosslinking may result either through the polymerization of monomers with functionality greater than 2 (by condensation mode) or by covalent bonding between preformed polymer molecules accomplished by irradiation, sulfur vulcanization, or miscellaneous chemical reactions. Crosslinking restricts chains from sliding past one another and generates elasticity in an amorphous polymer. It makes a polymer more resistant to heat, light, and other physical agencies, giving it a high degree of dimensional stability, mechanical strength, and chemical and solvent resistance. The effects of crosslinking on the physical properties of the polymers are primarily influenced by the degree of crosslinking, the regularity of the network formed, and the presence and absence of crystallinity in the polymer. For crystalline polymers there may be a reduction in crystallinity with a low degree of crosslinking as it hinders chain orientation, and the polymer may become softer, more elastic, and have a lower melting point. Crosslinking changes the local molecular packing and leads to a decrease in free volume. This is reflected in an enhancement of glass transition temperature. Improvement in creep behavior also results from crosslinking as it restricts the viscous flow. Hence for property synergism crosslinking is unavoidable in the present arena of polymers.

A macroscopically homogeneous mixture of two or more different polymers may be defined as a polymer blend. The blending of polymers provides a means of producing new materials, which combine the useful properties of all the constituents. Most of the polymers are immiscible. Hence the combination of two arbitrary polydispersed polymers (quasi-binary mixture) exhibits neither a true two-phase nor a single-phase structure. It may appear to have a two-phase morphological pattern whereby both phases contain a different and finite concentration of each polymer or it may have multiphase structure. In both the cases, no definite interface exists between the two adjacent phases. The relative concentration of the constituents in different phases depends upon their intensity of mixing and solubility parameters. The ultimate behavioral pattern of polyblends depends upon extent of phase separation, nature of the predominant phase, character of the dispersed phase, and interaction between the component polymers. Most useful blends have the characteristics of uniform dispersion, i.e., fine discrete particles of one component being uniformly dispersed through the matrix of the other component.

It would be advantageous if the size and size distribution of the domains of the dispersed phase remain practically unaltered over long periods of time, from the conditions of blending to the conditions under which the blend is to be performed. The two main objectives of making polymer blends are to maintain a proper cost performance ratio and reinforcement of properties, i.e., compensation of the demerits of a polymer by blending it with a second one that is rich in the aspects lagging in the initial one. The technology of blending is now advancing at a rapid pace. New blends are constantly being developed and characterized, and blends with metallocene resins, liquid crystal polymers, and biodegradable resins are now available. Blending technology is also proving to be of use in plastics recycling.

A composite, as its name suggests, is made by combining two or more dissimilar materials in such a way that the resultant material is endowed with properties superior to any of its components. These components neither take part in the chemical reaction nor do they dissolve or completely merge with one another. Nevertheless, they remain strongly bonded together while maintaining an interface between one another and act in concert to give a much improved performance. In polymer composites, polymers that are the major components contain different reinforcing materials such as glass fiber, carbon fiber, silica, or mica as minor components. Reinforcement results in increased tensile, flexural, compression, and impact strength; rigidity; size stability; fire retardancy; corrosion resistance; electrical properties; and processibility. Polymer composites have versatile applications from the construction sector to commodity materials.

The two predominant modes of polymer modifications—grafting and cross-linking—are primarily considered in this book. The basic theories and exploration of these techniques in different sectors of human life are described in the following chapters.

2

BASIC FEATURES AND TECHNIQUES

AMIT BHATTACHARYA AND PARAMITA RAY

INTRODUCTION

"Physiosorption," "grafting," and "crosslinking" are the techniques by which the association of monomers and polymers is described. The term physiosorption signifies that it is related to physical attractive forces. The process is a reversible one and is achieved by the end functionalized polymers on to the solid surface or self-assembly of polymeric surfactants, where "grafting" can be described as the covalent attachment process and irreversible. Grafting can be accomplished by either "grafting to" or "grafting from" approaches. In "grafting to" approaches, functionalized monomers react with the backbone polymer to form the grafted one. On the other hand, "grafting from" is achieved by treating a substrate with some method to generate immobilized initiators followed by polymerization. High grafting density polymer also can be accomplished using this technique. The schematic presentation of all the processes is depicted in Figure 2.1A. The "crosslinking" is the association of polymers through a chemical bond. In most cases, the crosslinking is irreversible. It may be intra- and intermolecular (Figure 2.1B).

TECHNIQUES USED

The grafting and crosslinking reactions can be performed by different pathways such as chemical or radiation. The important modes of reactions are discussed in the following.

Polymer Grafting and Crosslinking, Edited by Amit Bhattacharya, James W. Rawlins and Paramita Ray

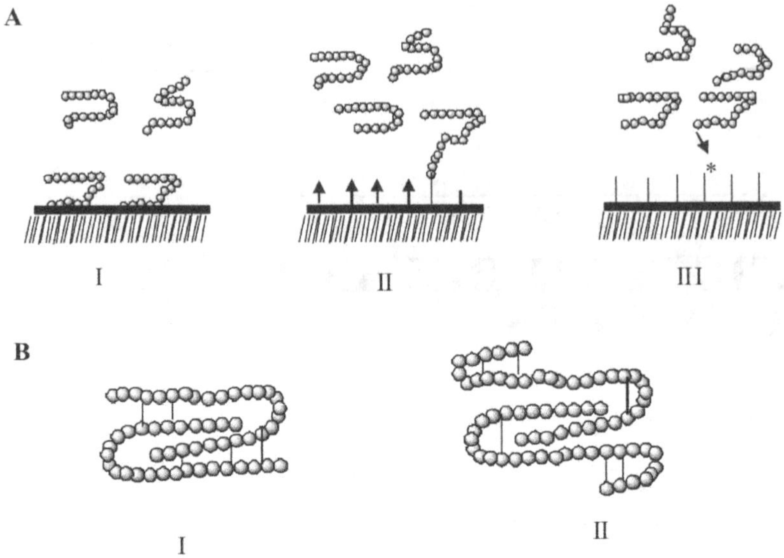

Figure 2.1. A. Schematic diagram of (I) physiosorption, (II) grafting to, (III) grafting from. B. Schematic diagram of (I) intermolecular crosslinking and (II) intramolecular crosslinking.

Chemical Routes

By Redox Reaction The redox reaction is the conventional one to produce the free radicals. There are different redox reagents in which radicals can be generated and relayed to the polymer so that the grafting reaction occurs. The features of the chemical reactions are as follows:

Features:

- Simple to carry out and no constraints in location point of view.
- Feasibility exists at room temperature and also in the aqueous solution.
- Extent of grafting can be controlled by tuning the reaction variables [viz. mixture composition (monomer/initiator), reaction time, and temperature].

The probable reaction pathways are described below.

(I) With Fenton's reagent (Fe^{2+}/H_2O_2) A HO• is generated by the reaction of Fe^{2+} and H_2O_2

$$H_2O_2 + Fe^{2+} \rightarrow HO\bullet + OH^- + Fe^{3+}$$

$$HO\bullet + |{-}H \rightarrow |\bullet + H_2O$$

(|–H is the polymer and M is the monomer)

The hydroxyl radical has the capability to abstract the hydrogen atom from the polymeric substrate, producing the free radical on it (|•). The monomer molecules that are in the vicinity of the polymeric substrate become the acceptor of the radicals, resulting in chain initiation. Thereafter, they become the free-radical donors to the neighboring molecules. In this way, the grafted chains usually grow. These grafted chains terminate by coupling to yield the grafted copolymer.

Initiation: $|\bullet + M \rightarrow |-M_1\bullet$

Propagation: $|-M_1\bullet + M \rightarrow |-M_2\bullet$

$$|-M_2\bullet + M \rightarrow |-M_3\bullet$$

$$\ldots$$

$$|-M_{n-1}\bullet + M \rightarrow |-M_n\bullet$$

Termination: $|-M_n\bullet + |-M_m\bullet \rightarrow$ Graft copolymer

Fenton's reagents are used in grafting, but in the crosslinking reaction, they are not used very much. However, crosslinking reaction from the bimolecular reaction of Fe^{2+} and H_2O_2 depends upon the free-radical concentration at once [1].

(II) With Fe^{2+}/Persulphate Persulphate and a ferrous ion act as a source of $SO_4^{-\bullet}$

$$S_2O_8^{2-} + Fe^{2+} \rightarrow SO_4^{-\bullet} + Fe^{3+} + SO_4^{2-}.$$

There are different views regarding the activity of $SO_4^{-\bullet}$. One is that $SO_4^{-\bullet}$ reacts with water to form OH•, subsequently producing free-radicals on the polymeric backbone

$$SO_4^{-\bullet} + H_2O \rightarrow HSO_4^- + OH\bullet$$

$$HO\bullet + |-H \rightarrow |\bullet + H_2O$$

(|–H is the polymer and M is the monomer)

The propagation and termination steps are similar to the previous example.

An alternate view is that $SO_4^{-\bullet}$ reacts directly with the polymeric backbone to produce the requisite radicals

$$SO_4^{-\bullet} + |-OH \rightarrow HSO_4^- + |-O^\bullet$$

However, Misra et al. [2] established that during grafting of vinyl monomers onto wool/cellulose, OH• is more reactive than the $SO_4^{-\bullet}$.

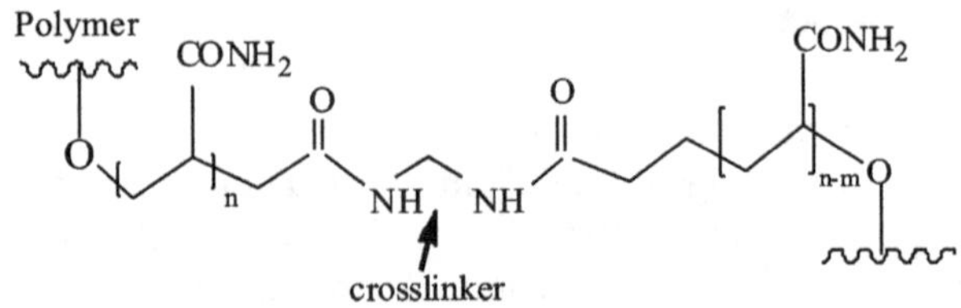

Figure 2.2. Involvement of the crosslinker in the crosslinked polymer.

(III) With Persulfate and Reducing Agent Persulfate and a reducing agent (e.g., sodium bisulphite, thiosulphate, or Ag^+) is also the source of $SO_4^{-\cdot}$, which subsequently produces OH•. $SO_4^{-\cdot}$ has the ability to initiate the polymer free radical directly or indirectly (as in the above example).

Persulfate/Reducing agent M: $S_2O_8^{2-} + Ag^+ \rightarrow SO_4^{-\cdot} + Ag^{2+} + SO_4^{2-}$

Persulfate/thiosulphate: $S_2O_8^{2-} + S_2O_3^{2-} \rightarrow SO_4^{-\cdot} + S_2O_3^- + SO_4^{2-}$

Persulfate/bisulphate: $S_2O_8^{2-} + HSO_3^- \rightarrow SO_4^{-\cdot} + HSO_3^{\cdot} + SO_4^{2-}$

The $SO_4^{-\cdot}$ and $HSO_3^{\cdot}$ produced in the reaction generate OH• in an aqueous medium (as shown earlier), which forms the free radical on the polymer.

Persulfate-initiated crosslinking occurs in the presence of different crosslinking agents (viz. N, N′ methylene bis acrylamide, NNN′N′tetramethylethyl ene diamine) [3, 4]. The crosslinking mechanism is similar to grafting, except for the crosslinker participation. The involvement of the crosslinker is shown above (Figure 2.2).

(IV) With Fe^{2+}/Hydroperoxides Hydroperoxides and Fe^{2+} comprise another important redox system, with free radicals generated by the interaction between them via thermal decomposition [5]. The activity of a tertiary butyl hydroperoxides —Fe^{2+} system is caused to the formation of a t-butoxy radical arising from one electron transfer between t-butyl-hydroperoxide (TBHP) and Fe^{2+}.

$$\text{t-BuOOH} + Fe^{2+} \rightarrow \text{t-BuO}^{\cdot} + OH^- + Fe^{3+}$$

The resulting t-BuO• may participate in hydrogen abstraction to generate OH• and the macroradical on the polymeric backbone:

With monomer (M): t-BuO• + M → t-BuO-M• → t-BuO-M_n-M•

With polymer (|—H): t-BuO• + |—H → t-BuOH + |•

With water: t-BuO• + H_2O → t-BuOH + OH•

(V) By Direct Oxidation Unlike the above system, free-radical sites are generated on a polymeric backbone by direct oxidation of the polymeric backbone by certain transition metal ions (e.g., Ce^{4+}, Cr^{6+}, V^{5+} and Co^{3+}) [6–8]. The metal ions with low oxidation potential are the proper choices for the reaction. The proposed mechanism for such processes has been ascribed to the intermediate formation of a metal ion polymer complex of chelate type (I). Such complex formation is not restricted to all polymers especially for cellulose types, where there is a possibility glycol groups may participate in the reaction. The mechanism is as follows:

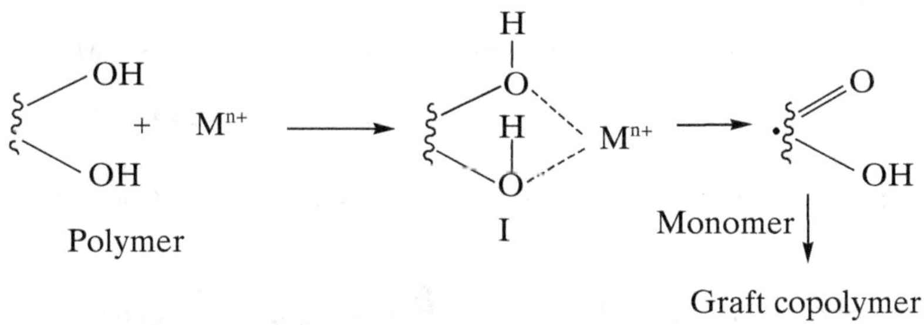

Grafting through Ce (IV) initiation is feasible in an aqueous medium, but for styrene grafting it is not the proper choice because of its low solubility in water.

Crosslinking through Ce (IV) initiation is possible in the presence of different crosslinking agents. The mechanistic way is similar to the above (through persulfate) [9].

MnO_4^- in the acid medium reduces to Mn^{3+} ions to Mn^{2+}. These highly reactive ions are responsible for initiating graft copolymerization.

$$| -H + Mn(III) \rightarrow \vdash H - Mn(III)\,[\text{complex}]$$

$$| -H + Mn(III) \rightarrow |\bullet + Mn(II) + H^+$$

$$|\bullet + M \rightarrow |M\bullet$$

$$Mn(III) + M \rightarrow M\bullet + Mn(II) + H^+$$

$$|M\bullet + nM \rightarrow \vdash M_{n+1}\bullet$$

$$|M_{n+1}\bullet + Mn(III) \rightarrow |M_{n+1} + Mn(II) + H^+$$

$$|\bullet + Mn(III) \rightarrow \text{oxidized product} + Mn(II)$$

(VI) Using Metal Chelates To avoid undesired reactions, controlling redox potential in the form of metal chelates (e.g., $Fe(acac)_3$, $Zn(acac)_2$, $Al(acac)_3$, $VO(acac)_2$:acac:acetyl acetonate) are useful in the grafting reaction. The metal ion forming complex with the monomer decomposes to the free radical, which undergoes a chain transfer to the polymer. The formation of the free radical in the polymer through monomer is sketched below [10, 11], and the grafting reaction is the same as shown in the earlier example.

$$Z^{n+}(acac)_n \longrightarrow Z^{(n-1)+}(acac)^{+}_{n-1} \qquad \dot{C}H\,(COCH_3)_2$$

$$\dot{C}H\,(CH_3CO)_2 + M \longrightarrow (CH_3CO)_2CH_2{-}M_n{-}M^{\bullet}$$

$$\vdash H + (CH_3CO)_2CH_2{-}M_n{-}M^{\bullet} \longrightarrow \,\vdash^{\bullet}$$

[Z: metal, (acac)=acetyl acetonate, M: monomer, $\vdash H$ polymer]

(VII) By Indirect Method The secondary free radicals are generated in the system and may ultimately induce the free-radical sites required for grafting [12].

$$Mn^{4+} + C_2O_4^{2-} \rightarrow Mn^{3+} + CO_2 + CO_2^{-\bullet}$$

$$Mn^{3+} + C_2O_4^{2-} \rightarrow Mn^{2+} + C_2O_4^{-\bullet}$$

V^{5+} in the presence of H_2SO_4 produces secondary $HSO_4^{\bullet}$, which can induce free-radical sites on the polymeric backbone. The oxidative reaction is as follows:

$$V^{5+} + H_2SO_4 \rightarrow HSO_4^{\bullet} + V^{4+} + H^{+}$$

Limitations: The proper selection of the metal ions or oxidative process is a factor, as the higher redox potential of the system has the possibility to react with the monomers that will lead to an increase in homopolymerization. To circumvent this situation, systems with low oxidation potential and polymers of strong reducing agents are preferred.

On heating: Apart from the redox reaction, certain organic compounds (azo, peroxides, hydroperoxides, etc.) [13] are capable of producing free radicals on heating and transferring them to the chain. It is seen that azoisobutyronitrile (AIBN) (I) is not as reactive as benzoyl peroxide (BPO), because of the resonance stabilization of the fragmentation.

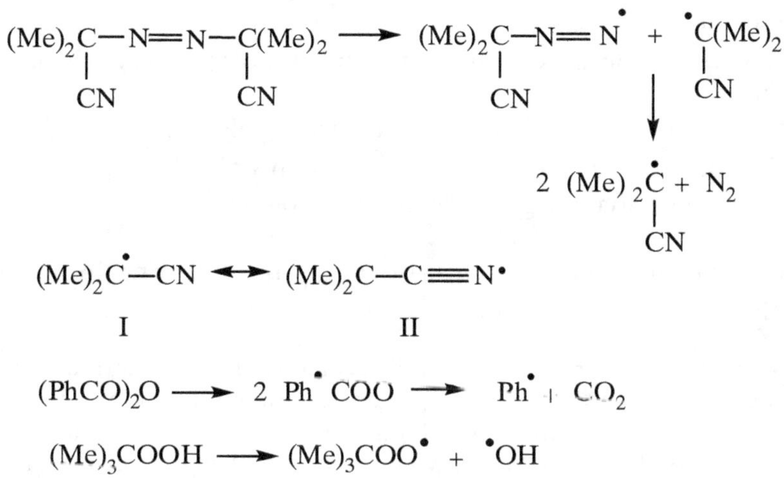

Crosslinking is also reported by using different crosslinking agents and initiators; the mechanism is similar to the grafting reaction [14–18].

Chemical pretreatment (e.g., ozonation, diazotization, xanthation) of the polymer backbone may also generate free radicals that provide sites for grafting as follows:

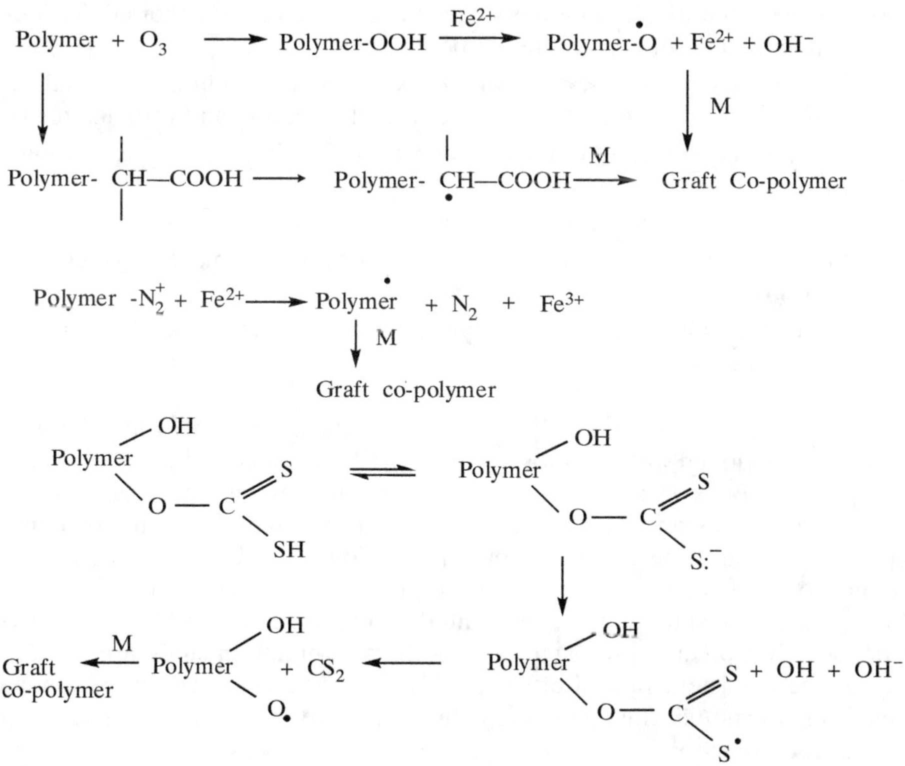

Limitations:

- The conventional free-radical graft polymerization needs continuous initiation.
- The termination step leads to an unreactive polymer.
- The grafted product is of broad molecular distribution.
- The time invariant degree of polymerization.

These limitations lead researchers to adopt the following living radical technique.

By Living Radical Formation In this technique, the chain termination step is avoided in two ways: either by increasing the rate of initiation with respect to the rate of propagation or by eliminating chain termination and transfer reactions. The polymer chains grow at a more constant rate, and their chain lengths remain very similar. Thus, it provides polymers that are able to grow whenever an additional monomer is supplied. The conditions are discovered for various types of reactions, e.g., atom transfer, nitroxide mediated, degenerative transfer.

Features:

- It proceeds until all monomers have been consumed. Further addition of a monomer results in continued polymerization.
- In these reactions $R_i << R_p$, where, as for the conventional free-radical grafting, $R_i > R_p$ (R_i and R_p are the rate of initiation and propagation).
- Dynamic equilibrium occurs between a propagating radical and dormant species.
- Mono and/or narrow dispersities of the products are possible.
- Predetermination of molar mass of the grafted/crosslinked polymer molecules is possible.
- Telechelic polymers are easily synthesized by using proper terminating reagents.

(I) Through Atom Transfer The key molecule is the one that contains halogen, positioned in the α-position with carbonyl, alkoxy-carbonyl, cyano, and phenyl groups. The reversible pseudohalogen homolytic transfer between a dormant species, an added initiator, or the propagating dormant chain end, (I-X) and a transition metal complex in the lower oxidation state (M^{n+}/L), occurs to form the propagating radicals (I*) and the metal complex in the higher oxidation state with a coordinated halide ligand [$X\text{-}M^{(n+1)+}/L$]. Activation, deactivation, and propagation, as well as termination steps, are present but as the persistent radical effect (PRE), the radical termination is diminished, and the equilibrium is strongly shifted toward the dormant species [19]. The general reaction is as follows:

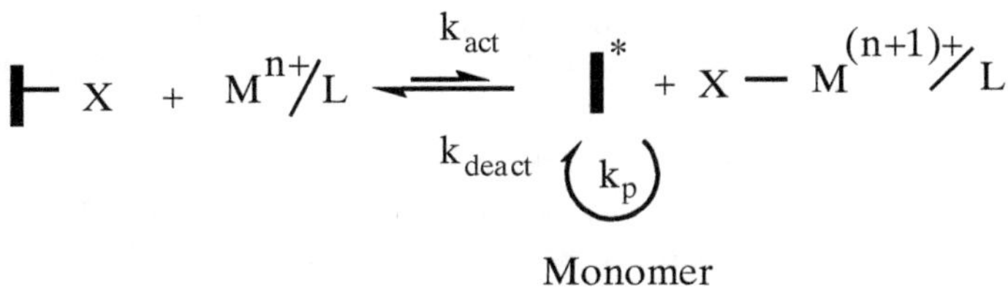

M^n - transition metal, L - complexing ligand, ⊢ Polymer chain, X = Br or Cl

Features:

- Scope for choosing both the initiator and catalyst complex.
- It can occur in solution as well as suspensions.

Polymers prepared by other processes can be functionalized at the terminus or along the backbone and incorporated into an ATRP as a macroinitiator/ macromonomer, or simultaneously through use of both a macroinitiator and a macromonomer to improve incorporation of the macromonomers into the polymer leading to well-defined block and graft copolymers [20–25]. Examples related to the grafting of acrylates and styrene to the polymer based on macroinitiators are as follows (Scheme 2.1):

There may be one or multiple initiating sites, leading to chain growth in several directions, which also proves the potential for the preparation of

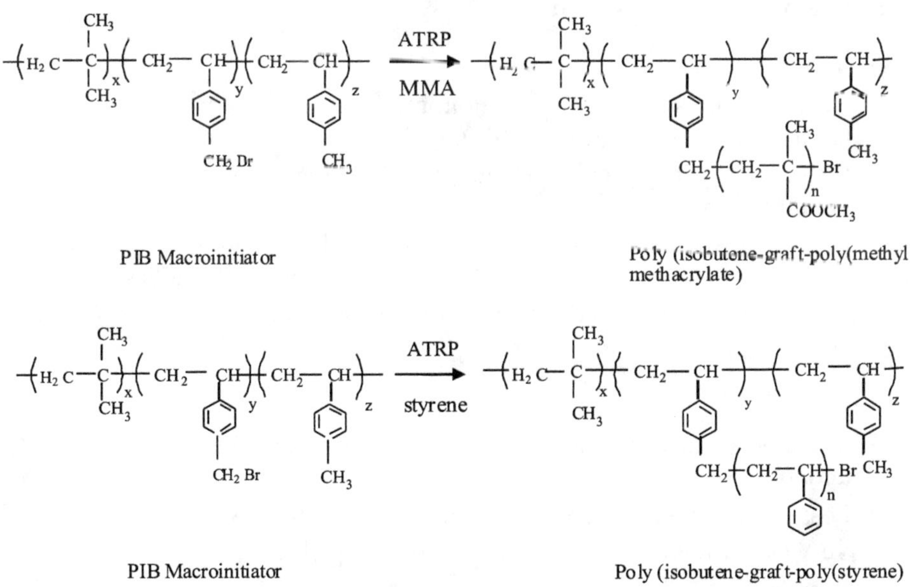

Scheme 2.1. ATRP reaction to form the graft polymer. Reproduced from *Macromol Chem and Phys*, **202**: 3392 (2001).

molecular brushes, i.e., high grafting density polymer, using a "grafting from" approach. The synthesis of poly (MMA-grad-BPEM)-graft-n-BuA [26] is sketched in Scheme 2.2.

Surface grafting by controlled radical polymerization on the polymeric substrate, colloidal particles and inorganic materials are also possible [27–29]. The examples are displayed in the Schemes 2.3 and 2.4.

Limitations:

- Selection of suitable ligands is critical.
- The reactions preferably conduct in the absence of oxygen, or otherwise some reducing agents [e.g., Cu (0), Sn (EH_2), or ascorbic acid] are essential.

(II) Nitroxide Mediated Free-radical nitroxides [viz. 2,2,6,6-tetramethyl-1-piperidinoxyl (TEMPO)] are key molecules. The reactions are based on reversible capping of the active polymeric chain end radical with a nitroxide leaving group. The dormant chains are generated by reversible deactivation of growing chains through covalent bond formation. At first, the nitroxide free radical, as well as active growing chain (through homolytic cleavage), forms at high temperature. After that, the reaction occurs with the preformed macromonomers [30–32]. The reaction feature shows that preparation of dendritic polymers [33–35] is possible through this technique. Examples related to the mixture of styrene and p-chloromethyl styrene is presented in the Scheme 2.5. In this case the polymerization mechanism is switched from nitroxide mediated to atom transfer "living free-radical" conditions.

Features:

- Reactions are acceptable with monomers, e.g., styrene, butadiene, methyl methacrylate, styrene sulfonate (aq).
- Reaction feasibility exists when using commercial initiators as well as initiator formed in situ.
- Possibility to introduce stable free-radical adduct to preformed structures including dendrimers, functionalized polymers, and telechelics.
- Narrow molecular weight distribution up to moderate conversions.
- No Trommsdorff (Gel) effect.

Limitations:

- Monomers (alkyl alkylates, vinyl acetate, vinyl chlorides) do not usually undergo the reaction.
- Loss of molecular weight control at high conversions.
- Very slow rates.

(III) Degenerative Transfer In this mode, the grafting reaction occurs through the chain transfer reagents (alkyl iodides, thiol compounds, and unsaturated

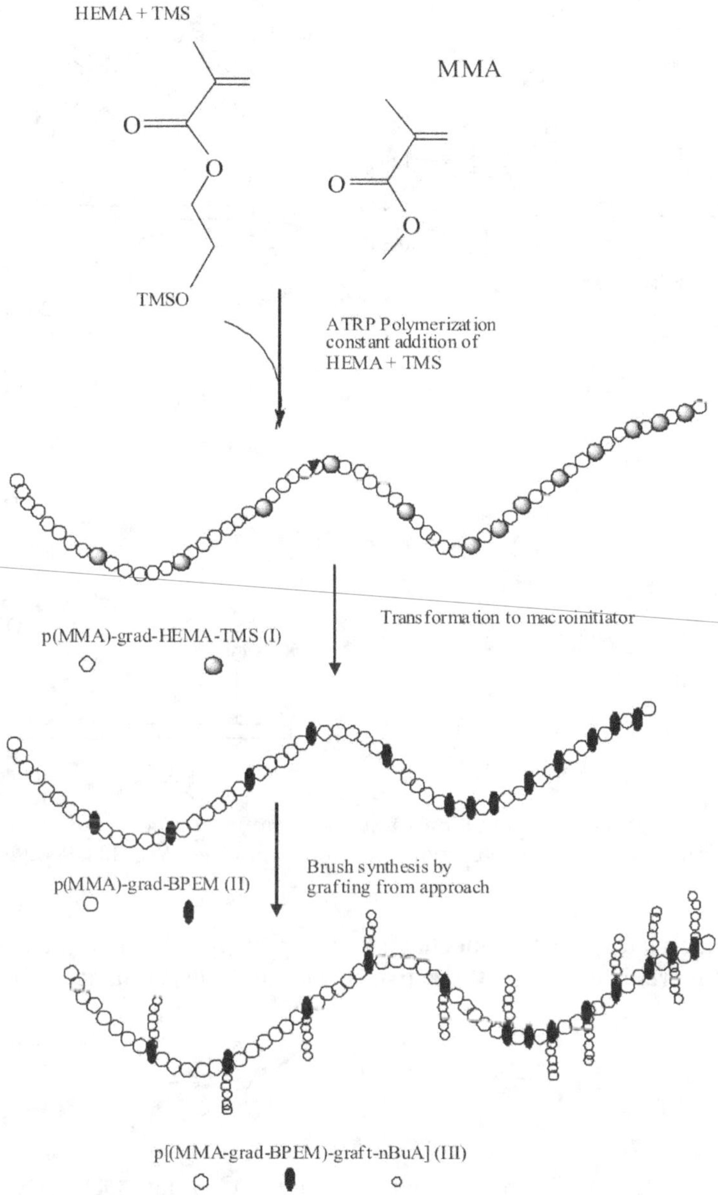

Scheme 2.2. (I) Subsequent synthesis of the macroinitiator precursor. (II) Macroinitiator. (III) Macromolecular brush copolymers. Reproduced from *Macromolecules*, **35**: 3387 (2002).

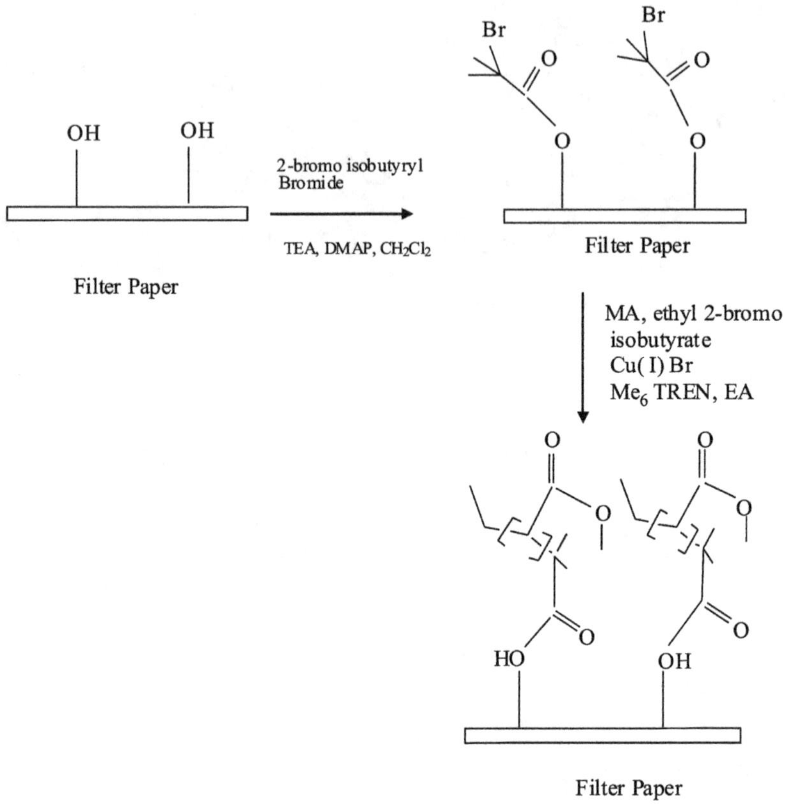

Scheme 2.3. Schematic presentation of surface grafting on cellulose via controlled living radical polymerization. Reproduced from *J Am Chem Soc*, **124**: 900 (2002).

polymethacrylates). In this mechanism, the polymer (P_x) is attacked by the propagating radical to form the active species and the dormant species.

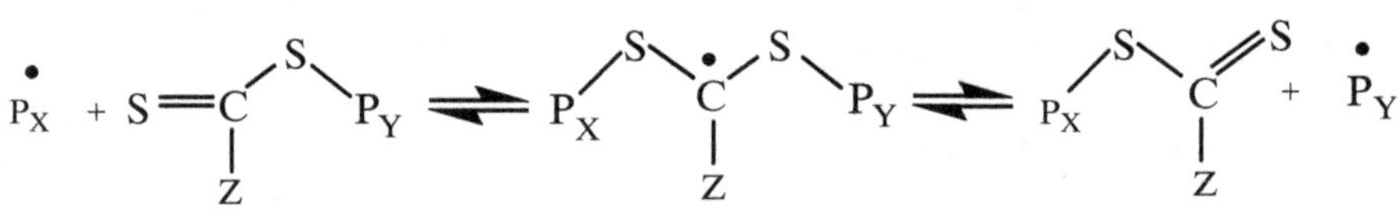

P_X and P_Y are polymeric chains, Z=C_6H_5 and CH_3

This process is also called a reversible addition-fragmentation chain transfer [36, 37]. The synthesis of poly(ethylene-co-vinyl acetate-g-methyl methacrylate) (EVA-g-PMMA) poly(ethylene-co-vinyl acetate-g-styrene) (EVA-g-styrene) based on free-radical polymerization of methyl methacrylate intitiated by AIBN, in the presence of mercapto-modified EVA as chain transfer agent (Scheme 2.6), is the example used in this regard [38, 39].

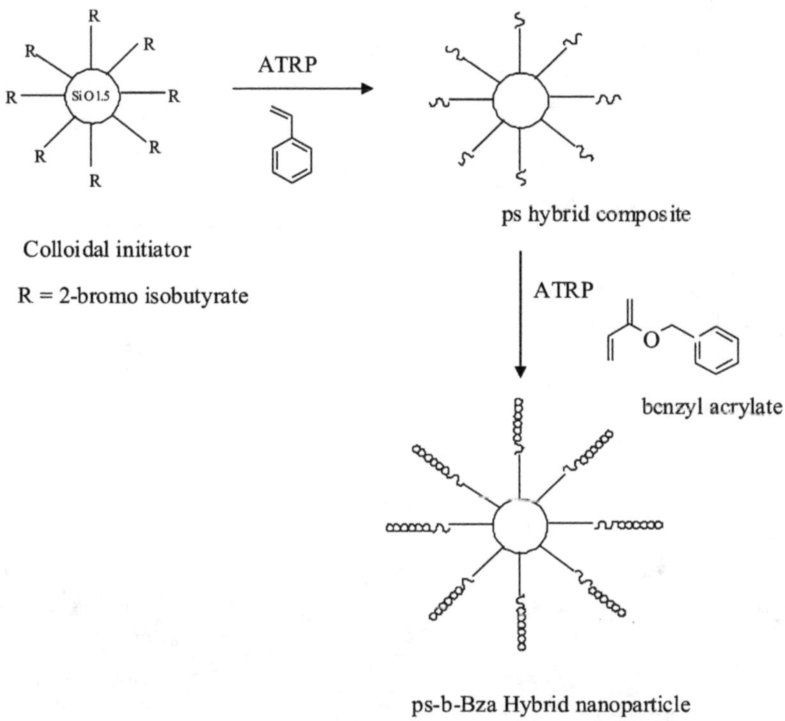

Scheme 2.4. Synthesis of 2-bromo isobutyrate functional nanoparticles and ps-b-p-Bza hybrid particles using microemulsion and ATRP processes. Reproduced from *Macromolecules*, **36**: 5094 (2003).

Crosslinking of monomers/macromonomers is also possible through ATRP. It is usual practice, in the presence of crosslinking agents (such as ethylene glycol dimethacrylate or dimethacrylate); free radicals produced in the reaction combine to form the crosslinked polymer [40–43]. The preparation of star polymers, through macromonomers, is shown in Scheme 2.7 [44]. The macroinitiator and macromonomer methods are featured. In the macroinitiator method, linear initiator is used, and in macromonomer method an initiator (such as ethyl bromo propionate) with the macromomonomer is used. In the macroinitiator there is the possibility of formation of broad polydispersity polymers.

Advantages:

- High grafting efficiency and effective control in the grafting points.

Limitations:

- The intermediate radical should not behave as damagingly as an initiator or a radical trap.

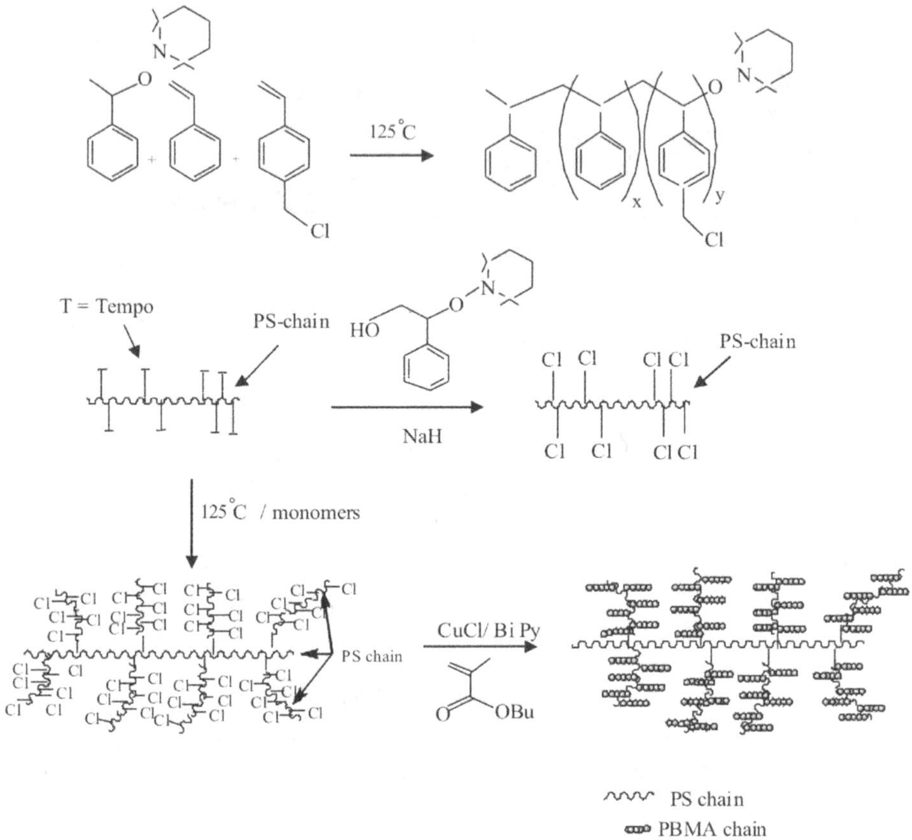

Scheme 2.5. Preparation of dendritic polymers. Reproduced from *Chem Rev*, **101**: 3661 (2001).

Enzymatic Technique The graft and crosslinking reactions with the help of enzymes (i.e., biological proteincious catalyst) are the features of this technique. Most of the grafting reactions by using oxido reductase enzymes (horseradish peroxidase, Tyrosinase, etc.) are focused in the literature [45–48]. They are capable of forming free radicals (e.g., phenol/amine derivatives) by abstracting electrons or by supplying molecular oxygen to the molecules. The active species can then either condense each other or react with the nucleophiles. The general mechanism is shown in Scheme 2.8.

It shows that there are the nonenzymatic reactions followed by enzymatic ones. Experimental evidence supports the hypothesis that quinone can undergo at least two different types of reaction with amines, i.e., Schiff bases and/or Michael-type adducts.

The uses of oxidative enzymes in the grafting reaction to the biopolymers are extensively studied for natural polymers rather than synthetic polymers

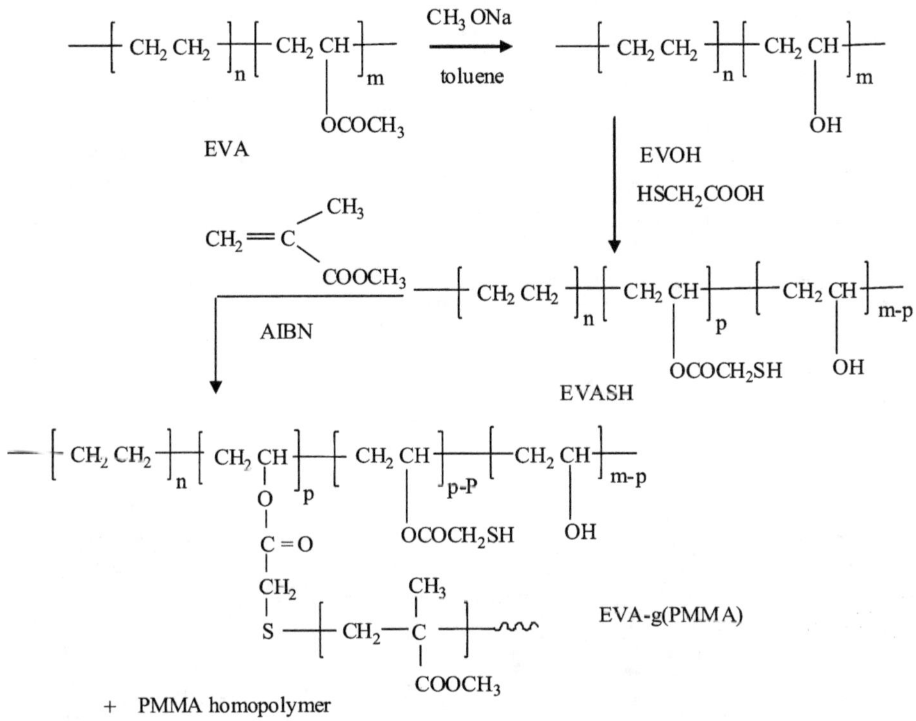

Scheme 2.6. Synthetic pathway for preparation of poly(ethylene-co-vinyl acetate-g-methyl methacrylate) (EVA-g-PMMA). Reproduced from *Macromolecules*, **29**: 416 (1996).

[49, 50]. Only few reports are available in case of synthetic polymers. At the first step, generation of radical occurs through enzymatic reaction (viz. HRP/H_2O_2) from the phenolic compounds, and the phenoxy radical propagates through the polymeric chain and monomers similar to the redox reaction [51].

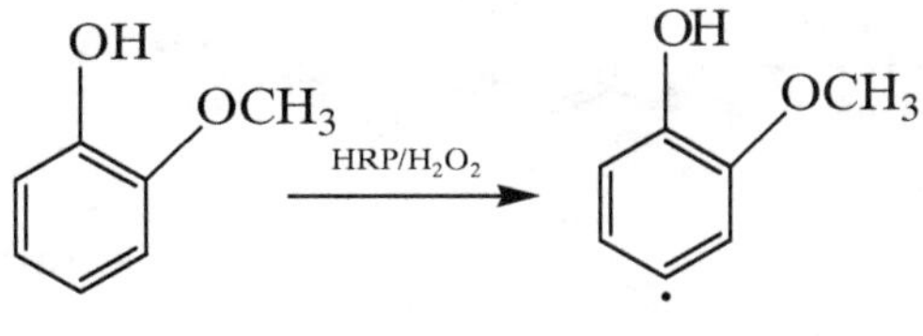

Features:

- It occurs in milder reaction conditions and lowers the hazardous use of harsh chemicals.
- Initiation needs the enzyme to form the free radicals and is followed by nonenzymatic reactions.

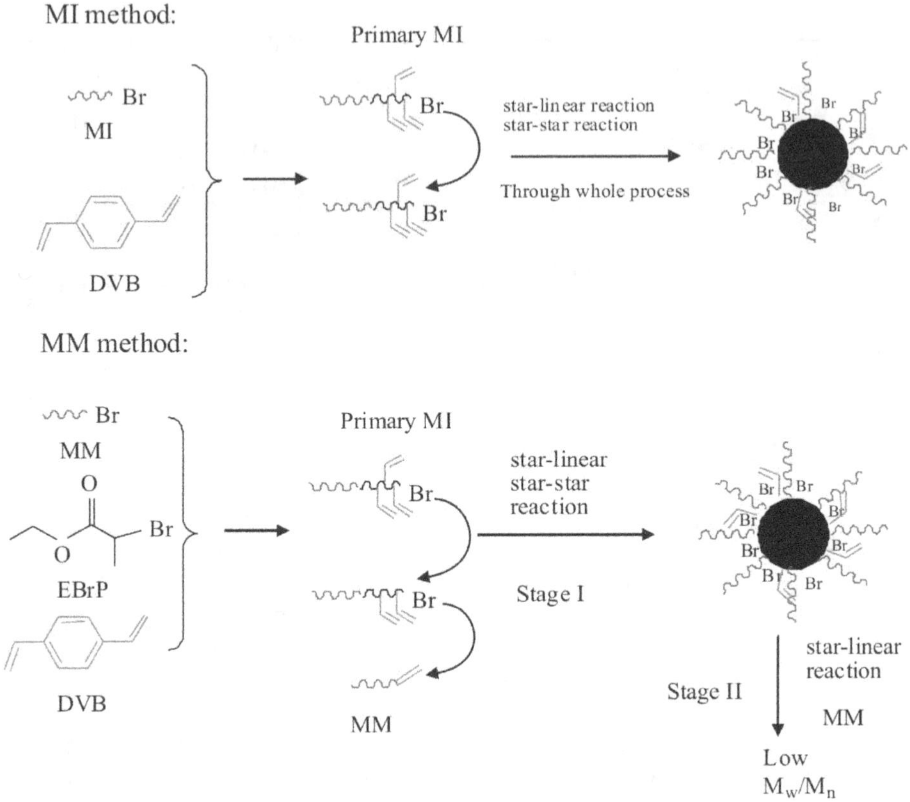

Scheme 2.7. Comparison of star synthesis by the MI method and the MM method. Reproduced from *J Am Chem Soc*, **128**: 15111 (2006).

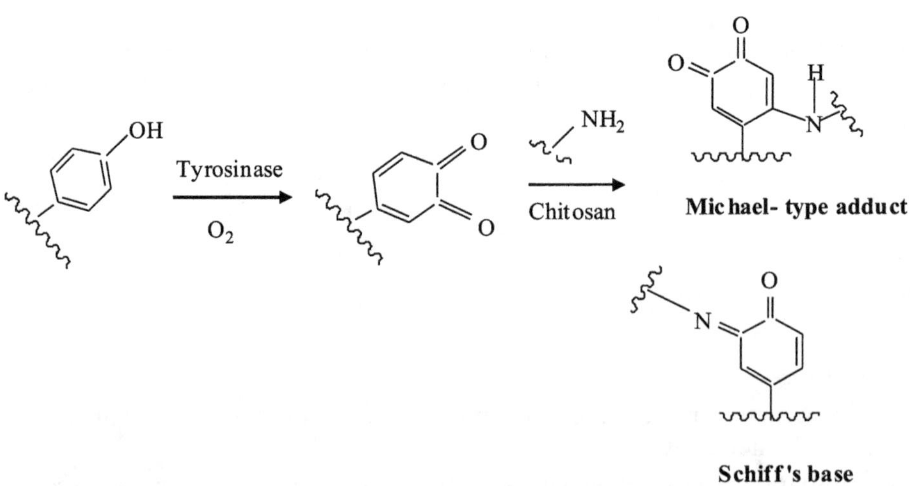

Scheme 2.8. General mechanism by enzyme.

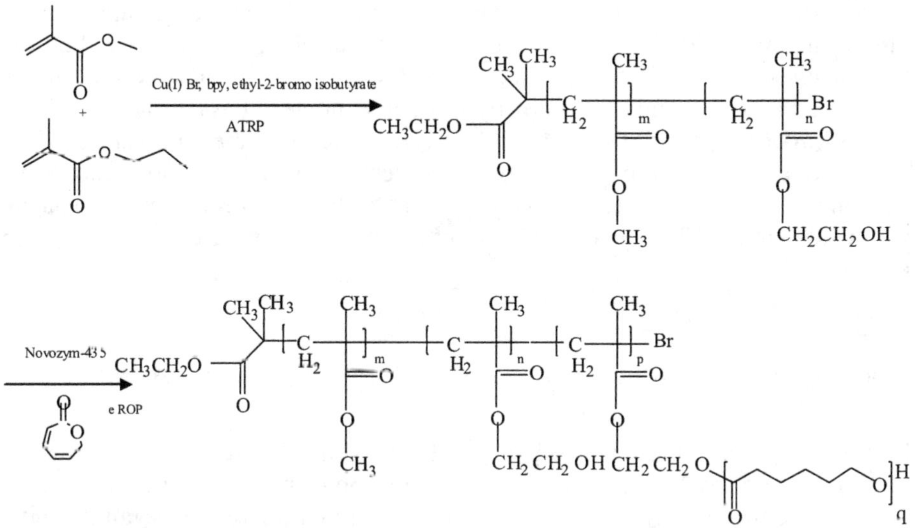

Scheme 2.9. Combination of ATRP and enzymatic ring opening polymerization in sc CO_2. Reproduced from *Macromolecules*, **39**: 9080 (2006).

- Enzyme selectivity and specificity makes the reaction simple and offers the potential for better controlling macromolecular structure without the need of protection or de-protection steps.
- It can be coupled with the other polymerization methods.

The flexibility of the enzymatic reaction is that it can combine with chemical reactions. Synthesis of well-defined polymethacrylate g-polycaprolactone copolymers by the combination of ATRP and enzymatic ring opening polymerization in sc CO_2 is the example (Scheme 2.9) [52].

Multiple sites in the quinone from the phenol moiety by enzymatic way have the feasibility of crosslinking of polymer (viz. chitosan) [53]. Phenolic polymer crosslinking can occur by using vanadium-dependent bromoperoxidase [54]. Chemoenzymatic crosslinked polymers through macroinimers or formation of telechelics are also possible [55].

Limitations:

- Expensive.
- It cannot be performed at high temperature.

Crosslinking Through Polycondensation

(I) Bulk Polymerization A specific class of polymers formed through condensation reactions are known as condensation polymers. Such polymerization, known as step-growth polymerization, proceeds through a step-by-step

succession of the elementary reactions between two functional groups such as amine, alcohol, acid, or isocyanate, etc. Each of the steps involves the elimination of two reactive sites and results in a new linking unit between a pair of molecules with the elimination of byproducts. The very basic criterion of the formation of polymers by such a method is that the monomers must be at least bifunctional similar to the formation of polyester, polyamide, polyimide, polycarbonate, polyether, etc., with the elimination of different byproducts as water, alcohol, HCl, or NaCl; for example, the reaction of diol with dibasic acid.

$$HO(CH_2)_xOH + HOOC(CH_2)_{x'}COOH \longrightarrow HO[-(CH_2)_xO\overset{\overset{O}{\|}}{C}(CH_2)_{x'}\overset{\overset{O}{\|}}{C}O-]_yH + H_2O$$

Step-growth reaction may also proceed by polyaddition reaction without the elimination of any byproducts like the formation of polyurethane, polyepoxy, or polycyanurate polyurea; for example, the reaction of diisocyanate with diols.

$$HO(CH_2)_xOH + OCN(CH_2)_{x'}CNO \longrightarrow [-O(CH_2)_xOCONH(CH_2)_{x'}NHCO-]_y$$

Condensation polymers may be formed by the reaction between two monomers, two oligomers, a monomer and oligomer/macromolecule, or two macromolecules. Hence a gradual enhancement in the molar mass of the product is inevitable but the reactivity remains equal regardless of molecular weight. There is always a probability of reaction between two x-mers as there is for monomers. The structure of a condensation polymer is determined by the functionalities of the monomers and the molar ratio of the reactive sites. Two bifunctional monomers react to give a linear polymer but when polyfunctional monomers are involved, successively higher conversion leads to branch polymers and the probability of network formation enhances. The condensation polymer formed by the reaction of trifunctional monomer (glycerol, T) and a bifunctional monomer (adipic acid, B) may be presented as:

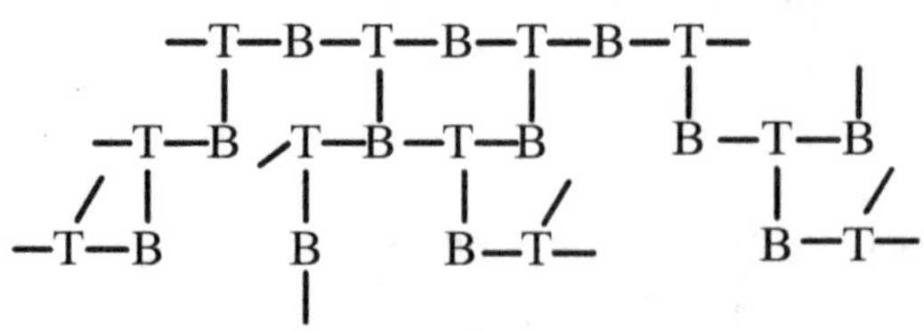

The structures of branched polymers are more complex than the linear polymers and widely differ in their properties.

Gelation and its significance: Three-dimensional polycondensation reactions are associated with an abrupt change from liquid to solidlike behavior.

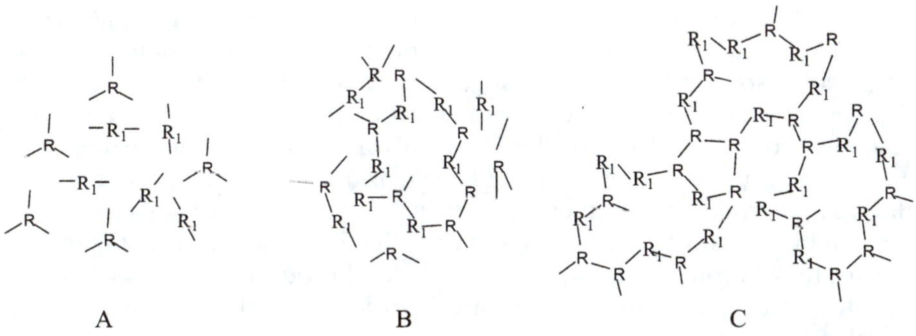

Scheme 2.10. Stages of network formation in step-growth copolymerization. A. Sol stage. B. Mixture of sol and gel. C. Pure gel stage.

Such a phenomenon is known as gelation, which is the formation of essentially infinitely large polymer networks in the reaction mixture and occurs when one of the growing molecules reaches a mass so large that it interconnects every boundary of the system [56]. Gelation is one of the most important criteria of polyfunctional polycondensation, which results in thermosetting polymers (Scheme 2.10).

The network formation with the progress of condensation polymerization between multifunctional monomers has two different stages, i.e., pregel and postgel stage. The pregel stage (Scheme 2.10A) contains a homogeneous system, whereas the postgel stage is heterogeneous, as the sudden onset of gelation at this stage divides the reaction mixture into two parts: the gel, which is insoluble in all nondegrading solvents, and the sol, which remains soluble and can be extracted from the gel (Scheme 2.10B). Beyond the gel point, the amount of gel increases at the cost of sol, and the mixture rapidly transforms from a viscous liquid to an elastic material of infinite viscosity (Scheme 2.10C). In gelation, the weight average molecular weight of the achieved polymers becomes infinite, whereas the number average molecular weight is insignificant. Prediction of gelation is highly important for practical applications of thermosetting polymers. Depending upon the final applications (as adhesives, surface coating, decorative laminates, molded products, etc.), it is necessary to maintain a specific viscosity of the precursors. Hence it is highly necessary to arrest the polymerization at a particular conversion before the gel point. The time to gel may be predicted from the knowledge of polymerization kinetics, reaction temperature, and conversion of the reactive groups at the gel point. Crosslinking of few commercial polymers through polycondensation is as follows:

Phenolic resins: Polymers formed by the reaction of formaldehyde with phenols and substituted phenols cover a wide volume in the arena of thermosetting polymers formed by multidimensional polycondensation.

Formaldehyde reacts with phenol in both acidic (molar ratio $CH_2O/PhOH$ 0.8) and alkaline medium (molar ratio $CH_2O/PhOH$ 1.5–3.0) to produce novolac and resol, respectively, which are oligomeric precursors of the final resin. On average, novolac contains five to six benzene rings per molecule, while resol contains two to four. Both novolac and resols are soluble and fusible in nature and defined as an **A** stage resin. On extension of their molecular weight through polycondensation they pass through a rubbery stage where they are not soluble but swollen by many solvents; this is defined as a **B** stage resin. Finally, they become an insoluble infusible hard product, i.e., a **C** stage resin. The **B** and C stage resins of resol are known as resitol and resit [57]. The novolac reactions are as follows:

$$HCHO + H_2O \rightleftharpoons {}^{+}CH_2OH + H_2O$$

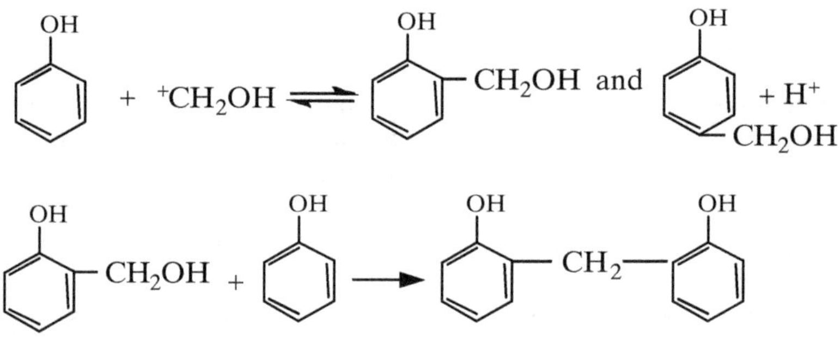

As novolac contains no reactive methylol group, further chain extension to form a network structure needs the addition of a hardener, mostly hexamethylenetetramine (HMTA). Benzoxazine and o-hydroxy benzylamine are the major intermediates formed at the initial stage of crosslinking [58–60]. The methylene linkage necessary for chain extension and crosslinking is formed from these initial intermediates. An optimum balance among the concentration of acid, formaldehyde, and heating rate is necessary to achieve the desired properties in novolac resin [61].

For resol, the phenoxide ion is the active species

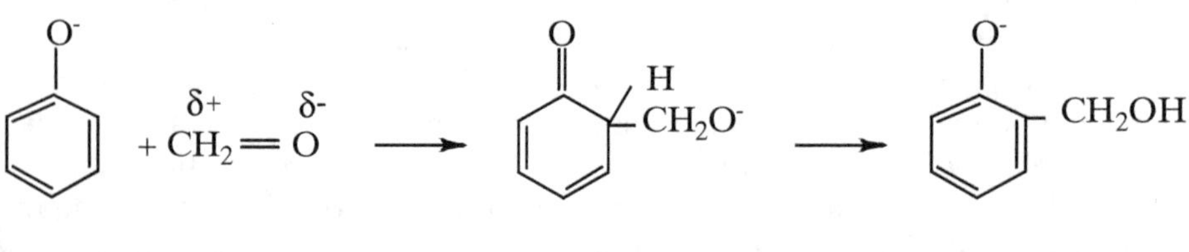

Due to the presence of excess formaldehyde, hardening of resols may be carried out just by heating at a temperature of 130 °C to 200 °C without any hardener, and the resulting macromer contains bridges like $-CH_2-O-CH_2-$ or $-CH_2-$ [57].

The reactions are

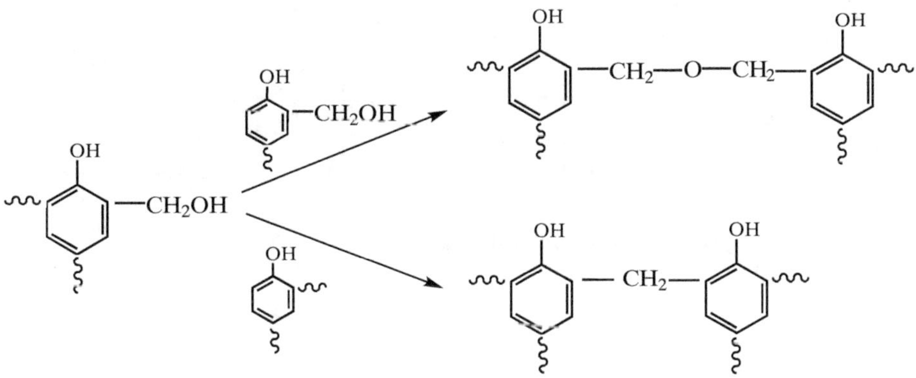

A molar ratio of formaldehyde:phenol 1.3-1.4 is found to have the highest total methylene bridge concentration, i.e., the highest crosslink density [62].

Amino resins: Amino resins are formed by the reaction of amines or amides with aldehydes. Among them, the urea formaldehyde and melamine formaldehyde resins have significant commercial importance. The reaction of urea with formaldehyde can yield a number of products whose content depends on the process factors: component molar ratio, pH, temperature, etc. [63]. Urea formaldehyde (UF) resins are prepared by a two-stage reaction. The first stage is the reaction between formaldehyde and urea under neutral and mildly alkaline conditions to produce mono- or dimethylol urea, the ratio of which depends upon the urea-to-formaldehyde ratio. At the same time tri- and tetramethylol urea may also exist in the solution. These monomeric building blocks contain reactive terminal groups.

$$H_2N{-}\overset{\overset{\displaystyle O}{\|}}{C}{-}NH_2 + 2H{-}\overset{\overset{\displaystyle O}{\|}}{C}{-}H \longrightarrow HOCH_2NH{-}\overset{\overset{\displaystyle O}{\|}}{C}{-}NHCH_2OH$$

The second stage is the heating of methylol-substituted urea in acidic condition, which results in condensation of the monomeric units with the formation of a polymer chain along with byproducts, which is reflected by a gradual increase in viscosity. Linear polymers from mono- and dimethylol ureas have the following structures:

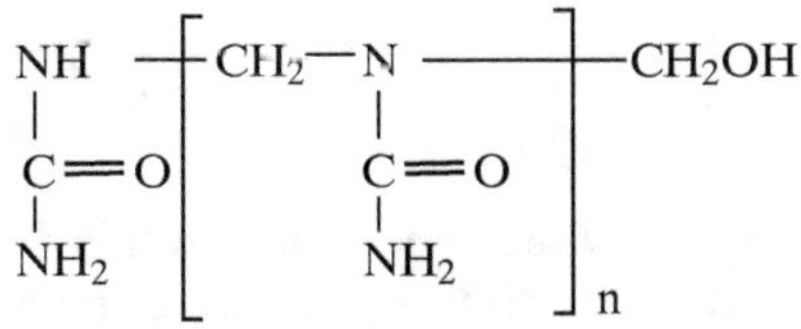

and

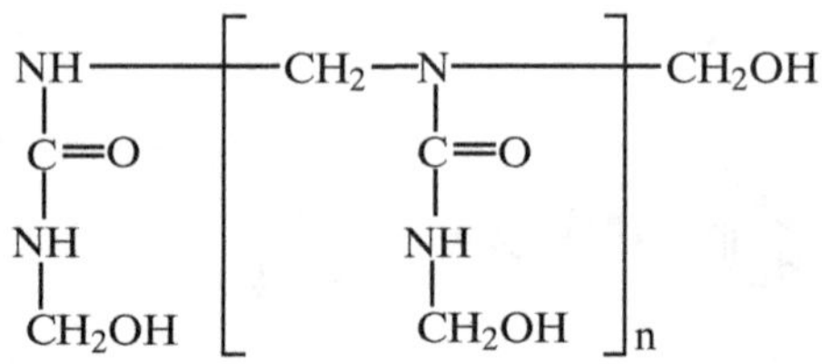

Further heating produces an insoluble and infusible product.

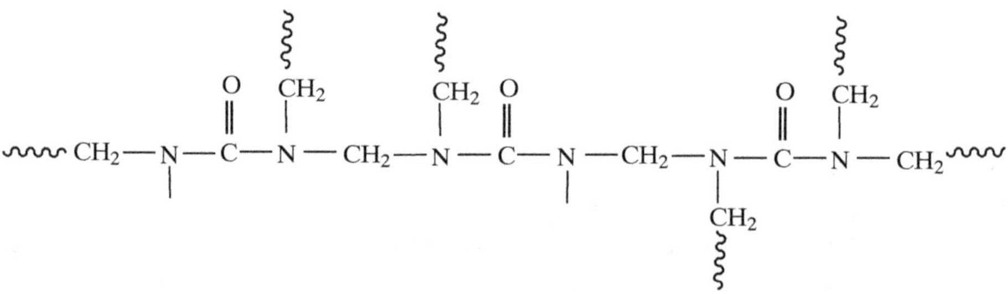

It is observed that in the curing of UF resin (urea: formaldehyde 1:2 and 1:1.8), in addition to intermolecular condensation of methylol and imine groups, intramolecular interactions of methylol groups also occur, which results in the formation of a urone ring in the polymer chain. Increase in the proportion of urea brings about more complete usage of the functional groups in curing and results in an increase in network density [64]. The complex structure of UF resins is due to the high functionality of urea and formaldehyde (4 and 2, respectively), which results in the formation of large sets of structure.

Reaction of melamine with formaldehyde at a temperature of 80 °C to 100 °C produces a mixture of water-soluble methylol amines. The methylol content in the mixture depends on the melamine formaldehyde ratio and reaction conditions. The derivatives may possess a maximum of six methylol groups per molecule.

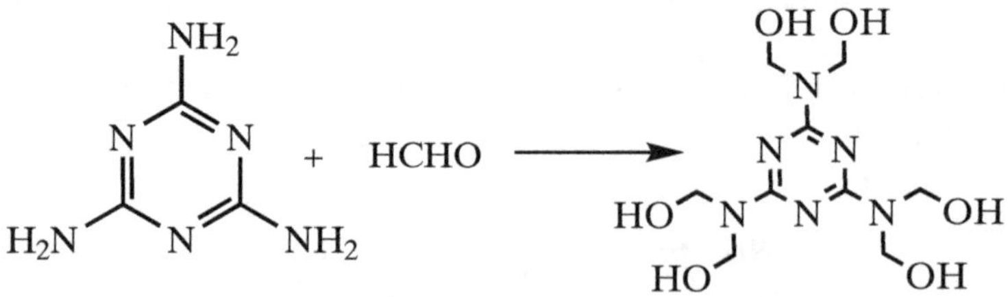

Condensation of these hydroxy methylol compounds results in the formation of oligomers with a methylene or dimethylene ether bridge. On further heating beyond a temperature of 100 °C, a network structure is formed because of the multifunctionality of melamine. The resinification is strongly dependent on pH. Side group activated species through protonation are the most reactive ones, whereas ring protonation results in deactivation [65]. MF resins that contain very low level of –NH and/or >NCH_2OH are functionally quite reactive in cocondensation reaction with polyol coreactants [66].

Polyurethane resins: Polyurethanes are formed by the reactions among polyhydroxy and polyisocyanate compounds.

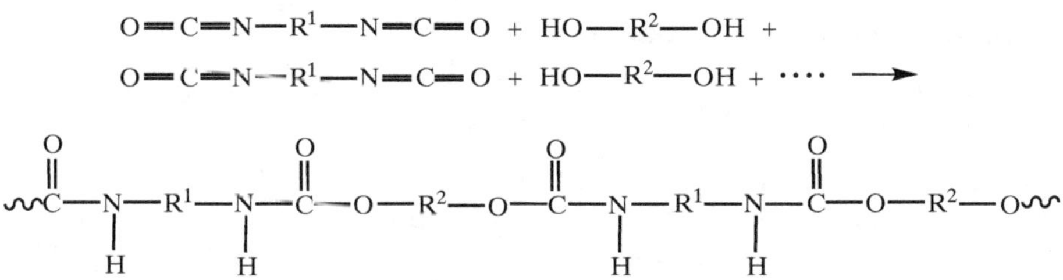

If the functionality of the –OH or –NCO bearing compounds is more than 2, the resultant polymer may be either branched or crosslinked. Crosslinking may also be achieved by secondary reactions. For example, reaction of isocyanate with water gives an amino group that immediately reacts with excess isocyanate to form a urea linkage. This reaction is accompanied by the evolution of carbon dioxide, which acts as a blowing agent. Further reaction of the urea group with the isocyanate leads to crosslinking via a biuret group.

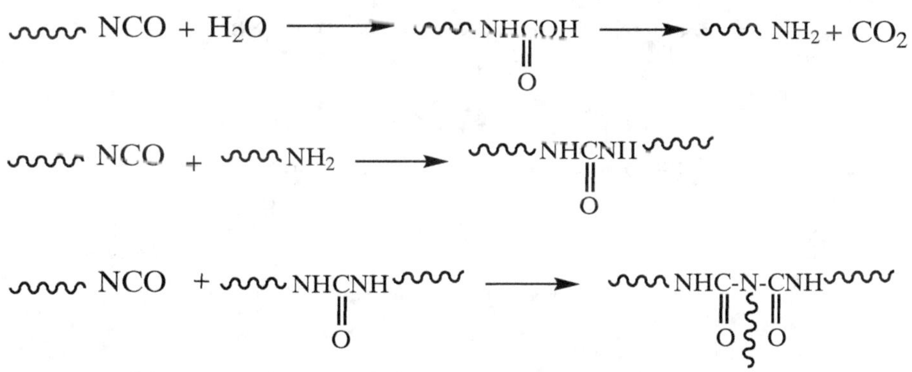

Because of the wide flexibility in the selection of polyol and polyisocyanate, polyurethanes have a wide versatility in their applications as foams, elastomers, adhesives, sealants, and fibers. A trifunctional polyol and a short diol and diisocyanate may react to form crosslinked PU elastomers by a one-shot reaction-injection-molding process. Such crosslinked PU elastomers contain both hard and soft blocks as follows:

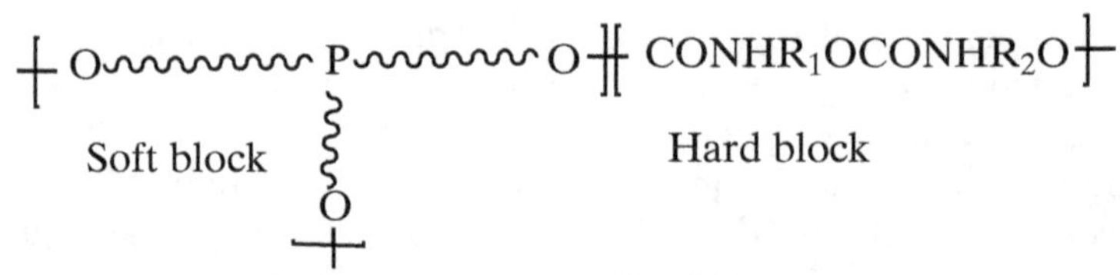

Diamines may also be used as short chain extender for PU:

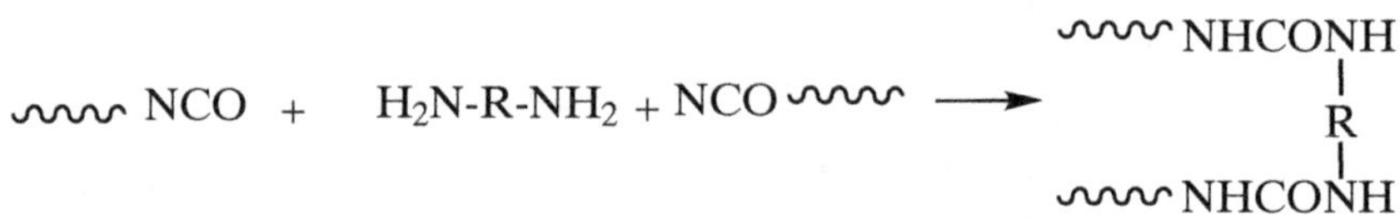

Currently, much research is being conducted on the crosslinking of polyurethane resins. Partially modified hyperbranched polyester based on bis-hydroxy ethyl propionic acid may be used as a crosslinker of polyurethane to yield a product of high temperature stability [67] (Scheme 2.11).

Hexamethoxy methyl melamine has been found to be an ideal crosslinker for phosphated polyurethane dispersion of nanoparticle size [68]. Such crosslinking improves the dynamic mechanical properties and solvent and chemical resistance of polyurethane dispersion. A hexanediol diacrylate-diazetidine complex may be used as a curing agent for aqueous PU dispersion containing

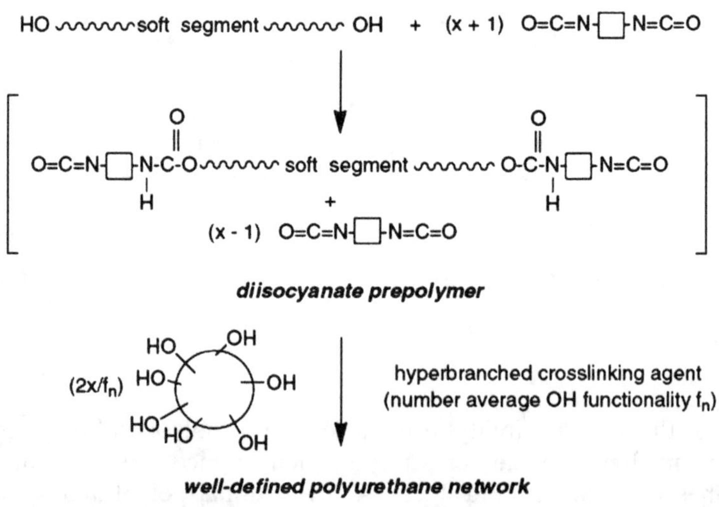

Scheme 2.11. Ideal reaction scheme for the synthesis of well-defined polyurethane networks based on hyperbranched polyester resin. Reproduced from *React Funct Polym*, **66**: 1462 (2006).

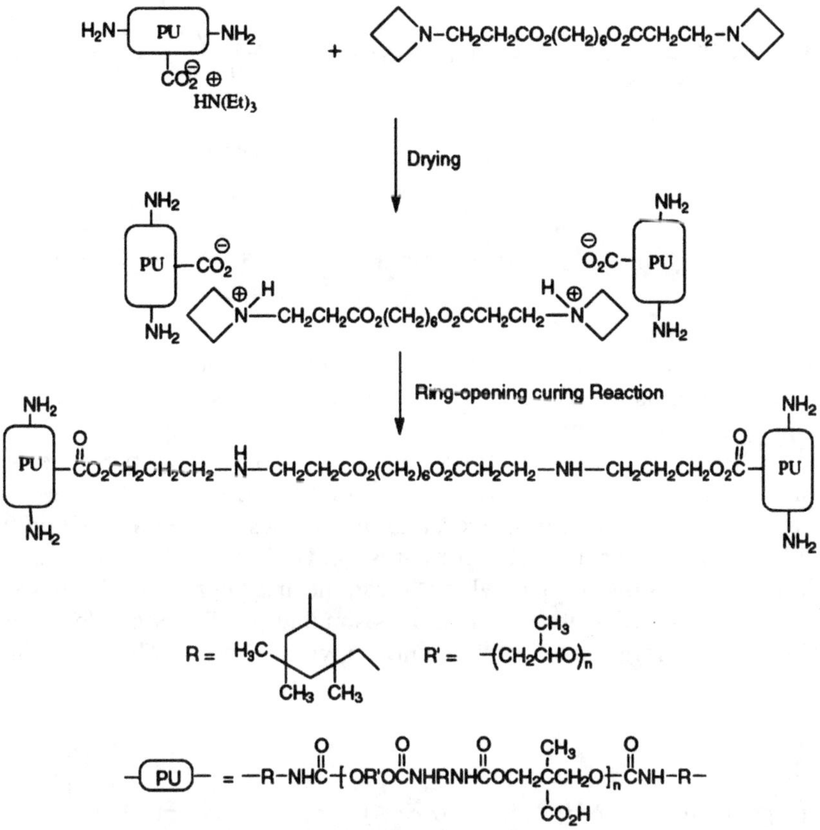

Scheme 2.12. Self-curing of aqueous-based polyurethane with a hexanedioldiacrylate-diazetidine complex. Reproduced from *React Funct Polym*, **67**: 299 (2007).

carboxyl group. There is a vast improvement in the gel content and mechanical properties of cured PU resin in comparison with the original one [69] (Scheme 2.12).

Epoxy resin: Epoxy resins are thermosetting epoxide polymers containing more than one 1,2 epoxy group per molecule. Epoxide resin formed by the reaction of bisphenol A and epichlorohydrin in the presence of sodium hydroxide is the first and till now the most important commercial resin and contains mainly low molecular weight diglycidyl ether of bisphenol A (DGEBPA) along with small quantities of high molecular weight polymers. High molecular weight epoxy resin may be prepared by proper adjustment of the ratio of bis phenol A and epichlorohydrin and reacting them in more strongly alkaline conditions as it favors the reaction of epoxide groups with bisphenol A.

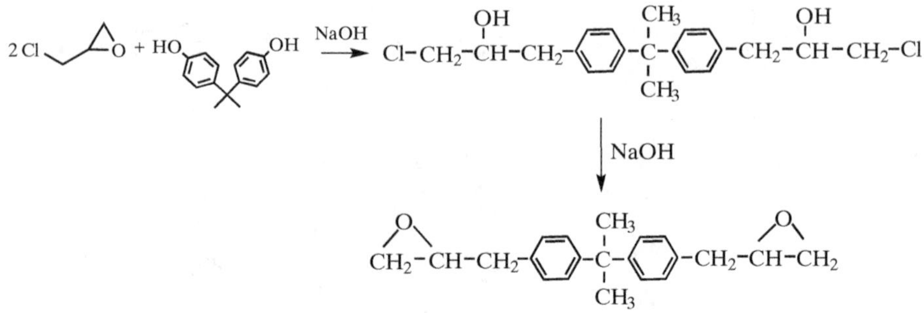

Epoxide or hydroxyl groups are the crosslinking sites for epoxy resins. The epoxy group is highly reactive because of the presence of ether linkage separated from it by methylene group. Hence any compound containing active hydrogen may readily attack the epoxy groups. Two types of curing systems are available for epoxy curing: polyfunctional crosslinking agents and catalytic systems. Amines, acids, and anhydrides [70–72] act as polyfunctional crosslinking agents of epoxy resins. Highly reactive amines such as diethylene triamine or triehylene tetramine may cure the glycidyl ether at room temperature, whereas when a reasonable pot life is needed dimethylaminopropyl amine or diethylaminopropyl are used. The crosslinking reaction is

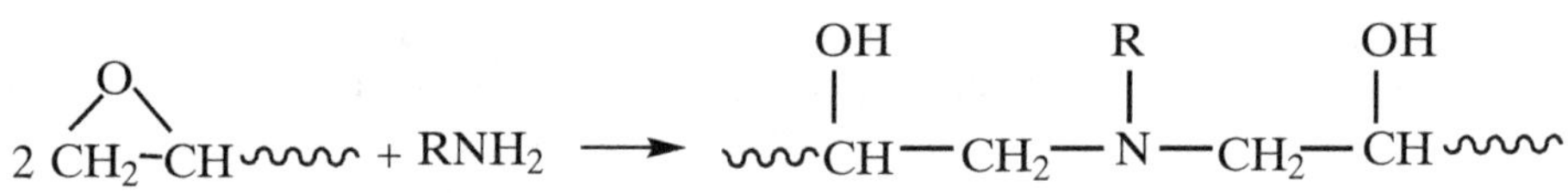

Anhydrides (phthalic anhydride, hexahydrophthalic anhydride, maleic anhydride, pyrometallic dianhydride, etc.) are preferred over acids, as acid-based crosslinking releases more water, leading to foaming of the product. In catalytic crosslinking an ionic complex is formed, which may open up a new epoxy group and generate another ion; thus the crosslinking proceeds.

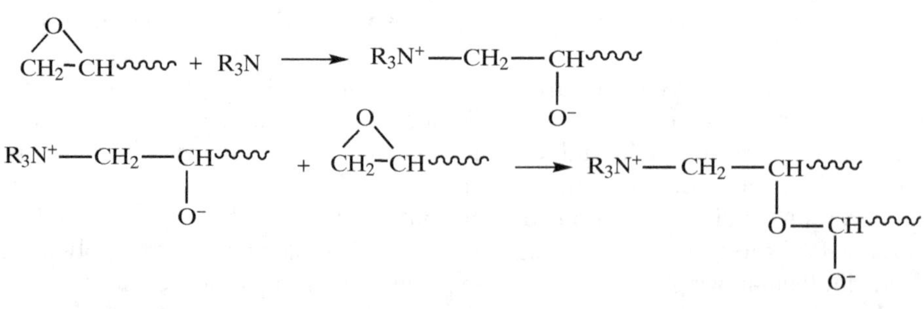

Polymers, containing –OH or –SH groups, may cure epoxy resin; therefore phenolic and amino resins [73] may be used as crosslinkers of epoxy

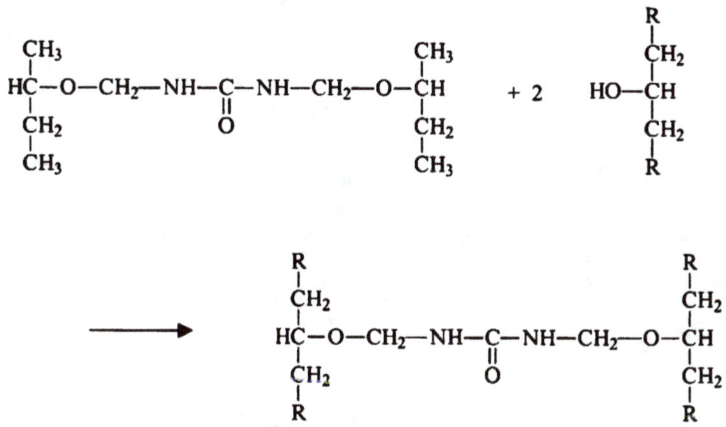

Scheme 2.13. Crosslinking reaction between urea formaldehyde and epoxy resin. Reproduced from *J Electron Spectrosc Relat Phenom*, **121**: 233 (2001).

(Scheme 2.13). Such crosslinkers enhance the flexibility of the cured resin and broaden the spectrum of applications.

(II) Interfacial Polycondensation A special classification of condensation polymerization is interfacial polycondensation (IP). In this process, two multifunctional reactive monomers dissolved in two mutually immiscible liquids may meet and react (copolymerized) by polycodensation reaction only at the interface of the solutions to obtain a high molecular weight polymer. A very fast polymerization occurs between the two very reactive monomers and the polymer precipitates quickly forming a thin dense film. Reactions between difunctional monomers result in linear polymers, but if the functionality of any one of the monomers is greater than 2, network polymers result. As the reaction continues, the thickness of the polymer film formed at the interface gradually increases, and this gradual enhancement of the thickness hinders the diffusion of the monomers. Because of the limited supply of reactants, the interfacial polymerization reactions are self-inhibiting, and thus an extremely thin film down to 50 nm thickness may be achieved [74]. However, a continuous removal of the polymeric film from the interface maintains the reaction. Interfacial polymerization is a well-known method for the preparation of a variety of polyamides, from multifunctional aliphatic and aromatic amines in water phase and acid chlorides such as trimesoyl chloride, isophthaloyl chloride, and terephthaloyl chloride in organic phase. Interfacial polymerization has become a well-established method for the preparation of thin films and the most significant commercial exploitation of this process is in the area of thin film composite reverse osmosis and nanofiltration membranes [75–77].

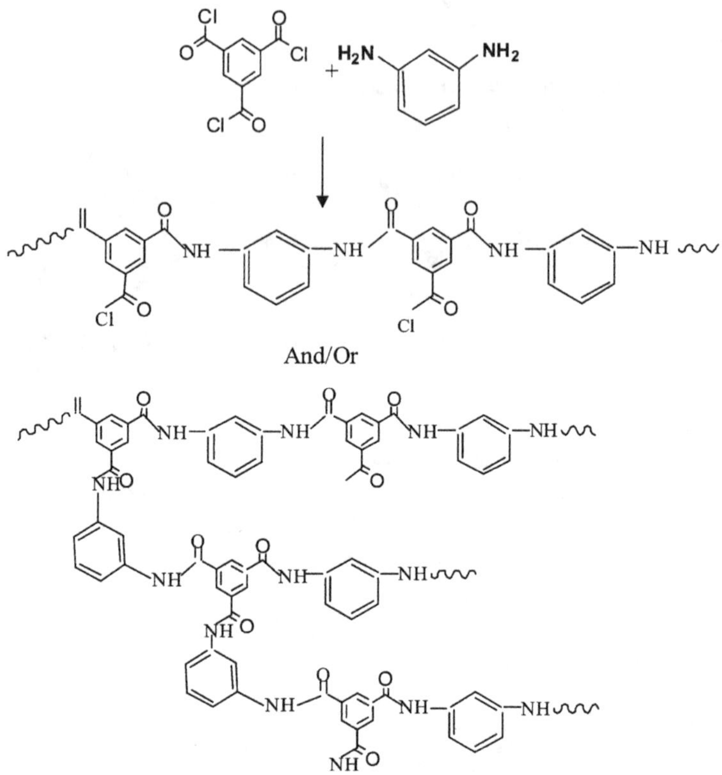

Scheme 2.14. Network formation through interfacial polymerization.

It has been established that the interfacial polymerization takes place in the organic phase [78]. The highly unfavorable partition coefficient of acid chloride limits its availability in the aqueous phase; hence the amine has to cross the water organic interface and diffuse through the already formed polyamide layer to make reactive contact with the acyl chloride on the organic solvent side of polyamide layer. The growth of the polyamide layer takes place on that side of the interface.

Reactions of diamine with trimesoyl chloride may lead either to linear chain formation, where the third acyl chloride group remains unreacted and may hydrolyze to the carboxylic acid group, or the third acyl chloride group may react with another diamine molecule to produce chain branching or a crosslinked network structure [79, 80]. The probable structures are given in Scheme 2.14.

Interfacial polymerization may be diffusion controlled or diffusion-reaction controlled. The polyamide formation from diacid chloride and diamine is diffusion-controlled polymerization [81], whereas polysulfonamide formation from disulfonyl chloride and diamine is a reaction and diffusion controlled IP [82].

The composition and morphology of the polymeric film formed at the interface depends on several variables such as the concentration and partition coefficient of the reactants, reactivity ratios where blends of reactants are employed, solubility of the nascent polymer in the solvent phase as it is formed, the overall kinetics and diffusion rates of the reactants, presence of byproducts (such as hydrogen chloride in the case of reaction of amine with acid chloride), hydrolysis or potentially competitive byreactions, crosslinking reactions and post reaction, or treatment of the resulting interfacial films [83]. The reaction rate constant and diffusion coefficient strongly affect the thin film thickness and growth rate. The thickness rapidly increases with time in the early stage of polymerization and then levels off with time. Hence, lengthening the polymerization time in an attempt to increase the film thickness is ineffective. Increase in the concentration of the monomer in the aqueous phase increases the growth rate and also the maximum thickness of the film, whereas increase in the concentration of the monomer in the organic phase increases only the growth rate but not the maximum thickness. Therefore, increase in the monomer concentration is more effective than lengthening the polymerization time to obtain a polymeric film of desired thickness at the interface [84]. As in the IP process, a thin film is formed instantaneously at the interface between the two phases; hence the process may lack strict requirement of reactant purity. The process permits the minimization of defects formed inside the thin film due to two effects: self-sealing and self-termination as a result of the growing resistance of the film [85].

Interpenetrating Polymer Networks An interpenetrating polymer network (IPN) is an intimate combination of two polymers both in network form, at least one of which is synthesized and crosslinked in the immediate presence of the other [86]. The basic difference between a chemical polyblend and an IPN is that there are no induced covalent bonds between the two polymers; i.e., if an IPN is an intimate combination of network of monomer I and monomer II, then during its formation monomer I has reacted only with the molecules of monomer I, as does II.

Classification of IPNs by Chemistry Simultaneous IPN: IPNs formed by simultaneous polymerization of two monomers along with their crosslinkers and initiators by noninterfering modes.

Sequential IPN: IPNs in which polymer network I is synthesized first and then swollen in monomer II, its crosslinker, and initiator, polymerized II in situ.

Classification of IPN by Structure Full IPN: Any material containing two or polymers in which there are no induced crosslinks between the individual polymers.

Semi-IPN: Sequential IPN in which one of the polymers is crosslinked and the other is linear. If polymer I is crosslinked and polymer II is linear, it is known as semi-I-IPN and the reverse is known as semi-II-IPN.

Pseudo IPN: Simultaneous IPN in which one of the polymers is in network form (i.e., crosslinked) and the other is linear.

Latex IPN: It is formed by the mixture of two kinds of latex particles, followed by film formation and crosslinking of both the polymers.

Thermoplastic IPN: It is a two-polymer IPN in which the individual polymers are thermoplastics. One may be a block copolymer like SBS rubber, and the other is typically a semicrystalline or glassy polymer. Polymers may contain physical crosslinks arising from ionic groups, crystallinity, or glassy domains.

Millar IPN: The IPN in which two polymers are chemically identical.

Gradient IPN: An IPN of nonuniform macroscopic composition and crosslink density, usually by nonequilibrium swelling in monomer II, and polymerizing rapidly before diffusion equilibrium takes place.

Semi and full IPNs are represented in Figure 2.3.

Advantages of IPNs One of the most important routes for the modification of the properties of polymers to meet specific needs is the physical/chemical combination of two or more structurally different polymers. The properties of such multicomponent polymeric systems depend on the properties of individual polymers and the ways in which they are combined. Generally, mixing of polymers results in a multiphase morphology because of their thermodynamic incompatibility. This is due to relatively small gain in entropy upon the mixing of polymers, as long chain length imparts contiguity restrictions. If the mixing is carried out at a low molecular weight level, and subsequent polymerization and crosslinking are carried out simultaneously, the phase separation may be kinetically controlled by permanent interlocking of the entangled polymer chains; i.e., covalent bonds would have to be broken to disentangle the individual polymer chains. Hence, the balance between thermodynamic and kinetic parameters controls the degree of mixing in IPNs. Thus, by introducing sub-

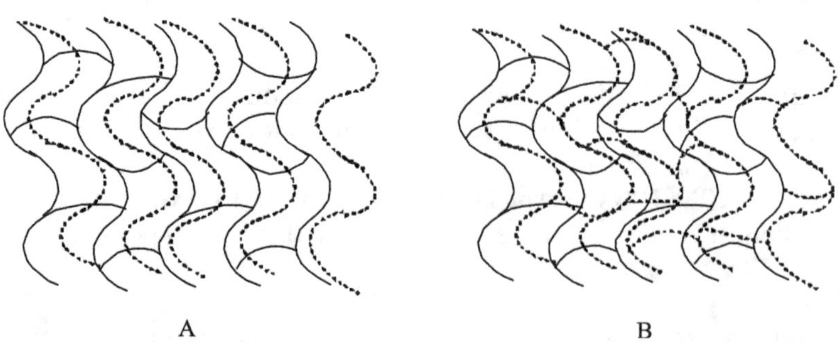

Figure 2.3. A. Semi–interpenetrating polymer network. B. Full interpenetrating polymer network.

stantial crosslinking of both the components before their significant diffusion, phase separation may be largely prevented and a high degree of mixing may be achieved in IPNs. The degree of phase separation, which depends mainly on the miscibility of the polymers, varies widely for different IPNs. For highly incompatible polymers, thermodynamic forces leading to phase separation are so powerful that phase separation occurs substantially before the kinetic ramification can prevent it. Hence, in such case marginal gain in phase mixing can be obtained. Phase separation may be avoided to a great extent if the polymers are compatible. However, to achieve complete phase mixing, complete miscibility, which is almost impossible, is not essential in IPNs, as permanent entanglement prohibits phase separation [87]. With moderate compatibility, intermediate and complex phase behavior may be observed. IPNs with dispersed phase domains have been reported, ranging from a few micrometers (incompatible) to a few hundred nanometers (intermediate) to those without a resolvable domain structure (complete mixing) [87]. IPN synthesis is highly significant and interesting because it is the only way of intimately blending two crosslinked polymers.

Synthesis and Property Achievements Simultaneous interpenetrating network (SIN): Among the different processes of IPN synthesis, the SIN technique is generally the best one to achieve a high degree of intermixing, as the compatibility of monomeric mixture is much higher than that of a polymeric mixture. One primary condition for SIN is that the choice of monomer is restricted to compounds, which do not react together in the polymerization conditions. Hence, formation of both polymers by radical polymerization is not possible. Generally, one network is formed by radical polymerization, and the other one is obtained by polyaddition reaction. The gel point of the two systems involved may be reached at the same time or may be shifted in time. In an extreme case, the formation of the first network is completed before the second monomer starts to react. If both systems reach gelation simultaneously, a very fine state of phase dispersion may be obtained. If one monomer system reacts faster, a larger domain size is obtained and the mechanical properties of the SIN improve. The three significant events that occur during the synthesis of SINs are gelation of polymer I, gelation of polymer II, and phase separation of polymer I from polymer II. These three events may occur in any time order, and the number of independent ways of carrying out such a polymerization is 3!, i.e., 6. Each is expected to produce different morphologies and as a consequence a specific set of physical and mechanical properties [88].

Studies were carried out on polyurethane-polymethyl methacrylate SIN [89] to determine how morphology develops from the initial mixture to the final IPN. This was done by starting the polymerization of MMA at various degrees of polymerization of PU, especially in the PU pre-gel region. When the time lag between the onset of the two reactions is short, macroscopic phase separation occurs as the PU network is incompletely formed. However, when the time lag increases, PMMA forms a more continuous network, which limits

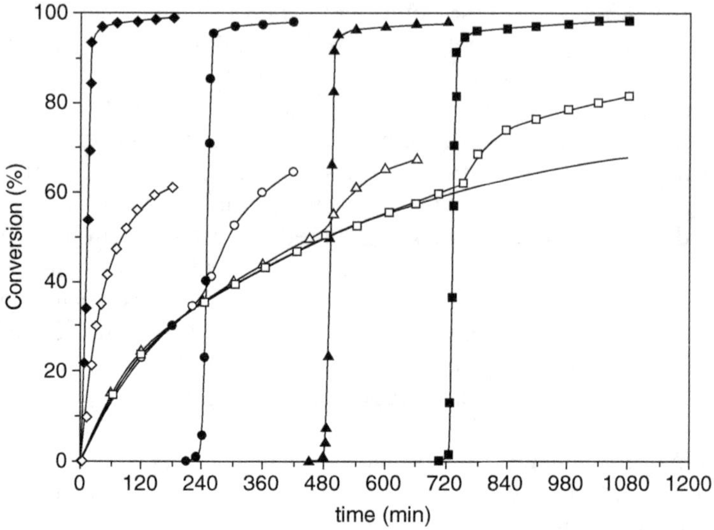

Figure 2.4. Kinetics of in situ delayed PU/PMMA IPN formation. (◆,◇) Simultaneous onset of both reactions; (●,○) initiation of free-radical polymerization delayed 4 h; (▲,△) time lag of 8 h; (■,□) time lag of 12 h. The full symbols refer to MMA polymerization and the open symbols to PU. Reproduced from *Polymer*, **46**: 3318 (2005).

the growth of phase separation to a close environment. Hence, by simply varying the lag time between the onset of the two reactions, quite different morphologies and, therefore, properties were obtained. Kinetics of PU-PMMA IPN formation at different onset conditions is shown in Figure 2.4.

Sequential interpenetrating network (Seq IPN): In sequential IPNs, the formation of undesired chemical bonds between separate networks is avoided because of the sequential synthesis. This broadens the probability of obtaining IPN from a wide range of starting materials. IPNs prepared by this technique are generally considered true IPNs if both the polymers are crosslinked. To obtain such an IPN, one polyfunctional monomer is first polymerized and crosslinked. The crosslinked polymer is then swollen in another monomer, which was previously mixed with its initiator and crosslinker. The monomer was then crosslinked inside the existing polymer network. Among the factors such as miscibility of the polymer components, composition, crosslink density, polymerization sequence, polymerization, and phase separation kinetics—which affect the phase morphology of sequential IPNs—the polymerization sequence is specifically important. If polymers I and II are immiscible, and the crosslink density of I is small, i.e., the number of monomer units between crosslinks is high, then during the polymerization and network formation of II, the growing chains of monomer II will push apart the already existing chains of I and result in phase-separated IPN. However, the positions of the chains of monomer I will be hardly affected by the growth of network II if network density of I is high. In such cases, network II will form by interpenetrating the existing network I, and a homogeneous IPN will result through forced compatibilization of both the polymers [90]. Thorough study on the sequential IPN of immiscible polymers [e.g., polymethyl acrylate (PMA) and polymethyl methacrylate (PMMA)] show phase separation at low crosslink densities detected by the occurrence of two clearly differentiated main dynamic mechanical relaxation processes corresponding to the two components. However, increasing the crosslink density [91] may compatibilize these two components. PMA/PMMA IPNs crosslinked with 10% diethylene glycol dimethacrylate shows a single dynamic mechanical relaxation process. It is shown that the key factor controlling the miscibility of sequential IPN is the crosslink density of the network that is polymerized first. In an IPN, there may be some accidental grafts between the networks, but the deliberately introduced crosslinks should always outnumber the accidental grafts. The difference between the experimental and calculated values of the refractive index and the swelling degree of in situ sequential polyurethane (PU)-polystyrene (PS) IPNs shows the presence of additional physical crosslinks for such system with different combinations of PU and PS [92]. The maximum entanglement was achieved for a 50:50 IPN composition. Sequential method results to the higher miscibility of the components compare to the simultaneous ones. Comparison of the reaction kinetics of sequential and simultaneous PU-PS IPNs shows that although for simultaneous IPNs both reactions start at same time, the formation of PU and PS are not simultaneous. The PU formation is found to be faster than the polymerization of styrene but gelation, which takes place at 70% conversion, occurs too late to prevent the polymerization-induced phase separation to extend. On the other hand, in the in situ sequential process, the formation of a PU network is complete and the polymerization of styrene can only proceed in the immediate vicinity of the preformed PU network due to topological hindrance. Therefore, higher miscibility of PU and PS results in a sequential technique [93]. The polymerization kinetics of sequential and simultaneous PU-PS full IPNs is shown in Figure 2.5.

An interesting range of properties may be achieved when an elastomer and plastomer are blended in different proportions for the preparation of an IPN [94]. A natural rubber polystyrene-based sequential IPN where the crosslinked NR phase was prepared first shows an intimate mixing of the two component phases, and a finer and more uniform distribution of PS in the NR phase was obtained [95]. The system showed a cocontinuous morphology above 50% PS, which results in an enhancement of different properties.

Elastomer Crosslinking Elastomers are made of long chain molecules with molecular weight vary in the range of 1.5–5 × 10^5. Uncrosslinked elastomers have limited commercial value because of their easy deformation under load and rapid dissolution in different solvents. To achieve the desired properties in the final products, it is necessary to crosslink the elastomers known as vulcanization.

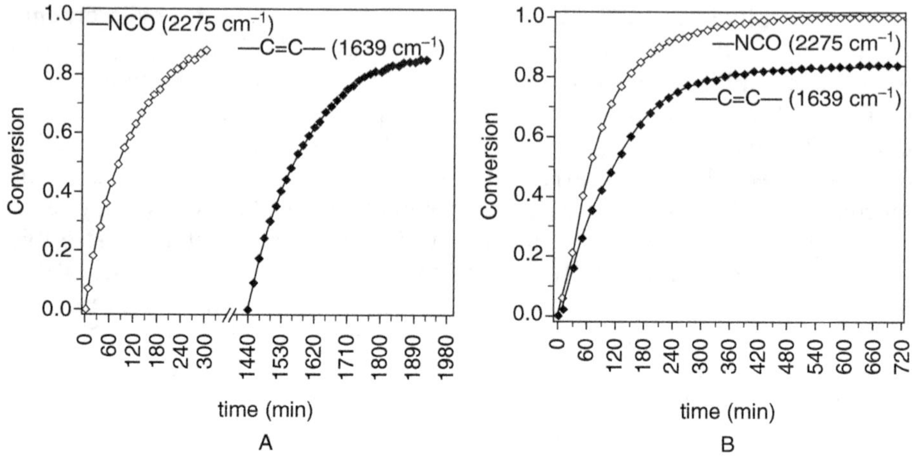

Figure 2.5. Kinetics of in situ (◇) PU; (◆) -PS full IPN formation. A. Sequential IPN. B. Simultaneous IPN. Reproduced from *Macromol Symp*, **216**: 179 (2004).

The term "vulcanization" refers to a chemical process in which the uncured long chain rubber molecules are tied together into a three-dimensional elastic network by the insertion of crosslinks [96]. Hence, by vulcanization a raw rubber is transformed from its thermoplastic state to an elastic rubbery or hard ebonitelike state. The three-dimensional network structure imparts strength, rigidity and elastiticity, improves solvent properties, and enhances the resistance to deformation of elastomers under heat and cold. Initially, vulcanization referred to the crosslinking of elastomer chains with the aid of sulfur but in due course different types of new crosslinking agents have been developed. The selection of crosslinking agents depends upon the type of elastomers and performance properties desired.

Curing of the Rubber Stock and Stages of Vulcanization The curing and vulcanization of rubber are synonymous. By plotting the change in a specific property such as tensile strength or modulus with the length of curing, the cure curve is obtained. The rate of vulcanization is an extremely important parameter in rubber industries as it controls the optimum cure time, which in turn determines the total batch cycles that may run per day in an industry. However, the rate of cure may be monitored by suitable choices of crosslinking agents and conditions. Vulcanization may occur in three different stages in an elastomer, which may be presented graphically by the generation of tensile modulus with time (Figure 2.6).

The stages are the induction period, i.e., the onset of vulcanization, a curing and crosslinking stage that consists of under and optimum vulcanization steps and a reversion or overcure stage [97]. In the induction period, i.e., in the prevulcanization stage, gradual thickening of the elastomer starts with no

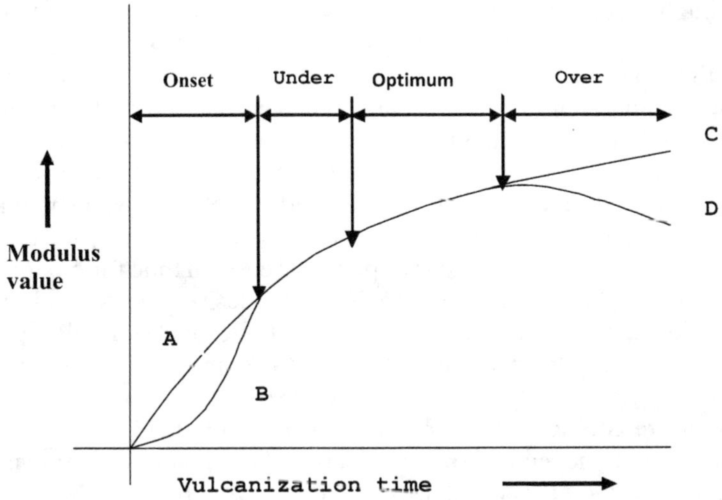

Figure 2.6. Stages of vulcanization. A. Compounds with rapid onset of vulcanization. B. Compounds with delayed onset of vulcanization. C. Compounds with increasing modulus. D. Compounds with reversion.

measurable crosslinking. During this period, rubber compounds still flow uninhibited in the mold. This duration determines the safety of a stock against scorching (premature vulcanization) during the various processing steps, which precedes the final cure. The choice of suitable crosslinking chemicals may shorten or lengthen the prevulcanization step. In most cases, the rapid vulcanization initiation is undesirable as it hampers the safe processing of the elastomer, but in some cases such as open curing, very fast vulcanization initiation is needed. The onset of vulcanization is followed by a crosslinking reaction at a rate dependent on the temperature and curative composition. This is the prime step of vulcanization, when most of the desired technological properties of the vulcanizate generate. All the properties may not attain their best values at a specific time, and there is no universal method for the selection of optimum cure time. Hence, for a specific formulation the time where the best combination of all the desired properties is achieved for a particular application is the optimum cure time. Further heating beyond the full cured state of the elastomer produces an overcured compound, which is evidenced either by continued stiffening reflected by a rise in modulus and a fall in elongation, or by reversion reflected by a fall in hardness, modulus and tensile strength, and a rise in elongation. Overcure is unavoidable in the vulcanization of most of the products. The surface of a thick product becomes overcured until its interior reaches a full cured state. Any curing shorter than the optimum is known as undercure.

Scorching is the term used for premature vulcanization. It takes place due to improper selection of compounding and processing time and temperature.

For thorough mixing of the compounding ingredients with the rubber stock, it is necessary to subject the rubber stock to a very high temperature for a definite time period in the mixing mill. If vulcanization starts during this period, scorching occurs. Once scorching has occurred, the rubber stock cannot be processed further and its molding to its final shape is impossible. Hence, it is desirable that vulcanization should not start until the processing is complete, i.e., it should start when required, accelerate when needed, and must stop at the right time. Scorch resistance is the time elapsed before the vulcanization starts. Scorch resistance provides enough time for mixing, storing, and molding of the rubber mixture into the desired shape and size. This can be achieved by the incorporation of a delayed action accelerator, which defers the onset of vulcanization and makes the compounding and processing operation safe.

Temperature coefficient: The rate of vulcanization is highly dependent on temperature. With the increase in temperature, the vulcanization rate increases. The ratio of the rate of reaction at two temperatures 18 °F (10 °C) apart is the temperature coefficient of the rate of that reaction. The temperature coefficient of vulcanization from 200 °F to 400 °F is ~2 [98].

Vulcanization Modes and Mechanisms The choice of the vulcanization mode depends primarily upon the nature of the elastomer to be crosslinked. Most of the elastomers are unsaturated in their nature, and for such polymers the predominant mode of crosslinking is sulfur vulcanization. Such a system consists of pure sulfur or a combination of sulfur and metal oxide (usually zinc oxide) and/or one or more organic promoters, known as accelerators, along with activators or retarders, and results in the formation of mono- and polysulphide linkages between the polymer chains. The crosslink density depends upon the time and temperature of cure and the concentration of curatives as well as of cure sites. Sulfur vulcanization is not suitable for crosslinking of saturated elastomers, which may be crosslinked by peroxides or high-energy radiation.

Vulcanization with Sulfur Sulfur is the prime material for crosslinking the elastomers. The dose varies between 0.25 and 5.0 parts per 100 parts of rubber (phr) for soft rubber products, whereas for hard rubber compounds like ebonite, the dose remains in between 25.0 and 40.0 phr. The vulcanization temperature may vary within 150 °C ± 30 °C. In addition to the polysulphide linkages results in intermolecular crosslinking, sulfur vulcanization leads to many undesirable side reactions such as formation of intramolecular cyclic structures that do not contribute to chain interlocking. Most of the evidence favors the free-radical mechanism for sulfur vulcanization. Thermally induced sulfur vulcanization is autocatalytic in nature, and the polysulphide products formed in the reaction splits to active thiyl (RSH•) and polysulfenyl (RS_X•) radicals, which act as catalysts. The reaction steps are shown in Scheme 2.15 [99, 100].

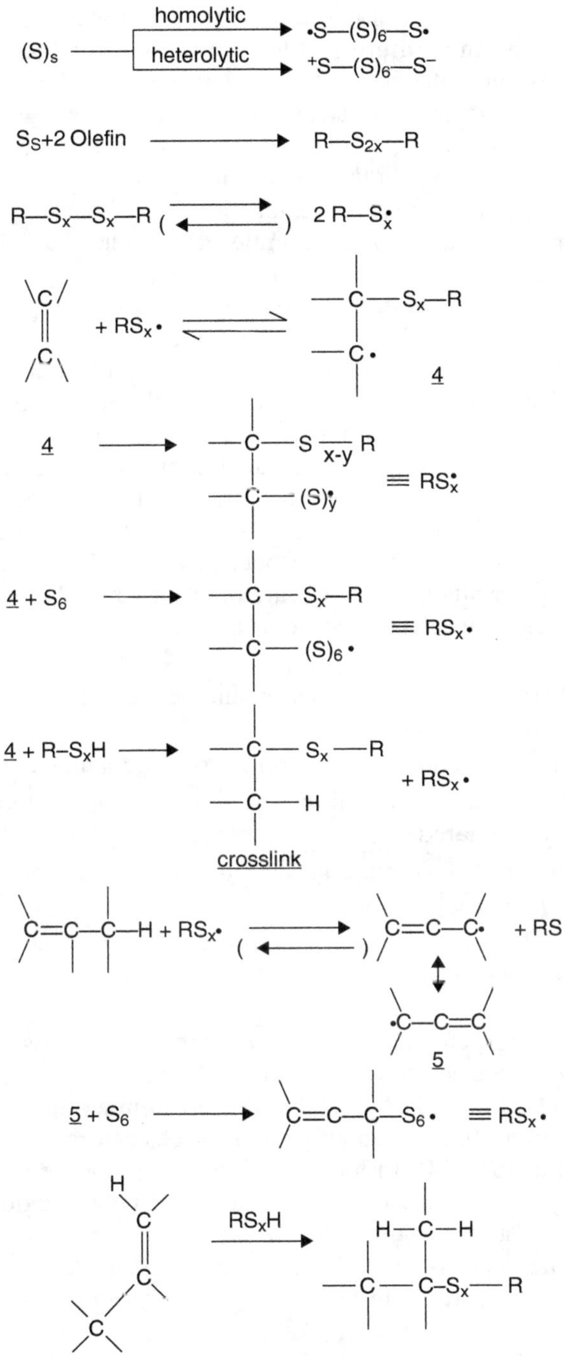

Scheme 2.15. Free-radical mechanism of sulfur vulcanization. Reproduced from *Prog Polym Sci*, **22**: 475 (1997).

However, the use of sulfur alone is an inefficient process, as it needs a higher dose of sulfur, high temperature and long duration of vulcanization, and the predominant existence of polysulphide linkages rather than mono- or disulphide, which results in unsatisfactory strength and poor aging resistance of the vulcanizates. Sometimes, instead of elemental sulfur, sulfur donors like dithio-di-morpholin, N-oxy diethylene dithiocarbamyl -N′ oxydiethylene sulphenamide, etc., may be used for vulcanization. They liberate sulfur at the vulcanization temperature and sometimes may act as vulcanization accelerators.

Modified Sulfur Vulcanization Systems To overcome the disadvantages of elemental sulfur vulcanization, a modified sulfur vulcanization system that consists of sulfur, accelerator, and activator has been introduced. The crosslink density of a vulcanized rubber is dependent on the nature of the elastomer, accelerator type and composition, time and temperature ratio, and the accelerator sulfur ratio [101]. Combined vulcanization systems for elastomers may be classified as conventional, semi-efficient, or efficient depending upon the accelerator–sulfur ratio. It is 0.1–0.6, 0.7–2.5, and 2.5–12.0 for conventional, semi-efficient, and efficient systems, respectively [102]. Enhancement of accelerator-sulfur ratio results in a higher rate of crosslinking [101].

The advantages of accelerator systems are

1. Accelerate the crosslinking reaction and hence reduce the cure time and make the process economic.
2. Reduce the cure temperature and improve the aging stability.
3. Reduce the amount of organic sulfur and hence unwanted side reactions are mostly eliminated.
4. Combination of two or more accelerators results in synergism and properties improve [103].
5. A broad plateau may be achieved in the stress heating time curve, indicating processing safety.

Depending on their chemical nature, accelerators are classified as acidic or basic. Important acidic accelerators are as follows:

Thiazole accelerators: They have highest economic importance in rubber compounding and are known as all-purpose accelerators. Mostly used are 2-mercapto-benzthiazole (MBT) and dibenzthiazyl disulphide (MBTS). These accelerators impart safe processing with a broad vulcanization plateau and excellent aging resistance of the vulcanizates. When combined with secondary basic accelerators such as guanidine, thiourea results in a higher crosslink density of the vulcanizates with an excellent synergism in product properties [104].

Benzthiazole sulfenamide accelerators: Cyclohexyl-, tert-butyl, and dicyclohexyl-benzthiazole sulfenamide are examples. These accelerators impart a great processing safety as they start the vulcanization in retarded fashion compared to MBT or MBTS. For molded articles they provide a long flow time but a very rapid vulcanization finish. The addition of a secondary accelerator may synergistically affect the vulcanization speed [105], and a favorable combination of flow time and cure time may be achieved.

Dithiocarbamate accelerators: Zinc, sodium or other metal dithiocarbamate are the examples. These are ultra-accelerators and to maintain processing safety, they are used in combination with other slow-acting accelerators. They may be used as sole accelerators in latex compounding [106], where the processing takes place at room temperature or in open curing. Proper combinations with secondary accelerators allow their use even for transfer or injection molding.

Xanthate accelerators: They are the derivatives of xanthic acid and are the fastest-acting accelerator. Hence they are used only in special cases of solid rubber. As they have a very short vulcanization plateau, a low cure temperature of 80 °C–110 °C is applied for them [107].

Thiuram accelerators: Tetramethyl thiurammonosulfide (TMTM), tetraethyl thiuramdisulfide (TETD), dimethyl diphenylthiuram disulphide are the prime examples. Thiurams can be used for a wide range of sulfur accelerator ratio [108]. Combination of thiurams as a secondary accelerator with rapid accelerators such as xanthate and dithiocarbamate, imparts a retarding effect without affecting the crosslink density of the vulcanizate, whereas the curing rate is enhanced with an accelerator such as thiazole [109].

The important basic accelerators are as follows:

Guanidine accelerators: Mostly used are diphenylguanidine (DPG) and di-o-tolyl-guanidine (DOTG). These are secondary accelerators as they result in unfavorable vulcanization plateaus, have long scorch and vulcanization times, and products have poor aging resistance. In combination with primary accelerators such as mercapto, they show excellent synergistic action [106].

Aldehyde amine accelerators: These are the condensation products of aldehydes and amines. The principally used amine is aniline. They widely vary in their activities and are relatively less important accelerators. Butyraldehydeaniline is a very rapid accelerator and leads to products of high crosslink density [97].

Amine accelerators: These are secondary accelerators and rarely used alone. Cyclohexyl amine and dibutylamine are mostly used for the activation of dithiocarbamate accelerators.

Activators: Zinc oxide has a pronounced effect on sulfur vulcanization. It improves the action of many accelerators and acts as an activator. Organic fatty acids (such as stearic acid) improve the action of ZnO. Hence, in sulfur vulcanization, stearic acid accelerates the action of ZnO, which accelerates the actions of accelerators, which accelerates the action of sulfur. The addition of ZnO enhances vulcanization efficiency and vulcanizate properties and reduces the vulcanization time. Introduction of nano-ZnO [110] results in lesser consumption of ZnO but higher improvement in the properties.

accelerator + ZnO

↓

Zinc-accelerator complex (I)

↓ S_8

Acc-S_x-Zn-S_x-Acc or polysulfidic accelerator (II)

↓ rubber

rubber-S_y-Acc (III)

↓

rubber-S_y-rubber (IV)

↓

rubber-S_z-rubber (V)

Scheme 2.16. General scheme for sulfur vulcanization. Reproduced from *J Appl Polym Sci*, **95**: 1388 (2005).

Mode of reaction: The mechanism of vulcanization is fairly complex. It is mainly dependent on the components present, temperature, and duration of the process. It has been proposed that the first step is the formation of an active accelerator complex (I) by the combination of accelerator and activator [111], which on subsequent reaction with sulfur produces an active sulfurating species (II). Zn^{++} ions play an important role in the formation of such active species. This reactive species reacts with the allylic hydrogen atom of rubber unsaturations to form a rubber bound intermediate, designated as a crosslink precursor (III). Reaction of such two intermediate moieties results in polysulfidic crosslinks (IV). The initial crosslinks may eventually desulfurate to form shorter crosslinks (V) or degrade to cyclic sulphides. The probable reactions are given in Scheme 2.16.

Vulcanization Without Sulfur

PEROXIDE CROSSLINKING Besides sulfur and accelerators, peroxides are also widely accepted today in rubber vulcanization. They are specifically used for rubber products requiring high thermal resistance. Peroxides are mostly used for curing saturated elastomers (i.e., ethylene propylene rubber, silicone rubber, chlorinated polyethylene, ethylene vinylacetate copolymer, etc.), but may also be applied for crosslinking of unsaturated elastomers (i.e., natural rubber, styrene butadiene rubber, nitrile rubber, etc.). The degree of crosslinking in peroxide vulcanization depends upon the radical yield from respective peroxide as well as the reactivity of the rubber. Peroxide vulcanization involves three different stages [112, 113]. These three stages are homolytic cleavage of peroxide to form two alkoxy radicals, abstraction of the H- atom from a polymer chain by the alkoxy radical, and coupling of two radicals on adjacent polymer chains to form a C–C bond. Hence, in peroxide vulcanization neither the peroxide nor the byproducts from the vulcanization process are part of the crosslinks, and the chains are interlinked by C–C bonds. The reaction mechanism is as follows:

$$\mathrm{ROOR} \xrightarrow{\text{Heat or light}} 2\mathrm{RO}\bullet$$

$$\mathrm{RO}^{\bullet} + \mathrm{PH} \longrightarrow \mathrm{ROH} + \mathrm{P}\bullet$$

$$2\mathrm{P}\bullet \longrightarrow \mathrm{P-P}$$

For unsaturated elastomers, the primary radical formed by the decomposition of peroxide abstracts the α hydrogen to the double bond [114, 115]. In peroxide vulcanization it is possible to apply a high vulcanization temperature without reversion. It is a rapid crosslinking process and the products possess good electrical and thermal properties. However, peroxides are expensive, and the products possess low tensile, tear strength, and poor abrasion resistance. Peroxides may also react with other compounding ingredients such as antioxidants, plasticizers, or resins, etc. Important commercial peroxides are diacetyl peroxide, dibenzoyl peroxide, dicumylperoxide, di-tert-butyl peroxidetc [116, 117].

RADIATION CROSSLINKING Crosslinking of elastomers may also be carried out by high-energy radiation such as X-ray, proton, electron, and neutron beams. Such crosslinking does not involve use of any curative. Irradiation generates a high concentration of free radicals by dehydrogenation in the rubber molecules, which combine to form C–C crosslinks as in the case of peroxide crosslinling. It is a physically induced chemical reaction, which is easier and preferable for continuous curing technologies and is used for many applications in the rubber industry [100]. Halogenated compounds, nitrous oxide, sulfur monochloride, and bases such as amine and ammonia act as sensitizers in radiation crosslinking by reducing the required radiation dose and time [118].

Thermal curing involves a long cure time, high-energy consumption, evolution of volatile toxic byproducts, and creates residual stresses in the material, whereas electron beam curing of rubber is a fast nonthermal process that utilizes highly energetic electrons at controlled doses for curing of variety of components for aerospace, automotive, and consumer applications [119]. Apart from crosslinking, electron beam radiation may also result in oxidation and degradation of a network structure [120]. The crosslinking efficiency of a rubber under an electron beam depends on its molecular structure. In photocrosslinking, the rubber absorbs light (near the ultraviolet and/or visible regions) whose wavelength is identical to the absorption band of photosensitive groups present in the rubber molecules. The absorbed energy usually generates radicals, which can initiate the crosslinking reaction [100, 121]. Photosensitive groups include cinnamoyl, cinnamylidene, acryloyl, diazo and dithiocarbamate, and double bonds.

Apart from the modes described above, unsaturated elastomers such as natural rubber, styrene butadiene rubber, or nitrile rubber may also be crosslinked by phenol formaldehyde resin [122, 123]. Ethylene thiourea has been established as a suitable crosslinker for specialty rubbers such as polychloroprene and epichlorohydrin rubbers [124, 125]. Fluororubbers may be crosslinked by polyamines, polyols, etc.

In Supercritical CO_2 It would be an incomplete discussion without mentioning the reactions in supercritical CO_2. The handling of common solvents is simple and the reaction process is smooth, but sometimes they are not environmentally friendly. Moreover, solvent removal from the final product creates additional wastes. To avoid these, supercritical CO_2 is the alternative choice. As supercritical fluid technology in recent years has made extraordinary progress in terms of commercial applications, its value in polymeric processes is obvious.

CO_2 is rather nontoxic, nonflammable, inexpensive, and volatile, as well as readily available. It can be easily separated from other components and has a readily accessible critical point (T_c 31.1 °C and P_c 7.38 MPa). In addition, for both Sc and liquid CO_2, the dielectric constant varies from 1.2 to 1.5. Low viscosity and surface tension makes the mass transfer speed is much faster than any other conventional liquid. The density of sc CO_2, and thus its solvent strength, is continuously tunable from gaslike to liquidlike by changing temperature or pressure [126]. All these advantages make the CO_2 an excellent reaction media for polymerization.

Features:

- Green solvent
- Good soluble and swelling agent
- Tunable strength depending upon pressure and temperature

The solubility and swellability of the polymers and small molecules in sc CO_2 makes the grafting reaction between monomer and substrate are has much accessible [127–132].

The free-radical mechanism controls the grafting and crosslinking reaction is similar to the conventional one.

Limitations: High-energy consumption and investment costs

Using Radiation Routes

Using High-Energy Radiation The use of high-energy radiation for the preparation of graft and crosslinked polymers began in early 1960s and since then, it has been investigated in great depth. The main radiation sources are for high-energy γ radiation and ion beam. Since radiation interacts with organic and other molecules to produce free radicals, cations, and anions, all three types of paths are possible.

Features:

- Simple and accurate, easy to control and clean process.
- Low energy consumption and of manpower saving process.
- Additives or catalysts are not needed to initiate.
- Local heating can be avoided as in the chemical method.
- Absorption of energy by the polymer initiates free-radical process unlike in the chemical method, where purity of the initiators is the matter of concern.
- Zero activation energy process for initiation, as it is a temperature independent step.
- High penetrating ability makes it appropriate to solid substrate in grafting.
- Possibility exists to mosaic graft (grafting of different monomers in different portions on the same substrate) on the solid substrate, as it can direct the location of interest

Grafting proceeds in three main methods: preirradiation, peroxidation, and simultaneous radiation. The mechanisms are discussed below.

(I) Preirradiation Techniques In this grafting [133–137] the polymeric backbone (|-H) is first irradiated, usually in vacuo or in inert gas, to produce relatively stable free radicals, which are then reacted with a monomer (M), usually at elevated temperature. Since monomers are not exposed to radiation in this technique, the obvious advantage is that it is relatively free from the problem of homopolymer formation. However, the disadvantage of the technique is that if the polymer is a degrading type, the scission of the base polymer due to its direct irradiation, brings about the formation of block rather than graft copolymers. Also, if the substrate polymer does not have the ability to trap the radicals for a long enough time, subsequent grafting reaction will not follow.

$$|-\mathrm{H} \rightsquigarrow |\bullet + \mathrm{H}^{\bullet}$$

$$|\bullet \overset{\mathrm{M}}{\rightsquigarrow} |-\mathrm{M}$$

(|– H is the polymer and M is the monomer)

(II) Peroxidation Techniques Unlike the previous method, here the trunk polymer is subjected to high-energy radiation in the presence of air or oxygen. The result is the formation of hydroperoxides/diperoxides depending on the nature of the polymeric backbone and irradiation conditions [138–140].

The stable peroxy products are then treated with the monomer at higher temperature, where the peroxides undergo decomposition to radicals, which then initiate grafting. It is also seen that the peroxide method is also a preirradiation one with similar attendant advantages and disadvantages. The long storage time of intermediate peroxy trunk polymers before the final grafting step is an obviously a great advantage of this technique.

$$| - \mathrm{H} \rightsquigarrow | - \mathrm{OOH} \text{ or } | - \mathrm{OO} - |$$

$$| - \mathrm{OOH} \text{ or } | - \mathrm{OO} - | \rightsquigarrow | - \mathrm{O}\bullet + \bullet\mathrm{OH}$$

$$| - \mathrm{O}\bullet + \mathrm{M} \rightsquigarrow | - \mathrm{O{-}M}$$

(III) Simultaneous Method In this most commonly used method, the substrate is irradiated simultaneously along with the monomer [141–146]. The monomer may be present as vapor, liquid, or solution. The free-radical and ionic reaction mechanisms are presented below (considering the vacuum/inert gas medium).

As both the monomer and substrate are exposed simultaneously to the radiation source, the formation of active sites in both the species occur. Of course, further reaction is dependent on the generation of free radicals on the polymer/monomer in the reaction. If the monomer has high G (M•) (radiation yield) value, homopolymer formation is favored over grafting. In practice, the radiation doses required to achieve graft in the simultaneous method are lower than the other two methods.

$$\vdash \mathrm{H} \rightsquigarrow |\bullet + \mathrm{H}^{\bullet}$$

$$\mathrm{M} \rightsquigarrow \mathrm{M}^{\bullet}$$

$$|\bullet + \mathrm{M}^{\bullet} \rightsquigarrow \vdash \mathrm{M}$$

$$\text{Undesired reaction: } \mathrm{M} \rightsquigarrow \mathrm{M}^{\bullet}$$

$$\mathrm{M}\bullet + \mathrm{M} \rightsquigarrow \mathrm{M}_n^{\bullet}$$

The reversible addition fragmentation transfer mechanism (discussed earlier) can also operate in the radiation initiated grafting technique [147, 148].

In the ionic mode, two different types (cationic and anionic) mechanisms can operate. In a similar approach to that presented earlier, the polymer is irradiated to form polymeric ions and then reacts to the monomer. The high reaction rate of the ionic mode needs small radiation doses to make the requisite grafting sufficient. The cationic grafting is initiated from the backbone as follows:

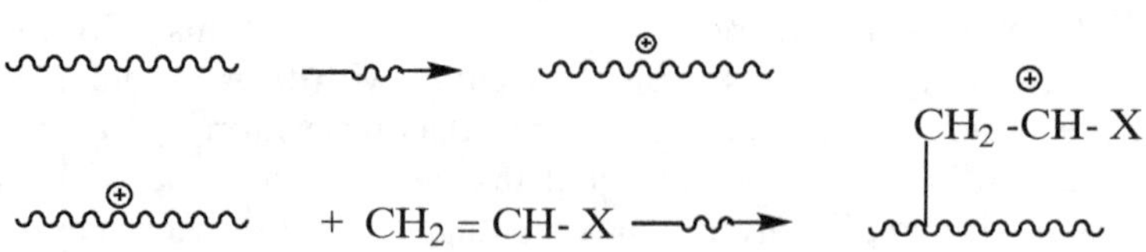

The alternate cationic grafting mechanism proceeds through monomer radical cation, which subsequently forms a dimer. Charge localization in the dimer occurs in such a way that the dimer radical cation reacts with the radical produced by the irradiation of the polymer to form the grafted polymer. The mechanism is sketched below.

$$\mathrm{CH\ X{=}\ CH_2} \rightsquigarrow \mathrm{[CH\ X{=}\ CH_2]}^{\bullet +}\ e^-$$

$$\mathrm{CH\ X{=}\ CH_2} + \mathrm{[CH\ X{=}\ CH_2]}^{\bullet +}\ e^- \rightsquigarrow \bullet\mathrm{CHX{-}CH_2-CH_2-{}^{+}CHX}$$

$$| \bullet + \bullet\mathrm{CHX{-}CH_2-CH_2-{}^{+}CHX} \rightsquigarrow |\mathrm{-CHX{-}CH_2-CH_2-{}^{+}CHX}$$

(|– H is the polymer and M is the monomer, $\mathrm{CHX{=}\ CH_2}$ is model monomer)

Analogous mechanisms involve the anion as the initiator in cases of anionic grafting.

Crosslinking also follows the same mechanistic way as grafting. It occurs between the same polymers as well as different polymers. The following reactions are expected to occur for the different polymers.

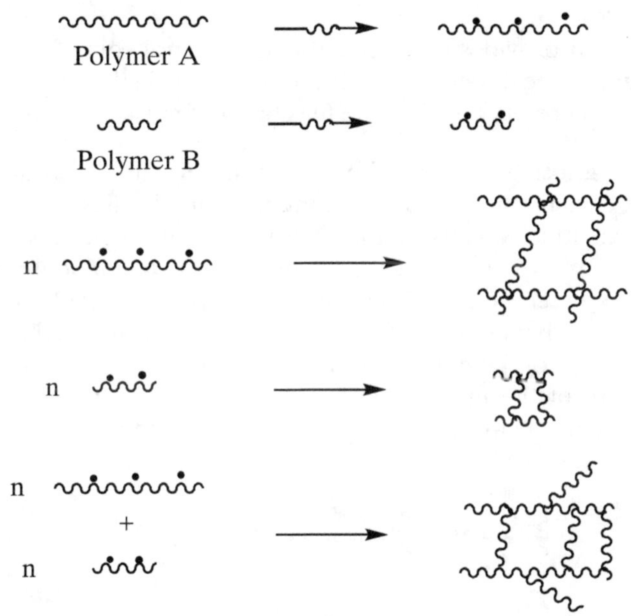

If one of the polymers is of the degrading type, one can still expect the formation of grafted structures such as

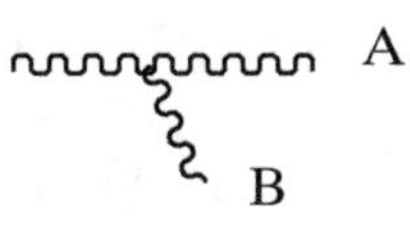

Limitations:

- Expensive technique.
- Length of irradiation time is long and setting up optimum conditions is a huge task.
- Possibility of scission exists by applying a higher than optimum dose.

Using Photoradiation Photografting and crosslinking are usually considered radical chain reactions where the initiation step is produced by a photochemical event.

Features:

- In general, the photoinduced reactions are of low activation energy with respect to chemical reaction and reaction rates are rapid.
- It can proceed under a low or room temperature and results in high monomer conversion and low monomer residue.
- It provides solvent-free formulations, which reduce the emissions of volatile organic pollutants.
- Controllable grafting on the polymeric surface without affecting the bulk properties.
- Moreover, it provides both spatial and temporal control of the reaction since light can be directed to locations of interest in the system and is easily shuttered on or off. For that, the feasibility of mosaic grafting also exists.

Photoinitiated grafting is achieved by two different pathways: directly, i.e., by photolysis of chromophoric groups on the macromolecules [149–150] or indirectly, i.e., by an attack on macromolecules by radicals or electronically excited molecules that are produced by the photolysis or solvent or additive molecules [151–155]. The direct method shows that apart from the graft copolymer, the unwanted homopolymer is also formed if radical Y• generated by the photolysis of side groups of the polymer (|-Y) is reactive towards the monomer. The chemical nature of the final product depends on the extent to which chain termination occurs via disproportionation versus combination. Actually, chain combination leads to crosslinks.

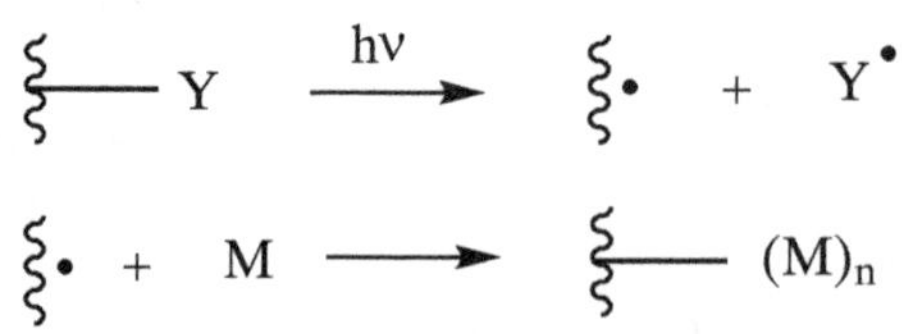

(Y: photo active group, M: Monomer)

The indirect method is illustrated in the routes for the cases of attack on polymers by free radicals (Route 1) and reaction of electronically excited photoinitiator molecules (PI) with lateral functional groups (Route 2), respectively. Actually, photoinitiator molecules are necessary where the monomers have low quantum yield.

Route 1:

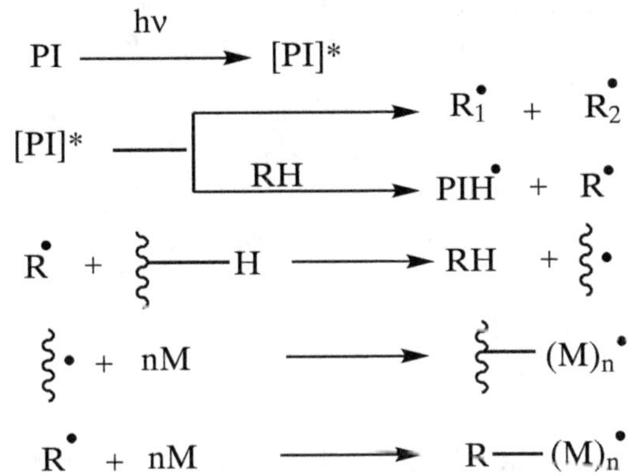

(PI: Photoinitiator, RH: Solvent, M: Monomer, R_1 nd R_2 are the fragments)

Route 2:

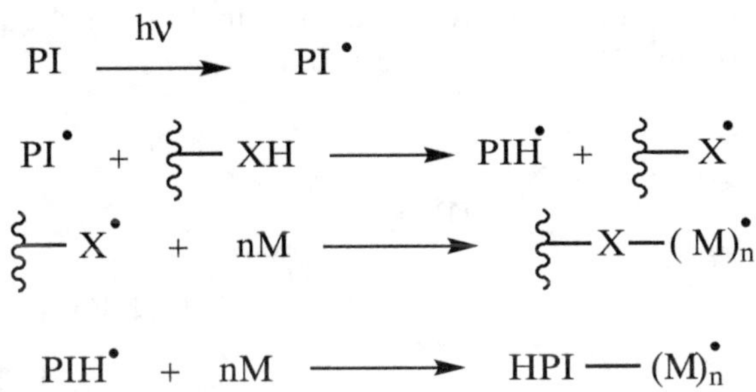

(PI: Photoinitiator, XH: Functional Moiety, M: Monomer)

The photochemical processes of the photoinitiator that result in a transformation of the photoinitiator molecule often take place in the triplet states (T). The transformation is mainly through cleavage processes, electron transfer reactions, hydrogen abstractions, isomerization, or rearrangement. Thus the primary radical is produced.

$$\underset{}{PI} \xrightarrow{h\nu} \underset{\text{Singlet}}{{}^{1}[PI]^{*}} \longrightarrow \underset{\text{Triplet}}{{}^{2}[PI]^{*}} \longrightarrow \underset{\text{Primary radical}}{R\bullet}$$

The produced primary radical will initiate the monomers to form graft polymers through the initiation, propagation, and termination reactions as in the usual free-radical polymerization.

The free-radical photointiators are in general of two types: α-cleavage photoinitiators and hydrogen abstraction photoinitiators.

1. α-cleavage photoinitiators involve a cleavage that takes place at the C–C bonds. This is a unimolecular one. The photochemical processes of this type of photoinitiators can be expressed as follows:

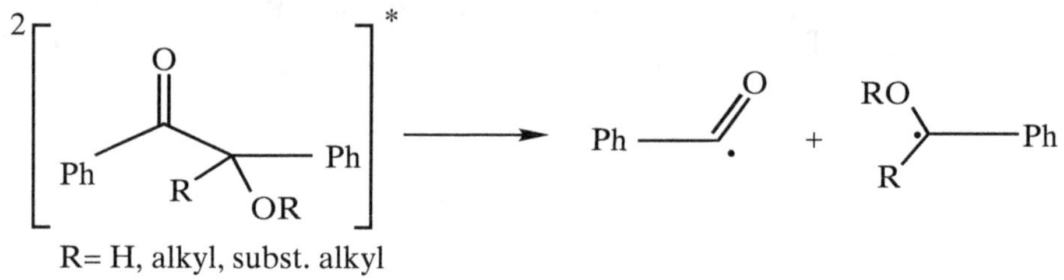

For example, benzoin ether, hydroxyalkyl phenyl acetophenones, phosphine oxide derivatives, etc.

2. Hydrogen abstraction photoinitiators consist of an initiator and a coinitiator, i.e., a bimolecular one. The coinitiator is generally an amine, alcohol, or thiol with an α-hydrogen and functions as the H-donor [156]. The initiation mechanism is expressed as follows:

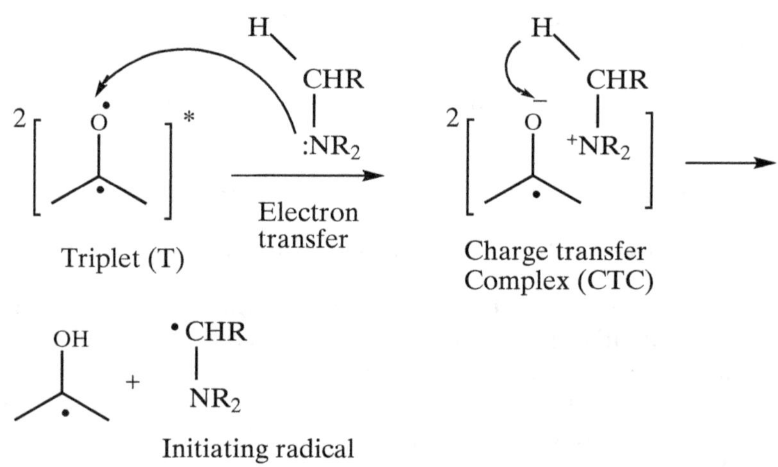

For example, benzophenones, thioxanthones, benzyls, etc. with the presence of alcohol or amines.

Solvent-initiated free-radical grafting can also occur when grafting occurs in a vapor phase or a mixture of aliphatic ketone/water and ethanol. According to photodecomposition and the self-initiation mechanism [157], the acetone molecule on UV light, is either split by a Norrish type I reaction and form acetyl and methyl radicals, or abstracts an H-atom from a nearby molecule or monomer or polymer.

However, this is only occurred when acetone is in the vapor phase. In the mixture of aliphatic ketone/water, there is the possibility of H-bond formation [158–160], and H-bond increases the energy of the excited state of ketone, thereby permitting it to abstract an H-atom from the polymer surface. Alcohol is used to promote photoinitiation efficiency.

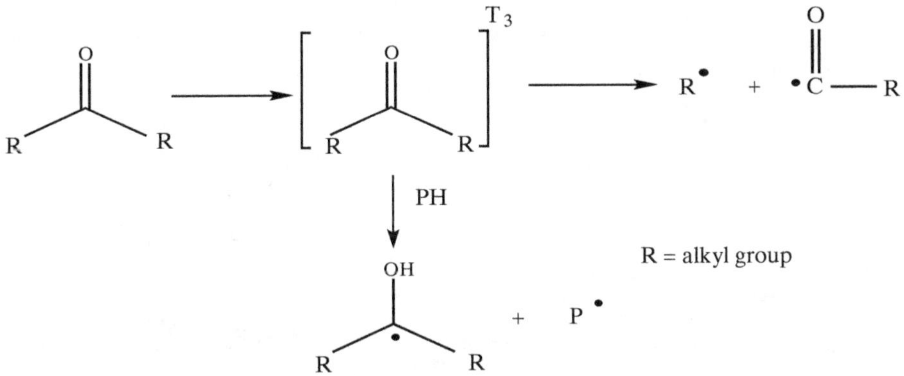

Photocrosslinking has also occurred through the free-radical mechanism as shown in grafting. The reaction also occurs through pendant photofunctional groups (Scheme 2.17) (azo, chalcone, cinnamoyl, acrylate, allyl ester moieties) of the polymers directly or indirectly [161–166]. Metal ion initiated photocrosslinking is also feasible in the indirect approach [167].

Limitations:

- Poor penetrating power of the light restricts graft copolymerization to the polymeric surface.
- Length of irradiation time and the set–up of optimum conditions as well as the selection of best photoinitiatior is huge task.
- Possibility of scission exists by applying above the optimum dose.

Using Plasma Radiation Plasmas are of neutral gaslike clouds or charged ion beams. They are typically formed by thermal, electrical, or light (UV-light or intense visible light from a laser) and ionizing a gas, stripping electrons away from atoms, thereby enabling the positive and negative charges to move freely. In the chemical sense, the interactions with electrons, ions are radical generation reactions. The radical generation reaction is either hydrogen abstraction or chain scission reactions. Plasma is an initiator to cause grafting. Essentially, radicals formed in the polymer chains by plasma reactions undergo similar reactions to that high-energy irradiation. The energy in plasma is lower than the usual conventional energy (e-beam, γ-irradiation). The effects of the plasma are milder than those of high-energy irradiation, because of the low level of high activated species, electrons, and ions. Thus, the plasma forms free

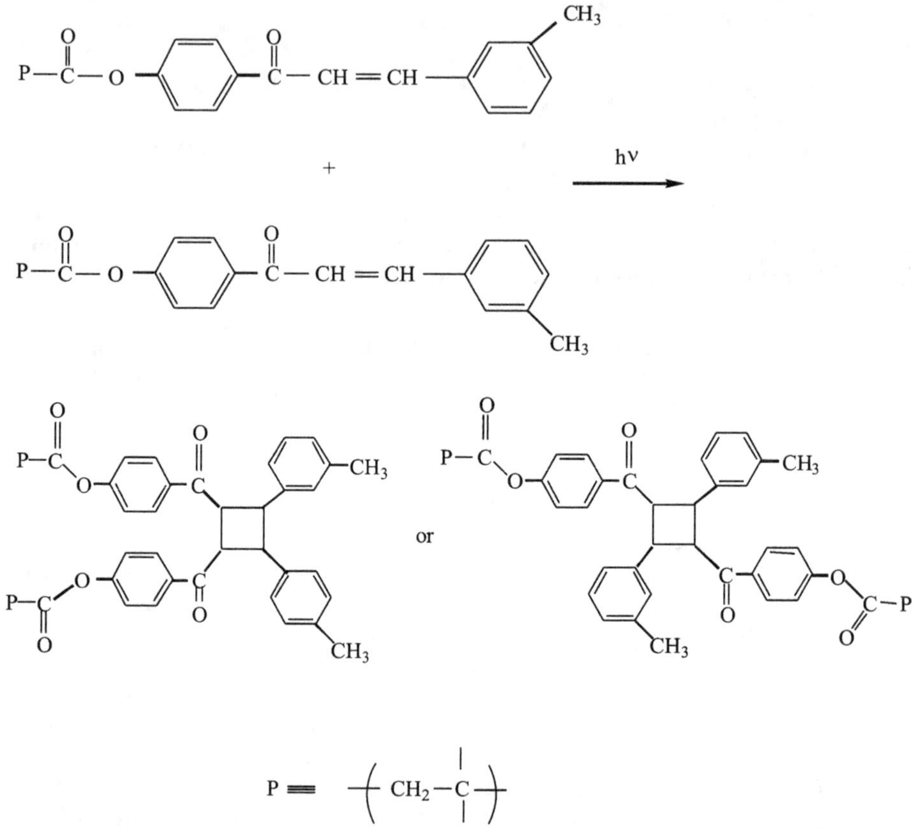

Scheme 2.17. Photocrosslinking in functionalized chalcone moiety. Reproduced from *React Funct Polym,* **49**(**1**): 77 (2001).

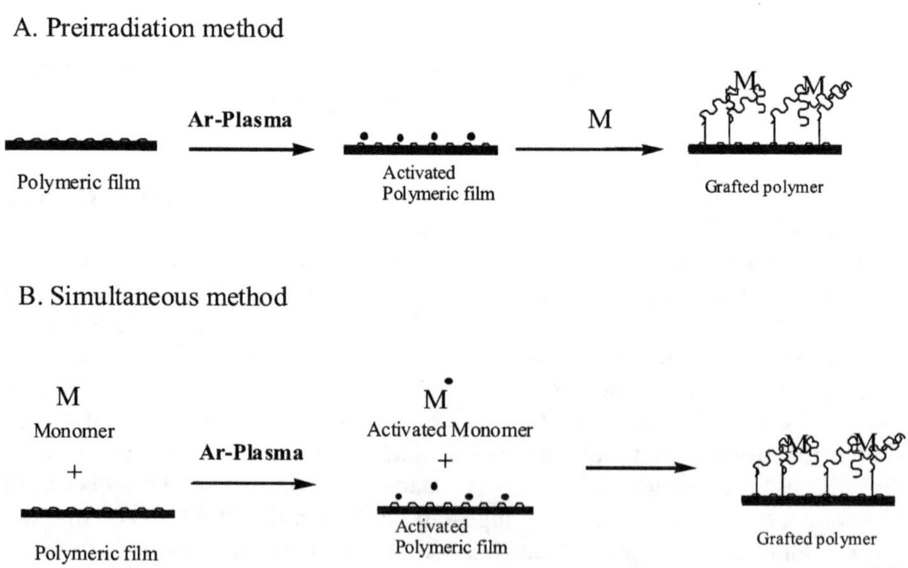

Scheme 2.18. Schematic diagram of the preirradiation and simultaneous methods.

radicals on the surface of the polymer rather than the ionization. As the changes are confined only to the depth of a few nanometers at the surface, the bulk properties (degree of polymerization and crystallinity) of the material are not very much influenced, with respect to high irradiation energy (γ-irradiation, accelerated electrons).

Features:

- The bulk properties will not change as the plasma induces vicinity of the surface.
- Surface grafting is more favored.
- It can be carried out without adding any other organics as a photosensitizer that is required in case of photoirradiation.
- Pore-filling polymerization can be carried out in this technique.

Radical sites are formed as shown in high-energy radiation techniques, i.e., preirradiation and simultaneous techniques [168] (Scheme 2.18). The preirra-

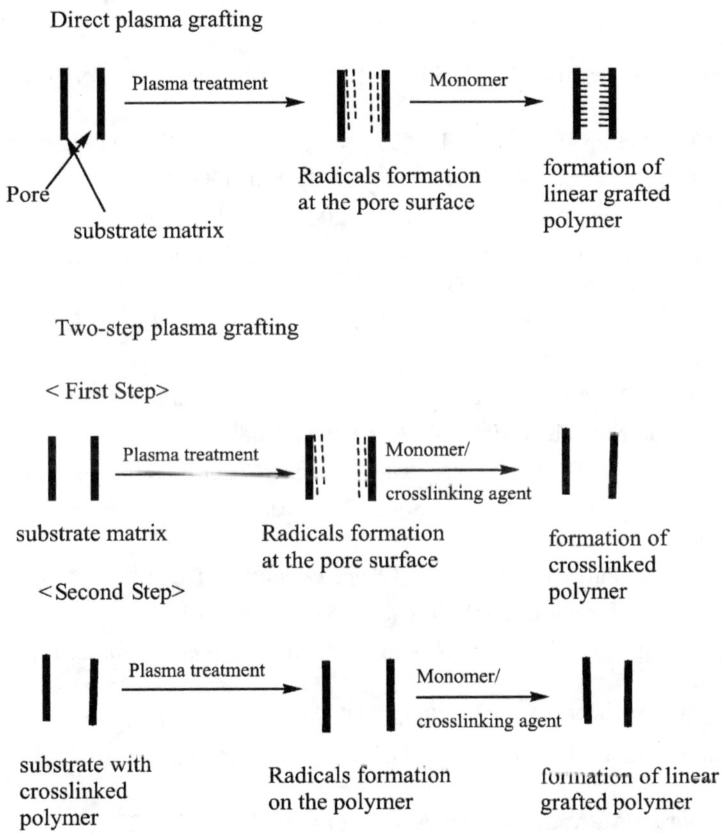

Scheme 2.19. Schematic diagram of pore-filling plasma-graft polymerization. Reproduced from *J Membr Sci* **95**(**1**):39 (1994).

diation technique is favored to minimize the formation of homopolymers. Polymers are first exposed to plasma to create the radicals in the vicinity of the polymer surface, and then contacted to the vapor of the adequate monomer or the aqueous or organic solution of the monomer.

Different plasmas are available, but for grafting, argon and oxygen plasma used the most.

In this context, the important technique to be addressed is pore-filling plasma graft polymerization [169–171]. The plasma treatment of the support matrix filled with monomer/monomer-crosslinker forms the grafted polymer. It can be a single step reaction or a two-step reaction. The schematic presentation is given below (Scheme 2.19).

Limitations:

- Expensive
- Restricted to surface grafting

REFERENCES

1. Barros JAG, Fechine GJM, Alcantara MR, Catalani LH. *Polymer*, **47**: 8414 (2006).
2. Misra BN, Mehta IK, Khetrpal RC. *J Polym Sci Polym Chem*, **22**: 2767 (1984).
3. Thomas TL, Mishra AK. *Eur Polym J*, **38(9)**: 1805 (2002).
4. Yener KM, Oguz O. *Polymer*, **44**: 5239 (2003).
5. Misra BN, Sood DS. in *Physicochemical Aspects of Polymer Surfaces*, ed, KL Mittal (New York: Plenum Press, 1981), p. 881.
6. Kurkuri MD, Lee JR, Han JH, Lee I. *Smart Mater Struc*, **15**: 417 (2006).
7. Bhattacharyya SN, Maldas D. *Prog Polym Sc*, **10**: 171 (1984).
8. Athawale VD, Rathi SC. *J M S—Rev Macromol Chem Phys*, C**39**(**3**): 445 (1999).
9. Naghash HJ, Massah A, Erfan A. *Eur Polym J*, **38**: 147 (2002).
10. Misra BN, Jassal JK, Dogra R. *J Macromol Sci Chem A*, **16(6)**: 1093 (1981).
11. Misra BN, Jassal JK, Dogra R, Sood DS. *J Macromol Sci Chem A*, **14(7)**: 1061 (1980).
12. Pradhan AK, Pati NC, Nayak PL. *J Appl Polym Sci*, **27(5)**: 1839 (1982).
13. Ibrahim N, Insegollu U, Sanli O. *Trk J Chem*, **27**: 403 (2003).
14. Chen Q, Guan Y, Min Zhang X, Peng YX, Xu J. *Chinese Chem Lett*, **12(11)**: 1029 (2001).
15. Huang JX, Yuan XY, Yu XL, Zhang HT. *Polym Int*, **52(5)**: 819 (2003).
16. Sanda F, Takata T, Endo TT. *J Polym Sci Part A: Polym Chem* **32(13)**: 2517 (1994).
17. Kampouris EM, Andreopoulos AG. *J Appl Polym Sci*, **34**: 1209 (1987).
18. Bremner T, Rudin A. *J Appl Polym Sci*, **49**: 785 (1993).

19. Wang JS, Matyjaszewski K. *Macromolecules*, **28**: 7901 (1995).
20. Janata M, Masar B, Toman L, Vlcek P, Policka P, Brus J, Holler P. *React & Funct Polym*, **50**: 67 (2001).
21. Deng G, Cao M, Huang J, He L, Chen Y. *Polymer*, **46**(**15**): 5698 (2005).
22. Hong SC, Pakula T, Matyjaszewski K. *Macromol Chem and Phys*, **202**: 3392 (2001).
23. Shinoda H, Matyjaszewski K. *Macromolecules*, **34**: 6243 (2001).
24. Hong SC, Jia S, Teodorescu M, Kowalewski T, Matyjaszewski K, Gottfried AC, Brookhart M. *J Polym Sci Part A: Polym Chem*, **40**: 2736 (2002).
25. Matyjaszewski K, Beers KL, Kern A, Gaynor SG. *J Polym Sci Part A: Polym Chem*, **36**: 823 (1998).
26. Boerner HG, Duran D, Matyjaszewski K, Da Silva M, Shieko SS. *Macromolecules*, **35**: 3387 (2002).
27. Edmondson S, Osborne VL, Huck WTS. *Chem Soc Rev*, **1**: 14 (2004).
28. Carlmark A, Malmstrom E. *J Am Chem Soc*, **124**: 900 (2002).
29. Pyun J, Matyzaszewski K, Kowalewski T, Savin D, Patterson G, Kickelbick G, Huesing N. *J Am Chem Soc*, **123**: 9445 (2001).
30. Miwa Y, Yamamoto K, Sakaguchi M, Shimada S. *Macromolecules*, **34**: 2089 (2001).
31. Hawker CJ, Mercerreyes D, Elce E, Dao J, Hedrick JL, Barakat I, Dubious P, Jerome R, Volksen W. *Macromol Chem and Phys*, **198**(**1**): 155 (1997).
32. Hawker CJ, Bosman AW, Harth E. *Chem Rev*, **101**: 3661 (2001).
33. Grubbs RB, Hawker CJ, Dao J, Frechet JM. *J Angew Chem Int Ed Engl*, **36**: 270 (1997).
34. Percec V, Barbolu B, Bera TK, Van der Sluis M, Grubbs RB, Frecht JMJ. *J Polym Sci Polym Chem*, **38**: 4776 (2000).
35. Tsoukatos T, Pispas S, Hadjichristidis N. *Macromolecules*, **33**: 9504 (2000).
36. Goto A, Fukuda T. *Prog Polym Sci*, **29**: 329 (2004).
37. Schellekens MAJ, Klumperman BJMS. *Rev Macromol Chem Phys*, C**40**(**2–3**): 167 (2000).
38. Moraes MAR, Moreira ACF, Barbosa RV, Soares BG. *Macromolecules*, **29**: 416 (1996).
39. Moreira ACF, Cairo, Jr. FO, Soares BG. *Eur Polym J*, **37**: 641 (2001).
40. Wang AR, Zhu S. *Macromolecules*, **35**: 9926 (2002).
41. Wang AR, Zhu S. *J Polym Sci Part A: Polym Chem*, **43**(**22**): 5710 (2005).
42. Yu Q, Zhang J, Cheng M, Zhu S. *Macromol Chem Phys*, **207**(**3**): 287 (2006).
43. Huang W, Baker GL, Bruening ML. *Angew Chem Int Ed*, **40**: 1510 (2001).
44. Gao H, Ohno S, Matyjaszewski K. *J Am Chem Soc*, **128**: 15111 (2006).
45. Kumar G, Smith PJ, Payne GF. *Biotech Bioeng*, **63**(**2**): 154 (1999).
46. Chao AC, Shyu SS, Lin YC, Mi FL. *Bioresource Tech*, **91**: 157 (2004).
47. Sampaio S, Taddei P, Monti P, Buchert J, Freddi G. *J Biotech*, **116**: 21 (2005).

48. Jayakumar R, Prabaharan M, Reis RL, Mano JF. *Carbohydrate Polym*, **62**: 142 (2005).
49. Zhao JC, Xie ZH, Guo ZA, Liang GJ, Wang JL. *Appl Surf Sci*, **229**: 124 (2004).
50. Zhao J, Fan G, Guo Z, Xie Z, Liang G. *Polym Bull*, **55(1–2)**: 1 (2005).
51. Kobayashi S, Hideyuk H. *Prog Polym Sci*, **289**: 1015 (2003).
52. Villarroya S, Zhou J, Thurecht KJ, Howdle SM. *Macromolecules*, **39**: 9080 (2006).
53. Kumar G, Bristow JF, Smith PJ, Payne GF. *Polymer*, **41**(**6**): 2157 (2000).
54. Berglin M, Delage L, Potin P, Vilter H, Elwing H. *Biomacromolecules*, **5**(**6**): 2376 (2004).
55. Heise A, Peeters JW, Xiao Y, Palmans ARA, Koning CE. *Polymer Preprints*, **47**(**2**): 224 (2006).
56. Pascault JP, Sautereau H, Verdu J, Williams RJJ. eds. *Thermosetting Polymers* (New York: Marcel Dekker Inc, 2002).
57. Brydson JA. *Plastics Materials* (London: Butterworth Scientific, 1982).
58. Zhang X, Potter AC, Solomon DH. *Polymer*, **39**(**2**): 399 (1998).
59. Zhang X, Solomon DH. *Polymer*, **39**(**2**): 405 (1998).
60. De Bruyn PJ, Foo LM, Lim ASC, Looney MG, Solomon DH. *Tetrahedron*, **53**(**40**): 13915 (1997).
61. Liu CL, Guo QG, Shi JL, Liu L. *Mater Chem Phys*, **90(2–3)**: 315 (2005).
62. Manfredi LB, de la Osa O, Fernandez NG, Vazquez A. *Polymer*, **40**(**13**): 3867 (1999).
63. Maslosh VZ, Kotova VV, Maslosh OV. *Russ J Appl Electrochem*, **76**(**3**): 483 (2003).
64. Pshenitsyna VP, Molotkova NN, Frenkel'MD, Tikhomirova YY, Gurman IM, Aksel'rod BYa, Potekhina YeS, Maizel'NS, Lagucheva YeS, Kochnov IM. *Polym Sci USSR*, **21**(**9**): 2145 (1979).
65. Berge A, Mejdell T. *Polymer*, **47**(**9**): 3249 (2006).
66. Jones FN, Chu G, Samaraweera U. *Prog Org Coat*, **24(1–4)**: 189 (1994).
67. Thomasson D, Boisson F, Reydet EG, Mechin F. *React Funct Polym*, **66**: 1462 (2006).
68. Mequanint K, Sanderson R. *Polymer*, **44**: 2631 (2003).
69. Wang SC, Chen PC, Yeh JT, Chen KN. *React Funct Polym*, **67**: 299 (2007).
70. Liu YL, Chen YJ, Wei WL. *Polymer*, **44**: 6465 (2003).
71. Glavchev I, Apetrova K, Devedjeiv I. *Polym Test*, **21**: 243 (2002).
72. Glavchev I, Petrova K, Devedjiev I. *Polym Test*, **21**: 89 (2002).
73. Watts JF, Abel ML, Perruchot C, Lowe C, Maxted TT, White RG. *J Electron Spectrosc Relat Phenom*, **121**: 233 (2001).
74. Zhao J, Wang Z, Wang J, Wang S. *J Membr Sci*, **283**: 346 (2006).
75. Yang F, Zhang S, Yang D, Jian X. *J Membr Sci*, **301**: 85 (2007).
76. Li L, Zhang S, Zhang X, Zheng G. *J Membr Sci*, **289**: 258 (2007).
77. Xu X, Kirkpatrick RJ. *J Membr Sci*, **280**: 226 (2006).

78. Morgan PW. *Condensation Polymers: By Interfacial and Solution Methods* (New York: Interscience Publishers, 1965).
79. Rao AP, Joshi SV, Trivedi JJ, Debmurari CV, Shah VJ. *J Membr Sci*, **211**: 13 (2003).
80. Roh IJ, Greenberg AR, Khare VP. *Desalination*, **191**: 279 (2006).
81. Janssen LJJM, Nijenhuis KT. *J Membr Sci*, **65**: 69 (1992).
82. Ji J, Dickson JM, Childs RF, Mccerry BE. *Macromolecules*, **33**: 624 (2000).
83. Petersen RJ. *J Membr Sci*, **83**: 81 (1993).
84. Ji J, Mehta M. *J Membr Sci*, **192**: 41 (2001).
85. Chen SH, Chang DJ, Liou RM, Hsu CS, Lin SS. *J Appl Polym Sci*, **83**: 1112 (2002).
86. Sperling LH. *Source-Based Nomenclature for Polymer Blends, IPNs and Related Materials*, Division of Polymer Chemistry Nomenclature Committee Document, 1984.
87. Sperling LH. *Introduction to Physical Polymer Science*, 2nd ed. (New York: Wiley Interscience, 1992).
88. Mishra V, Sperling LH. *Polymer*, **36(18)**: 3593 (1995).
89. Widmaier JM, Nilly A, Chenal JM, Mathis A. *Polymer*, **46**: 3318 (2005).
90. Fradkin DG, Foster JN, Sperling LH, Thomas DA. *Polym Eng Sci*, **26**: 730 (1996).
91. Sanchez MS, Ferrer GG, Cabanilles CT, Meseguer DJM, Pradas MM, Ribelles JLG. *Polymer*, **42**: 10071 (2001).
92. Chenal JM, Widmaier JM. *Polymer*, **46**: 671 (2005).
93. Widmaier JM, Chenal, JM. *Macromol Symp*, **216**: 179 (2004).
94. Das B, Gangopadhyay T, Sinha S. *J Appl Polym Sci*, **54**: 367 (1994).
95. Mathew AP, Packirisamy S, Thomas S. *J Appl Polym Sci*, **78**: 2327 (2000).
96. Coran AY. *Vulcanization*, Internal Technical Review, Monsanto Chemical Company, 1988.
97. Hofmann W. *Rubber Technology Handbook* (New York: Hanser Publishers, 1994).
98. Garvey BS. *An Introduction to Rubber Technology*, M Morton, ed. (New York: Reinhold Publishing Corporation, 1959).
99. Wolfe, Jr. JR in *Encyclopedia of Polymer Science and Technology*, Vol. 14; HF Mark, NG Gaylord, NM Bikales, eds. (New York: John Wiley and Sons, 1971).
100. Akiba M, Hashim AS. *Prog Polym Sci*, **22**: 475 (1997).
101. Cerveny S, Marzocca AJ. *J Appl Polym Sci*, **74**: 2747 (1999).
102. Quirk RP. *Prog Rubber Plast Technol* **4(1)**: 31 (1988).
103. Aprem AS, Joseph K, Mathew T, Alstaedt V, Thomas S. *Eur Polym J*, **39**: 1451 (2003).
104. Debnath SC, Basu DK. *J Appl Polym Sci*, **60**: 845 (1996).
105. Dutta RN, Debnath SC, Noordermeer WM. *J Appl Polym Sci*, **90**: 3835 (2003).
106. Sezdic JT, Jelencic J, Braver M, Frobe Z. *Eur Polym J*, **32(12)**: 1395 (1996).
107. Palaty S, Joseph R. *J Appl Polym Sci*, **78**: 1769 (2000).

108. Lawandy SFH, Halim SF. *J Appl Polym Sci*, **96**: 2440 (2005).
109. Giuliani BVMK, McGill WJ. *J Appl Polym Sci*, **58**: 1053 (1995).
110. Cao L. *Chem J Internet*, **45**: 4 (2002).
111. Heideman G, Datta RN, Noordermeer JWM, Baarle BV. *J Appl Polym Sci*, **95**: 1388 (2005).
112. Dluzneski PR. *Rubber Chem Technol*, **74**: 451 (2001).
113. Gonzalez L, Rodriguez A, Marcos-Fernandez A, Chamorro C. *Rubber Chem Technol*, **69**: 203 (1996).
114. Gonzalez L, Rodriguez A, DelCampo A, Marcos-Fernandez A. *Polym Int*, **53**: 1426 (2004).
115. Valentin JL, Rodriguez A, Marcos-Fernandez A, Gonzalez L. *J Appl Polym Sci*, **96**: 1 (2005).
116. Gonzalez L, Rodriguez A, Marcos-Fernandez A, Valentin JL, Fernandez-Torres A. *J Appl Polym Sci*, **103**: 3377 (2007).
117. Kranenburg JM, Duin MV, Schubert US. *Macromol Chem Phys*, **208**: 915 (2007).
118. Makuuchi K, Tsuda Y. *Nippon Gomu Kyokaishi*, **61**: 710 (1988).
119. Vijayabaskar V, Stephan M, Kalaivani S, Volke S, Dorschner H, Bhowmick AK, Wagenknecht U. *Radiat Phys Chem*. **77**: 511 (2008).
120. Banik I, Bhowmick AK, Raghavan SV, Majali AB, Tikku VK. *Polym Degrad Stab*, **63**: 413 (1999).
121. Bousquet JA, Fouassier JP. *Polym Degrad Stab*, **18(2)**: 163 (1987).
122. Jayaraman M, Frechet JMJ, Dias AJ, Wang HC. *J Appl Polym Sci*, **80**: 680 (2001).
123. Hofmann W. *Prog Rubber Plast Technol*, **30**: 1 (1985).
124. Sae-oui P, Sirisinha C, Thepsuwan U, Hatthapanit K. *Eur Polym J*, **43**: 185 (2007).
125. Matsumoto K, Matoba Y. *Nippon Gomu Kyokaishi*, **63**: 349 (1990).
126. Hayes HJ, McCarthy TJ. *Macromolecules*, **31**: 4813 (1998).
127. Sun D, Wang B, He J, Zhang R, Liu Z, Hanand B, Huang Y. *Polymer*, **45**: 3805 (2004).
128. Liu Z, Song L, Dai X, Yang G, Han B, Xu J. *Polymer*, **43**: 1183 (2002).
129. Kunita MH, Rinaldi AW, Girotto EM, Radovanovic E, Muniz EC, Rubira AF. *Eur Polym J*, **41**: 2176 (2005).
130. Li D, Han B, Liu Z. *Macromol Chem Phys*, **202**: 2187 (2001).
131. Li D, Liu Z, Han B, Song L, Yang G, Jiang T. *Polymer*, **43**: 5363 (2002).
132. Hou Z, Xu Q, Peng Q, Li J, Fan H, Zhang S. *J Appl Polym Sci*, **100**: 4280 (2006).
133. Hegazy EA, Ishigaki I, Desouki AM, Rabie A, Okamoto J. *J Appl Polym Sci*, **27**: 535 (1982).
134. Momose T, Kitazami T, Ishigaki I, Okamoto J. *J Appl Polym Sci*, **39(6)**: 1221 (1990).
135. Dworjanyn PA, Garnett JL. in *Radiation Processing of Polymers*, A Singh, J Silverman, eds. (Munich: Hanser Publishers, 1992.

136. Gupta B, Buchi FN, Scherer GG. *J Polym Sci Polym Chem*, **32**: 1931 (1994).
137. Bhattacharya A. *Prog Polym Sci*, **25**: 371 (2000).
138. Ishigaki I, Sugo T, Takayama T, Okamoto J, Machi S. *J Appl Polym Sci*, **27**: 1043 (1982).
139. Bozzi A, Chapiro A. *Rad Phys Chem*, **32**: 193 (1988).
140. Takacs E, Wojnarovits L, Borsa J, Papp J, Hargittai P, Korecz L. *Nuclear instruments and Methods In Physics Research B*, **236**: 259 (2005).
141. Misra BN. *J Polym Sci Polym Chem*, **23**: 307 (1985).
142. Bhattacharya SN, Maldas D, Pandey VK. *J Polym Sci Polym Chem*, **A24**: 2507 (1986).
143. Bhattacharya SN, Maldas D. *Prog Polym Sci*, **10**: 171 (1983).
144. Nho YC, Garnett JL, Dworjanyn PA, Jin JH. *J Korean Ind Eng Chem*, **3(3)**: 491 (1992).
145. Basu S, Bhattacharya A, Mondal PC, Bhattacharyya SN. *J Polym Sci Part A: Polym Chem*, **32**: 2251 (1994).
146. Liu Q, Zhu Z, Yang X, Chen X, Song Y. *Rad Phys Chem*, **76(4)**: 707 (2007).
147. Grasselli M, Betz N. *Nucl Instr Meth Phys Res B*, **236**: 201 (2005).
148. Barner L, Zwanefeld N, Perara S, Pham Y, Davis TP. *J Polym Sci Part A: Polym Chem*, **40**: 4180 (2002).
149. Taniguchi M, Pieracci J, Samsonoff WA, Belfort G. *Chem Mater*, **15**: 3805 (2003).
150. Yamagishi H, Cricvello JV, Belfort G. *J Memb Sci*, **105**: 237 (1995).
151. Irwan GS, Kuroda SI, Kubota H, Kondo T. *J Appl Polym Sci*, **83(11)**: 2454 (2002).
152. Kubota H, Fukushima Y, Kuwabara S, *Eur Polym J*, **33(1)**: 67 (1997).
153. Peng T, Cheng YL. *Polymer*, **42**: 6811 (2001).
154. He D, Ulbricht M. *J Mater Chem*, **16**: 1860 (2006).
155. Rohr T, Ogletree DF, Svec F, Frecht MJ. *Adv Func Mat*, **13(4)**: 264 (2003).
156. Mufotuoglu AE, Saner B, Demirel AL, Yaggi Y. *Design Monomers Polym*, **8(6)**: 645 (2005).
157. Wang H, Brown HR, Li Z. *Polymer*, **48**: 939 (2007).
158. Allmer K, Huk A, Ranby B. *J Polym Sci Polym Chem Ed*, **26**: 2099 (1988).
159. Coutinho K, Canuto S. *J Mol Struct (Theochem)*, **632**: 235 (2003).
160. Aloisio S, Francisco JS. *Chem Phys Lett*, **329**: 179 (2000).
161. Zhou X, Ren H, Chen H, Wang M. *J Appl Polym Sci*, **73**: 1913 (1999).
162. Balazi R, Nanjundan S. *React Func Polym*, **49(1)**: 77 (2001).
163. Selvam P, Victor Babu K, Nanjundan S. *Eur Polym J*, **41(1)**: 35 (2005).
164. Seo DK, Lim HS, Lee JY, Kim WG. *Mol Cryst Liq Cryst*, **445**: 323 (2006).
165. Yasuda N, Yamamoto S, Wada Y, Yanagida S. *J Polym Sci Polym Chem*, **39(24)**: 4196 (2001).
166. Martina M, Rene N, Helmut. K, Hartwig H. *Macromol Chem Phys*, **205**: 2429 (2004).
167. Lin Q, Ping Gao J, Wang ZY. *Polymer*, **44(19)**: 5527 (2003).

168. Gupta B, Hilborn J G, Bisson I, Frey P. *J Appl Polym Sci* **81**: 2993 (2001).
169. Kai T, Yamaguchi S, Nakao S. *Ind Eng Chem Res*, **39**: 3284 (2000).
170. Yamaguchi T, Yamahara S, Nakao S, Kimura S. *J Membr Sci*, **95**: 39 (1994).
171. Wenzel A, Yanagishita H, Kitamoto D, Endo A, Haraya K, Nakane T, Hanai N, Matsuda H, Koura N, Kamusewitz H, Paul D. *J Membr Sci*, **179(1–2)**: 69 (2000).

3

MECHANISM AND KINETICS

CHRISTOPHER M. FELLOWS

INTRODUCTION

The kinetics of a chemical reaction depends on the mechanism involved in the reaction. This makes it possible to probe the mechanism of polymer grafting and crosslinking reactions by considering their kinetics, and conversely, to predict the kinetics of these reactions from mechanistic information. The mechanisms by which graft and crosslinked polymer networks are generated may be classified on a physical basis, in terms of relative dimensions of the species reacting, or on a chemical basis, in terms of the chemical nature of the intermediate in the reaction.

Physically, these reactions may be divided into

(1) **Polymer/Polymer (P/P) Mechanisms:** Grafted and crosslinked polymers can be created by adding a linking agent to one or more preformed polymers. The best known instance of this class is the "vulcanization" process for preparing a crosslinked elastomer from viscoelastic natural rubber [1]. Physical stimuli such as shear or γ irradiation may also be used to effect grafting and crosslinking between preformed polymers, or polymers may be prepared with specific functional groups which react to form covalent bonds.

(2) **Polymer/Monomer (P/M) Mechanisms:** A polymer may be subjected to conditions such that reactive centers are generated on it from which new polymer chains may grow (a "grafting from" reaction), or to which new growing polymer chains may be attached (a "grafting to" reaction) [2].

Polymer Grafting and Crosslinking, Edited by Amit Bhattacharya, James W. Rawlins and Paramita Ray

(3) **Monomer/Monomer (M/M) Mechanisms:** Polymerization of a multifunctional monomer may be carried out in such a way that a highly crosslinked or branched product may be formed (e.g., the polymerization of multiply substituted acrylates [3, 4]).

This chapter will concentrate on the kinetics and mechanisms of grafted and crosslinked polymers prepared by P/M mechanisms. The kinetics of polymer networks prepared by M/M mechanisms is well described by conventional polymerization kinetics, while there are close analogies between the key kinetic events in typical P/M and P/P mechanisms.

Chemically, grafting and crosslinking reactions may be logically classified according to the nature of the intermediate involved. Perhaps the most important intermediates for grafting and crosslinking are free radicals. Of somewhat lesser importance, due to the typical chemistry of commodity polymers and the typical conditions under which grafting and crosslinking take place, are the chain reactions proceeding through a cationic, anionic, nonpolar intermediate (e.g., the Diels-Alder addition [5]) or through an enzyme-substrate complex (where the charge distribution in the intermediate is unclear). Grafted and crosslinked polymers can also be generated by step-growth mechanisms, through the reaction of a mixture of di-, tri-, or more highly functionalized monomers [6]. However, this chapter will focus on chain reactions.

RADICAL MECHANISMS

Radical intermediates can be generated in a number of ways (Scheme 3.1). Radicals may be introduced via the thermal or chemical decomposition of an initiator, by a redox reaction between two species, or by the excitation of a photoinitiator. Radical intermediates can also be introduced by exposure of a material to energies sufficient to cleave chemical bonds homolytically, using γ irradiation, photoirradiation, thermal treatment, or an electron beam. Where initiation is by irradiation, the radical flux generated will clearly be proportional to the flux of radiation (conventionally intensity, Φ, for lower-energy photons and radiation dose, G, for γ-rays). Preirradiation of a material by ionizing radiation is sometimes used to generate radical initiator species, typically peroxides, *in situ*.

Initiation is the first reaction of a chain carrier with a polymer or monomer molecule to generate a new covalent bond and regenerate the reactive center at a new location. This initiation may occur by addition, where the radical adds across a double bond, or by abstraction, where the radical removes a labile atom from a substrate. The thermodynamic driving force of these reactions is the replacement of a high-energy π-bond by an σ bond (addition) or the replacement of a weaker R–H or R–X bond by a stronger R′–H or R′–X bond (abstraction). Typically, a carbon-centered radical will be more reactive toward addition, and an oxygen-centered radical toward abstraction [7, 8], but

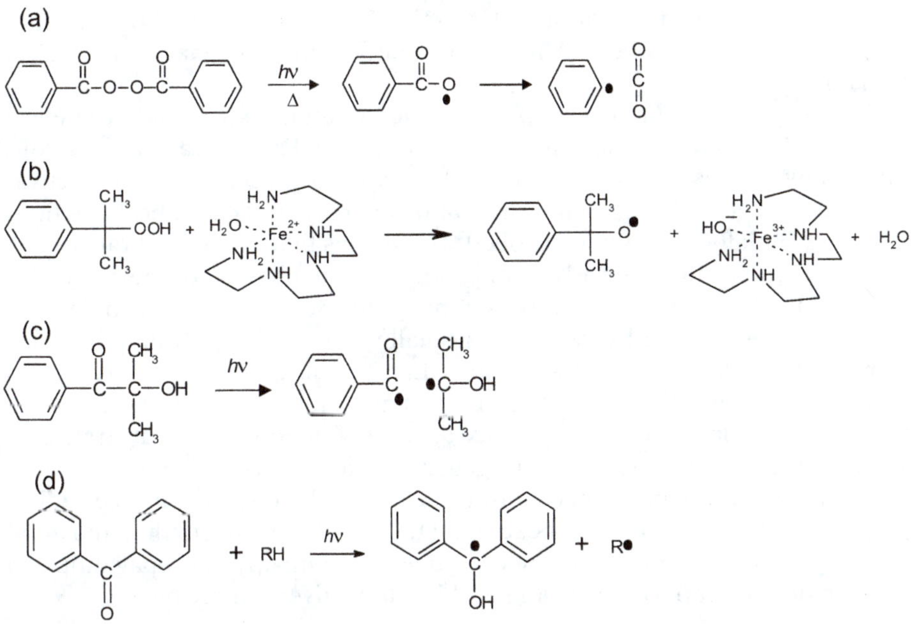

Scheme 3.1. Some initiation reactions: (a) decomposition of benzoyl peroxide; (b) generation of cumyloxy radical by redox initiation. Photochemical decomposition of a (c) PI1 and (d) PI2 photoinitiator.

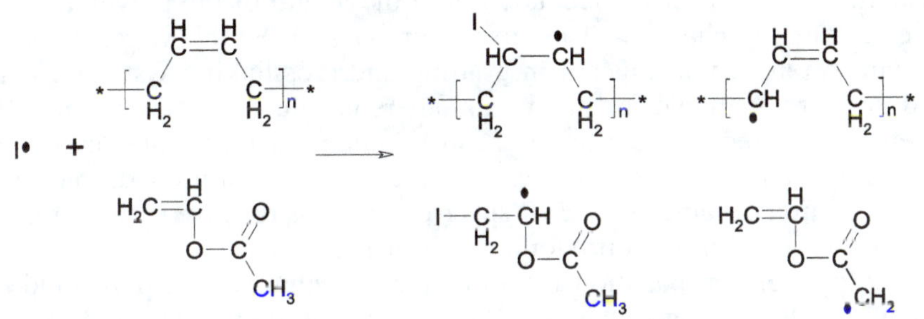

Scheme 3.2. Possible initiating radical reactions with poly(butadiene) and vinyl acetate.

it is difficult to make *a priori* predictions of rate coefficients for these reactions [9]. Some possible reactions are shown above for the grafting of vinyl acetate onto polybutadiene (Scheme 3.2). The initiating radicals may add to pendant or main-chain double bonds in polybutadiene, or to a vinyl acetate monomer, and will abstract chiefly from the most labile positions—the allylic hydrogens on polybutadiene and the acetoxy hydrogens in vinyl acetate. A single initiating radical species can generate a profusion of possible species in a typical reaction, and small populations of reactive defect structures in a polymer (e.g., terminal double bonds) may contribute disproportionately to radical formation.

The many possible reactions between the initiating radical and nonradical species will all have significant activation energy. Radical-radical reactions that remove radicals from the reaction, however, have very low or zero activation energy. Because of these low activation energies, radical-radical reactions will be essentially diffusion-controlled. This means that the overall rate of any radical reaction will be sensitively dependent on the physical properties of the system. Small radicals will relatively rapidly diffuse close enough to another radical to terminate and so will have a small window of opportunity in which to react with the nonradical substrates. Large radical species will last for a longer time, especially when there are no small radicals present to diffuse to them, and will have more opportunity to react with nonradical substrates.

Propagation is the next step in generating a graft copolymer by a free-radical polymer/monomer mechanism. The overall kinetics of any free-radical polymerization reaction will be dominated by this step, so that the total enthalpy of polymerization will be essentially the enthalpy of propagation and the rate of consumption of monomer, M, will be given simply by

$$-d[\mathrm{M}]/dt = k_{\mathrm{p}}[\mathrm{M}][\mathrm{P}^*], \qquad (1)$$

where k_p is the rate coefficient for propagation and [P*] is the concentration of polymer radicals. The coefficient k_p should not be referred to as a "rate constant," because it has a dependence on the length of the polymer radical (unimportant for chains >10 monomer units) and on the viscosity of the medium (often very significant in grafting and crosslinking systems). While good data are available for many monomers in dilute solution regimes, for more concentrated regimes there are little reliable data. In most cases, [P*] must be estimated indirectly rather than measured. Even if conditions allow the observation of radicals by ESR spectroscopy, less reactive radicals formed in side reactions will often predominate over P*.

Some possible propagation reactions for the vinyl acetate/polybutadiene system are shown below (Scheme 3.3). The propagations shown could conceivably continue without the need for further initiation until all the monomer is consumed, but the high and unselective reactivity of radicals makes this unlikely. Termination occurs by the pairing of two radicals, stopping both the kinetic chain reaction and the physical polymer chain. The physical chain may also be stopped by transfer, in which the polymer radical passes its reactivity to another species by abstraction or chain scission.

Considering first the "grafting from" case, a grafted polymer radical may lose its reactivity by encountering: (1) an abstractable atom, e.g., a relatively labile hydrogen on the polymer backbone; (2) a small radical species, e.g., a low-molecular-weight oligomer generated by addition of initiator to free monomer, or (3) a large radical species, e.g., another growing or a persistent

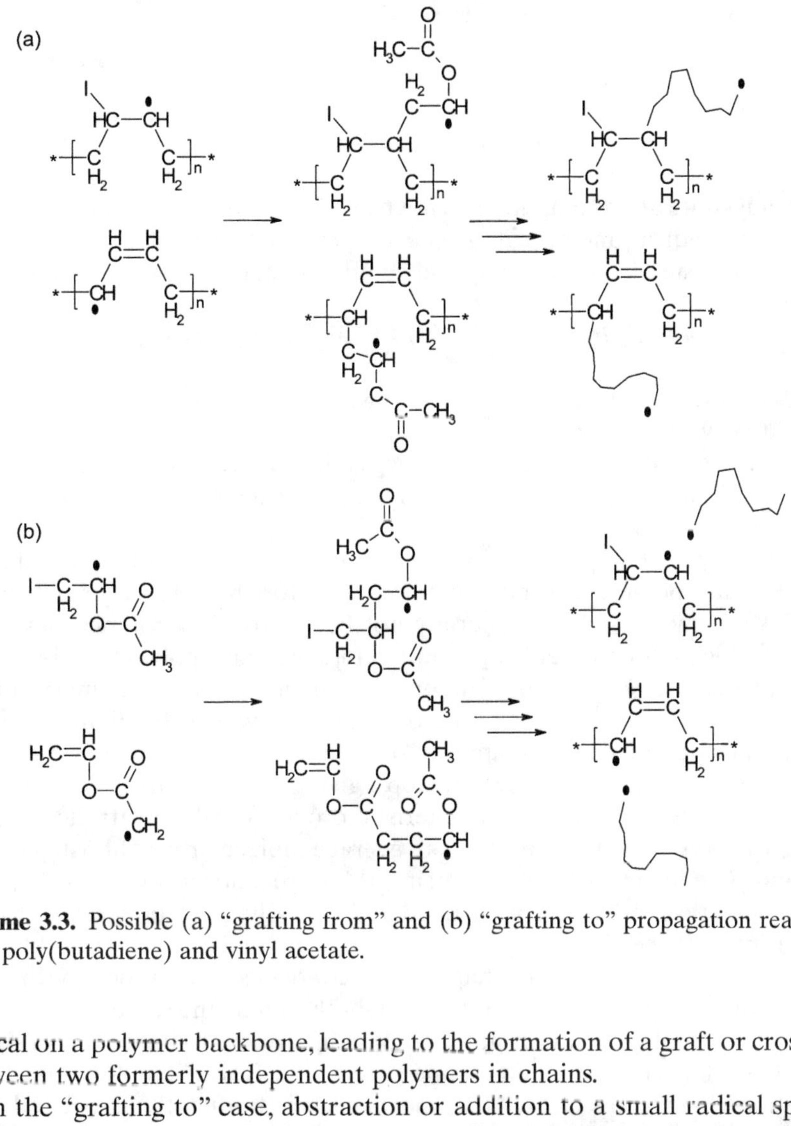

Scheme 3.3. Possible (a) "grafting from" and (b) "grafting to" propagation reactions with poly(butadiene) and vinyl acetate.

radical on a polymer backbone, leading to the formation of a graft or crosslink between two formerly independent polymers in chains.

In the "grafting to" case, abstraction or addition to a small radical species will not result in a graft copolymer, so conditions must be chosen such that the growing polymer radicals are most likely to terminate by encountering persistent radicals on the polymer backbone. As these termination reactions will be diffusion controlled, the termination rate coefficients k_t are very far from being constants: the larger the polymer chain, and the greater the mass fraction of polymer in the reaction, the lower k_t will be.

A powerful approximation is frequently used to estimate [P*]. The steady-state approximation is that over some relatively small time interval of interest, [P*] remains constant. Therefore, the rate of production of radicals can be set as equal to the rate of destruction of radicals.

$$R_i = k_t[P^*]^2. \tag{2}$$

Therefore,

$$-d[M]/dt = k_p[M](R_i/k_t)^{1/2}. \tag{3}$$

The steady-state approximation will not be valid for times less than the average lifetime of a radical, and here it may be appropriate to use the time-dependent expression arising from the analytical solution of the $d[P^*]/dt$ equation:

$$-d[M]/dt = k_p[M](R_i/k_t)^{1/2} \tanh\left((R_i/k_t)^{1/2}\, t\right). \quad [10] \tag{4}$$

As k_t is governed by physical factors that will be largely invariant to the chemistry of the system, it is the rate of initiation that is most significant in controlling the overall rate of a grafting polymerization. Varying the chemistry of the initiator and the initiation conditions is thus the primary means of controlling grafting and crosslinking.

The steady-state approximation can be applied to free-radical crosslinking of preexisting polymers to give an expression for the rate of generation of crosslinks. However, it will generally not be appropriate for photoinitiated systems, which undergo very rapid and inhomogeneous polymerization.

Controlled radical polymerizations (or living radical polymerizations) involve a reversible termination or transfer step, which distributes a single kinetic chain across a large number of physical chains. This allows the population of polymer chains to grow "in step," giving a much narrower molecular weight distribution. The key characteristic properties of a controlled radical polymerization are a linear increase in average molecular weight with conversion and a reduction of polydispersity as the reaction progresses. The distinctive kinetic features of controlled radical polymerization relate to the equilibrium between the active and inactive chain ends [11, 12]. The different mechanisms are described in Chapter 2. In controlled radical polymerization, the overall rate and length of a kinetic chain should be the same as in conventional radical polymerization unless the living radical polymerization opens up new mechanisms for chain stoppage (e.g., the Reversible Addition Fragmentation chain Transfer (RAFT) mechanism is frequently associated with retardation [13]).

IONIC CHAIN MECHANISMS

The same key steps of initiation, propagation, and termination will apply to ionic chain carriers, which, like radicals, may be generated by a number of mechanisms. Cations are suitable for polymerizing vinyl monomers bearing an electron-donating group and epoxides. Rather than being generated in pairs, cations will normally be generated singly, although this will depend on the

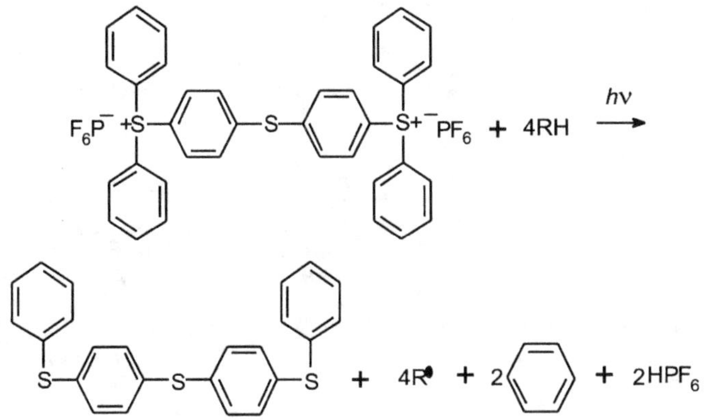

Scheme 3.4. Dissociation of a triarylsulfonium salt to generate a Brönsted acid, HPF_6, under photolysis.

initiator. For instance, the photolysis of triarylsulfonium salts (common cationic photoinitiators) in the presence of a hydrogen donor species generates two Brönsted acid initiating species (Scheme 3.4) [3]. This reaction is accompanied by the generation of free radicals, and in many interpenetrating network (IPN) syntheses, it is possible to independently polymerize one species by cationic polymerization and another by radical polymerization simultaneously using a single initiator.

To achieve grafting by the initiator illustrated above, the Brönsted acid species HPF_6 must add across a double-bond or epoxide functionality in the backbone polymer, as chain-termination processes will not give combinations of polymer chains. The termination in cationic chain polymerization will either be through a chain-transfer process, in which the polymer chain degrades or abstracts an anion to generate a low-molecular-weight cation, or by combination with a chain-terminating anion, A. In the steady state, therefore, the rate of cationic polymerization will be given by

$$d[\mathrm{M}]/dt = k_p[\mathrm{M}]\mathrm{R_i}/k_t[\mathrm{A}]. \quad (5)$$

Under "living" conditions where chain terminators are excluded, the number of cationic charge carriers will increase throughout the course of the initiation reaction, and no steady state will be achieved.

An obvious means of achieving grafted polymerization by cationic mechanisms is to use a polymer backbone with functionalities, which can readily be converted to carbocations. Another is to employ a "grafting to" mechanism where functionalities on the backbone can combine with polymer carbocations to generate a covalent bond (e.g., the grafting of poly(butyl vinyl ether) onto aminated cellulose [14]).

Monomers with electron-withdrawing substituents can undergo anionic chain polymerization. Anionic chain mechanisms are frequently associated

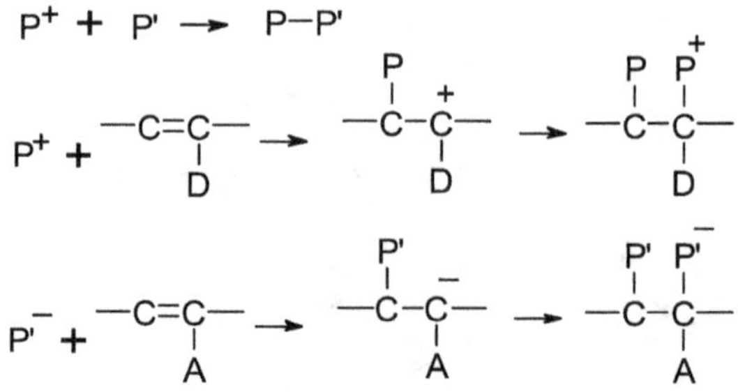

Scheme 3.5. Crosslinking pathways in ionic polymerization.

with living mechanisms in which the rate of polymerization will be given simply by

$$-d[\mathrm{M}]/dt = k_{\mathrm{p}}[\mathrm{M}]\,f[\mathrm{I}], \tag{6}$$

where f is the fraction of initiator converted to carbanions. Anionic chain polymerization may readily be adapted to "grafting from" processes by incorporating appropriate initiating groups on the backbone polymer, for instance, changing hydroxy groups on cellulose to alkali-metal alkoxides.

The overall kinetics for ionic chain polymerizations will closely parallel those of radical chain polymerizations. The expressions for "grafting from" and "grafting to" mechanism should generally be transferable, with appropriate allowances for different termination mechanisms. An important caveat is that ionic grafting and crosslinking mechanisms will rarely occur in the steady-state regime that provides analytical expressions for free-radical reaction kinetics. In ionic polymerizations, crosslinking will be achievable either through heterotermination or addition across an appropriate functionality on a substrate polymer (Scheme 3.5).

OTHER MECHANISMS

There is a great deal of literature about the preparation of crosslinked polymers by condensation mechanisms (e.g., the formation of phenol-formaldehyde resins) [15] and on the coupling of preexisting polymer chains by reaction with a low-molecular-weight species (e.g., sulfur vulcanization) [16]. These reactions will typically follow standard bimolecular kinetics, although the physical factors influencing interactions between polymers and leading to heterogeneous rates of reaction in space and time will still be applicable [17]. For instance, in highly viscous polymerization systems, the conventional condensation polymerization assumption that all chain ends are equally reactive irrespective of molecular weight will often not apply.

KINETICS OF FREE-RADICAL GRAFTING AND CROSSLINKING REACTIONS

Initiation: Controlling initiation is the key in controlling the kinetics of grafted and crosslinked polymers prepared by a radical mechanism. Initiation can be divided into three parts:

(1) Generation of the initial radical species.
(2) Reaction of these species to generate other radicals.
(3) Addition of either the initial radical, or its products, to a monomer, to begin the propagation reactions. In a P/P system, the equivalent step will be the reaction with the polymer that generates the radical site for crosslinking.

Step (1) will depend predominately on the nature of the initiator. For a given chemical initiator the dissociation rate can be controlled chiefly by varying the temperature, and a wide range of radical fluxes can be achieved by varying the initiator concentration. Where radicals are derived from irradiation, the intensity of incident radiation can readily be varied to control radical flux. Steps (2) and (3) will depend on the other species present as well as the initiator: substrate polymer, monomer, solvent, etc.

The distribution of the initiator, and hence the radical flux, can be expected to be inhomogeneous in time and space, complicating the kinetics of any radical process. Three general cases can be distinguished:

(1) An initiator initially homogeneously distributed. This would be the case for most single-phase reactions using a conventional thermally decomposable azo initiator (e.g., AIBN, 2,2′-azo *bis* isobutyronitrile), peroxy initiator (e.g., dicumyl peroxide) or alkyl halide initiator (e.g., 2-bromopropionate). The rate of generation of initiator-derived radicals [I*] will be the same everywhere in the system, and will be given by

$$-d[\mathrm{I}^*]/dt = fk_\mathrm{d}[\mathrm{I}], \tag{7}$$

where [I] is a function of time. The rate coefficent k_d can be considered to include the stoichiometry of the decomposition. While most initiators decompose to give two radicals, some may give one active and one persistent radical unreactive toward addition (e.g., the PI2 photoinitiator benzophenone, where the ketyl radical is unreactive [3]). The viscosity-dependent constant *f* accounts for any solvent-related "cage" effects, where two radicals held in close proximity combine to form an unreactive product, for example, the production of unreactive tetramethylsuccinonitrile (TMS) from the decomposition of AIBN. These reactions can consume a considerable fraction of initiator, and the strong influence of *f* is a main reason it is often necessary to abandon the kinetically simplest case of a homogeneously distributed initiator.

(2) An initiator entering the system from outside, giving a heterogeneous distribution in time and space. This can arise through direct irradiation to form radicals (e.g., with γ-rays or electrons [18]), and in heterogeneous systems where radicals are generated in one phase and migrate into the phase where grafting is to take place (e.g., emulsion polymerization [19]). The highest initiating radical density will be at the surface of the polymer to be grafted or crosslinked and will decline further from the source. This inhomogeneity depends on the nature of the initiating species. Electrons and γ-rays penetrate much more than small molecule fragments and can be used to initiate grafting in much larger samples than could be grafted by introducing radicals from outside.

The rate of initiating radical generation in these systems is

$$d[\mathrm{I}^*]/dt = \mu\Phi e^{-kx}, \tag{8}$$

where Φ is the flux of initiating particles or radiation entering the system, μ is the efficiency of the initiation, x is the distance from the surface, and k is a function of the resistance to propagation of radiation or diffusion of particles. Generally k will typically increase over time and become dependent on x as the system become less homogeneous.

An initiation quantum yield, Φ_i, can be defined in terms of the rate of initiation and the intensity of radiation, I_a:

$$\Phi_i = (d[\mathrm{I}^*]/dt)/I_a. \tag{9}$$

(3) Initiation by a two-component initiator where one component is originally distributed in the polymer to be grafted or crosslinked, and the other component enters from outside, giving a heterogeneous distribution of initiating radicals in time and space. Photoinitiated systems, where the photoinitiator PI is originally present throughout the system and light enters from outside to generate radicals, are of this kind.

A photoinitiator will give better control of the chemistry of initiation than exposure of the system to high energy radiation. However, while it is theoretically possible to continue irradiating a one-component system until the entire material is grafted to an acceptable extent, two-component systems are susceptible to exhaustion of the photoinitiator. The appropriate equation for the rate of radical generation will be

$$d[\mathrm{I}^*]/dt = [\mathrm{PI}]\Phi e^{-kx}. \tag{10}$$

Here the dependence of k on time and position is likely to be more complex than in case (2), and [PI] will depend both on time and position; e.g., depletion of photoinitiator in the surface volumes of the material may allow light to penetrate more deeply. Redox initiation systems, where an electron is transferred between chemical species to generate radicals, are often carried out in a similar fashion, with one half of the redox couple present at the beginning of the reaction and the other half added over time. The appropriate selection of a redox couple can localize the site of initiation: in the preparation of surface-grafted natural rubber latex particles, use of a hydrophobic hydroperoxide localized in the high-viscosity rubber phase, and complexed iron(II) localized in the aqueous phase generates radicals at the surface and leads to efficient grafting of the hydrophilic monomer [20].

As crosslinking or grafting increases, it will become more difficult to maintain relative homogeneity with a redox couple, even if the second half of the couple was originally able to diffuse rapidly throughout the polymer-containing phase. The general expression for initiation under these conditions will be

$$d[I^*]/dt = k_{12}[I_1][I_2], \tag{11}$$

where $[I_1]$ and $[I_2]$ are the time- and position-dependent concentrations of the two halves of the redox couple, and k_{12} is the bimolecular rate coefficient for the electron-transfer reaction between them.

REACTION OF RADICAL TO FORM INITIATING SPECIES

The initial radicals formed may undergo abstraction, addition or termination (Scheme 3.2).

Abstraction is an efficient way for a radical to introduce a reactive center to the backbone or side chain of a polymer. This reaction will depend on the H-abstracting capacity of the radical and susceptibility of the substrate to abstraction. Heteroatom-centered radicals are usually considered to be more effective in abstraction on electronic grounds, and a relatively persistent radical will usually be more reactive toward abstraction than addition on steric grounds. The strength of an R–X bond is correlated with the enthalpy change involved in removing X* to give the radical species, R*. As the reaction is strongly endothermic, the Hammond postulate states that the transition state should be productlike and hence dominated by the properties of R*. R–H bonds that give rise to tertiary radicals will be more favorable than bonds that give rise to secondary radicals, and secondary radical-forming reactions more favorable than primary ones. Other factors to be considered are the ability of heteroatoms to stabilize adjacent radicals by π-donation, and of hydrogens on neighboring atoms to do so by hyperconjugation. These principles allow qualitative estimation of the rate coefficients for grafting, k_{gr}, to different positions RH.

$$d[R^*]/dt = k_{gr}[I^*][RH]. \tag{12}$$

It is difficult to deconvolute experimental data on grafting reactions to make quantitative estimates of k_{gr}. Values may be estimated by carrying out

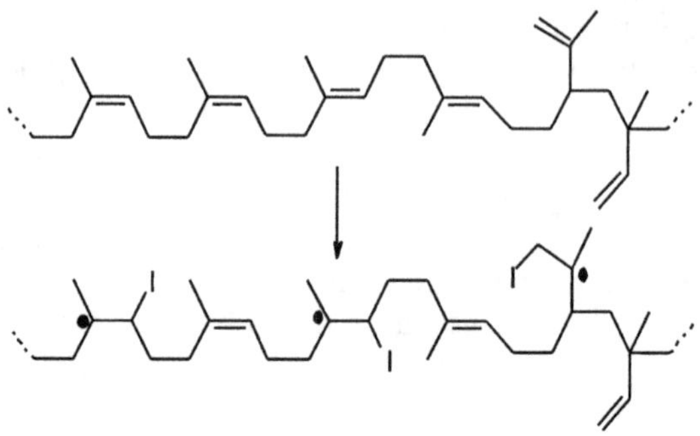

Scheme 3.6. Some addition reactions to polyisoprene. From left: addition to *cis*, trans, and vinyl isoprene residues.

analogous grafting reactions or computations on small-molecule models [9, 21], but these estimates will only be applicable under conditions such that I* is relatively free to diffuse. The larger I* is, the higher the weight fraction of polymer. The higher the degree of crosslinking, the more the system will approach the case where the grafting reaction is limited by diffusion. Typical values of activation energies of abstraction relevant to grafting and crosslinking are of the order 20–30 $kJmol^{-1}$ [22].

The addition to a double bond in the polymer is a common means for a small radical to generate a reactive center [23]. Common polymers based on a diene monomer, such as polyisoprene and polybutadiene, have both double bonds in the backbone and pendant double bonds to which addition may take place.

$$d[R^*]/dt = k_{ad}[I^*][-C{=}C-]. \tag{13}$$

Possible sites for addition to a random polyisoprene are shown below (Scheme 3.6). As each addition reaction will have a different rate coefficient, it will often not be possible to deconvolute these events.

ADDITION OF INITIATING SPECIES TO MONOMER/POLYMER

In P/M radical grafting or crosslinking reactions, the final step of initiation will be when the radical center R* adds to a molecule of a monomer. If R* is localized on the polymer backbone, this will give the "grafting from" process; if R* is a small molecule, this will give a potential "grafting to" process. This step is unlikely to be rate-controlling unless the previous steps of initiation give rise to persistent radicals.

PROPAGATION

A large number of reactions are possible once initiation is complete, and all events occurring before termination form part of the "kinetic chain." These reactions may include addition to the monomer and functionalities on the polymer, or chain transfer, typically to polymer or an additive.

In grafting polymerizations where consumption of monomer is the primary reaction a propagation rate can be defined as

$$-d[\mathrm{M}]/dt = k_p[\Sigma\mathrm{R}^*][\mathrm{M}], \qquad (14)$$

where ΣR^* is the sum of all radical species that can add to monomer. As [M] is initially fixed and can only decrease in the course of the reaction, $[\Sigma R^*]$ will be the main parameter controlling the kinetics. This assumes all R^* have the same reactivity, an assumption valid at degrees of polymerization greater than 10 in circumstances where there are few side reactions generating persistent radicals. If one or more comonomers are present, the rate coefficients for each of the possible propagation reactions and the possible change of the relative amounts of the comonomers over time must be considered. Monomers that are very unreactive toward homopolymerization, such as maleic acid or crotonic acid, will give very little propagation, and hence give very short grafts (~1 monomer unit). At very high viscosities, chemical control of propagation will be overwhelmed by physical constraints, and propagation will become controlled by the reactive diffusion of monomer to polymer radicals, giving a greatly reduced k_p value.

For P/P crosslinking reactions where addition reactions are the main mechanism of crosslinking, we may define propagation as the consumption of potential grafting sites, G.

$$-d[\mathrm{G}]/dt = k_p[\Sigma\mathrm{R}^*][\mathrm{G}]. \qquad (15)$$

TERMINATION

The kinetic chain of radical polymerization can only end by the pairing of two radicals to form nonradical species. As this reaction is essentially barrierless, termination will be rapid and effectively diffusion controlled under all conditions. As the polymer radicals are much larger than the monomer, with increasing viscosity k_t will decrease well before k_p begins to decrease. This is the physical basis for the rapid increase in the rate of free-radical polymerization with viscosity over the course of a reaction (the Trommsdorff-Norrish effect) [24]. An important consequence of the diffusion dependence of k_t for grafting and crosslinking reactions is that termination will most frequently occur not by the meeting of two large radical species but by the addition of a small mobile radical to a polymer-centered radical. "Grafting to" mechanisms will thus be unlikely except for reactions carried out in dilute solution.

Semiempirical expressions for the chain-length dependence of k_t have been determined [25, 26]. For termination between two equal chains of length i, the appropriate relations are

$$k_t^{i,i} = k_t^{1,1} \times i^{-es} \quad (16a)$$

for relatively short chains whose interactions are governed by centre-of-mass diffusion, and

$$k_t^{i,i} = k_t^{1,1} \times (i_{crit})^{-(es-e_L)} \times i^{-e_L} \quad (16b)$$

for longer chains ($>i_{crit}$) that interact by segmental diffusion. Expected values of e_S and e_L are 1/2 and 1/6, respectively. In most systems where radicals are continuously generated, the dominant termination mechanism will be center-of-mass diffusion between short and polymer radicals. In such systems, k_t is given by

$$k_t^{i,long} = 4\pi D_i p_i \sigma N_A, \quad (17)\ [19]$$

where D_i is the relative diffusion coefficient of a long radical and a radical of length i and p_i is the probability that they will terminate on encounter, defined as approaching within a distance σ of the order of an atomic diameter. The center-of-mass diffusion coefficient D_i for oligomeric radicals can be fit by an empirical scaling law dependent on the mass fraction of polymer in the reaction, x, where $D_1(x)$ is the diffusion coefficient of the monomer.

$$D_i = c_2 D_1(x) i^{c_0 + c_1(x)}. \quad (18)\ [27]$$

Few systems have been characterized well enough for Equations (17) and (18) to be used, and it is seldom that better than order of magnitude estimates of k_t are available for grafting and crosslinking reactions outside of dilute solution.

OVERALL KINETICS

The dependence of the rate of consumption of a monomer on the concentrations of various species provides a window into the mechanism of the reaction. Two important complicating factors should be noted: (1) competition between mechanisms giving rise to the grafted polymer and to the homopolymer, and (2) the difficulty of making measurements under conditions other than dilute solution.

Methods for following grafting include monitoring of monomer consumption by Fourier transform infrared spectroscopy (FTIR), nuclear magnetic resonance (NMR), ultraviolet/visible light (UV/Vis) spectroscopy, calorimetry, or use of a microbalance. In every case, however, it is necessary to verify that the "grafted" polymer is really grafted and not just entangled. A typical method suitable for grafted systems is Soxhlet extraction, in which a good solvent for the homopolymer is refluxed through the grafted material for an extended period of time [28]. In crosslinked systems, the rate of reaction will generally be expressed in the rate of crosslink formation, which is rarely experimentally accessible. Useful proxies for crosslink density are the gel fraction (the fraction of the material that is insoluble on extraction with a good solvent for the linear polymer analog) and the torque measured by an oscillating disk rheometer [29]. Under some circumstances it may be possible to determine the true concentration of crosslinks by NMR spectroscopy.

Chemical kinetics is typically concerned with the rate of consumption of a monomer ($d[M]/dt$), but it is often of interest to monitor the rate of consumption of graft or crosslinking sites ($d[G]/dt$). Fischer developed expressions for $d[G]/dt$ for the "grafting to" mechanism of styrene onto polybutadiene [30].

$$-\frac{d[G]}{dt} = ((k_{add} + k_{abs})[R^*] + (k_{add,P} + k_{abs,P})[P^*])[G], \tag{19}$$

where $[R^*]$ and $[P^*]$ is the concentration of small and polymer radicals, k_{add} and k_{abs} are the rate coefficients for the addition and abstraction reactions, respectively. From Equations (1) and (19), an instantaneous relative probability of grafting and chain extension can be calculated:

$$C_{gr} = \frac{-d[G]/[G]}{-d[M]/[M]} = \frac{(k_{add} + k_{abs})[R^*]}{k_p[P^*]} + \frac{k_{add,P}}{k_p} + \frac{k_{abs,P}}{k_p}. \tag{20}$$

When the rate of chain-initiating events is much less than the rate of propagation, the average degree of polymerization of the grafted chains, ν, can be incorporated into Expression (20) as follows:

$$C_{gr} = \frac{-d[G]/[G]}{-d[M]/[M]} = \frac{1}{\nu} \cdot \frac{k_{add} + k_{abs}}{k_i} + \frac{k_{add,P}}{k_p} + \frac{k_{abs,P}}{k_p}. \tag{21}$$

Huang and Sundberg derived general kinetic expressions for instantaneous graft efficiency, ϕ, defined as the amount of monomer incorporated into grafted chains as a fraction of total monomer polymerized [31]. These expressions, like those of Fischer, assume no molecular weight dependence of the rate coefficients and are, therefore, applicable only in dilute systems.

If grafting sites are generated by the reaction of primary radicals with the substrate and are not terminated directly before reaction with the monomer to an appreciable degree, then

$$\frac{1}{(1-\phi)^{1/2}}-1=\frac{k_{add}+k_{abs}}{k_i}\frac{[\mathrm{P}]}{[\mathrm{M}]}, \tag{22}$$

where k_i is the rate coefficient for the initiation reaction of the primary radical with the monomer. If grafting sites are generated by reaction of oligomeric or polymeric radicals with the substrate, then

$$\frac{1}{(1-\phi)^{1/2}}-1=\frac{k_{add}+k_{abs}}{(2k_t fk_d)^{1/2}}\frac{[\mathrm{P}]}{[\mathrm{I}]^{1/2}}. \tag{23}$$

Huang and Sundberg confirmed that in the grafting of styrene, benzyl methacrylate, and benzyl acrylate onto polybutadiene, the degree of grafting extrapolated to zero concentration was consistent with Equation (22) and not (23) when an initiator that gave carbon-centered radicals (AIBN) was used [32], but that Equation (23) applied for styrene and benzyl methacrylate when oxygen-centered radicals derived from benzoyl peroxide (BPO), and therefore more active in abstraction, were employed [33].

Zhou, Huang, and Liu have derived analytical expressions for the rate of polymerization in radical grafting systems where grafting is primarily by radical combination [34] or by addition across double bonds in the backbone [35]. The combination expression gives a good fit to the experimental data collected for the grafting of crotonic acid onto ABS rubber [34]:

$$\frac{d[\mathrm{M}]}{dt}=\frac{A[\mathrm{G}][\mathrm{M}]^2[\mathrm{I}]^{0.5}}{([\mathrm{M}]+B[\mathrm{G}])^2}, \tag{24}$$

where A is a constant given by $(2fk_d/k_t)^{1/2} \cdot k_p B$, and B is a constant given by k_{iP}/k_{iM}, where k_{iM} and k_{iP} are the coefficients for the reaction of initiator to monomer and polymer, respectively.

The addition expression gives a good fit to data obtained for the addition of undecylenic acid to ABS rubber [35]:

$$\frac{d[\mathrm{M}]}{dt}=\frac{A'[\mathrm{G}][\mathrm{M}]^n[\mathrm{I}]}{\left([\mathrm{G}]+B'[\mathrm{M}]^n\right)^2}, \tag{25}$$

where $n=\dfrac{\text{rate of P* growth}}{\text{rate of P* growth}+\text{rate of P* termination}}$, A' is a constant given by $fk_d/k_p\beta k_{iM}$ and B' by $k_{tr}/\beta k_{iM}$, where β is a constant expressing the degree to which monomer radicals are hindered by a solvent cage, and k_{tr} is the rate coefficient for chain transfer from polymer radical to monomer.

A kinetic model incorporating many of the features discussed here has been implemented by Vivaldo-Lima et al. and used to fit experimental data with good success, as shown in Figure 3.1 [36].

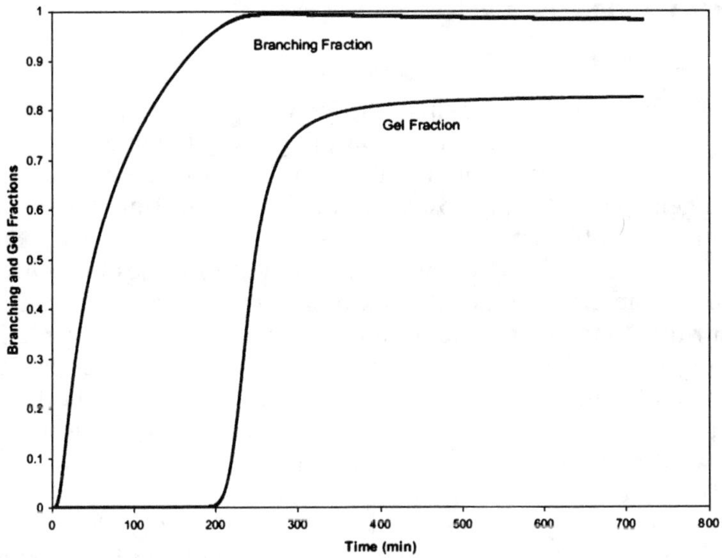

Figure 3.1. Model prediction of branching and gel fractions versus time for thermal crosslinking of a polybutadiene rubber of medium *cis* content, T = 170°C, 5 wt. % polymer in decalin. Reproduced with permission from *Macromolecular Theory and Simulations*, **14**: 539–553 (2005).

It can be seen that the model predictions are very far from the "linear" growth of branching and crosslinking density, with a sigmoidal gel-fraction curve giving a dramatic "gelation point" consistent with experimental results.

In homogeneous γ or photoinitiated systems where the details of the polymerization may be obscure, a polymerization quantum yield, Φ_P, can be defined in terms of the overall rate of polymerization and the intensity of radiation, I_a:

$$\Phi_P = (d[M]/dt)/I_a. \tag{26}$$

The ratio of this value to the initiation quantum yield Φ_I will give a first estimate for the expected degree of polymerization of grafted chains in the system.

Having outlined the chain-growth mechanisms for generating grafted and crosslinked mechanisms and the kinetic expressions that can be derived from these (mindful that they will usually only be strictly applicable in dilute solution), it is useful to survey the effect of varying different components of the system. Chemical effects (initiator, monomer, additives, etc.) will be considered first, followed by physical effects (primarily the viscosity of the system).

CHEMICAL EFFECTS

Effect of the Initiator

The initiator first generates the chain carrier, whether it is a γ ray, redox couple, or classical initiator. The most important thing about the initiator is the effective flux of radicals it can generate. This requires estimation of k_d and $f(x,t)$. In a homogeneous system, viscosity will often be high. Thus, the loss of radicals to futile recombination reactions $R^* + R^* \rightarrow RR$ will be large and f will be small. There are little good data on appropriate f values for concentrated solutions, but the effective radical flux in a viscous reaction mixture will be well below that found in a dilute solution. In many emulsion polymerizations, for example, f is small enough within monomer-swollen polymer particles that hydrophobic initiators are ineffective. In such cases, where only the initiator in the aqueous phase can contribute radical flux to initiation, it is necessary to consider what fraction of the initiator is partitioned into the aqueous phase. Thus in addition to k_d, the miscibility of the initiator and radicals generated from it with the polymer phase is of key importance. Pesetskii et al. investigated the grafting of itaconic acid onto PE and found that initiator solubility effects were more important than k_d values [37]. Furthermore, free-radical inhibitors soluble in polyethylene but insoluble in itaconic acid inhibited cross-linking but did not significantly affect grafting [38].

The kinetics of the grafting system will next depend on the possible fates of the primary radical: direct initiation, formation of a small radical by abstraction or addition, formation of a polymer-centered radical by abstraction or addition, or primary radical termination of a polymer radical. There are likely to be multiple possible abstraction sites on the polymer, and multiple species susceptible to abstraction, but in practice most of these pathways can be discounted. The relative rate coefficients for these processes and the relative concentrations of various species will determine what happens to the radical, but cannot yet predict the effectiveness of grafting. For example, if a large proportion of radicals add to the monomer, this may mean there is efficient grafting by a "grafting to" mechanism or that the polymer radicals formed are terminated in other ways and cannot contribute to grafting.

At a particular initiator concentration, a limit in the degree of grafting may be found, most likely indicating that the effective radical flux that is available to initiate grafting reaches a maximum and that additional initiator is expended in futile recombination reactions. In the grafting of *N*-vinyl carboxamido-2-methylpropanesulfonic acid onto polystyrene with BPO, for example, a maximum in grafting was found at a BPO concentration of 30 mM [39]. The more diffusion is inhibited, the lower this optimal radical flux will be. For example, the γ-initiated homopolymerization of vinyl imidazole was found to proceed at a maximum rate at a higher flux (0.26 kGray) than the graft polymerization of vinyl imidazole on cellulose (0.22 kGray) [40].

The failure to find expected differences in kinetics with variations in the initiator may indicate that the primary radical is not involved in initiation. In the heterogeneous (miniemulsion) grafting of a number of vinyl monomers onto alkyd resins reported by Tsavalas et al., AIBN, BPO, and potassium persulfate (KPS) all show very similar behavior [41]. It is likely that in these cases radicals could effectively be generated only in the aqueous phase, and grow into oligomers through reaction with the monomer in the aqueous phase before entering the hydrophobic alkyd/monomer particles to commence reaction. Hence only the aqueous phase concentration of these initiators, the appropriate effective dissociation coefficient fk_d, and their relative reactivities toward species present in the aqueous phase could affect the results.

Huang and Sundberg observed significant differences between BPO and AIBN for graft polymerization onto polybutadiene only for monomers where the polymer-derived radical was relatively unreactive [42]. The grafting of benzyl acrylate, which has a highly reactive polymer radical, gave similar results for both initiators. For styrene and benzyl methacrylate, which generate much fewer reactive polymer radicals, BPO generated significantly more polymer-centered radicals by H-abstraction than did AIBN. These polymer-centered radicals, relatively unreactive to addition, gave significant rate retardation for BPO in comparison to AIBN at the same radical flux. Similarly, in the crosslinking of polyethylene in the molten state with *t*-butyl cumyl peroxide and di-*t*-butyl peroxide, a much greater degree of crosslinking was obtained at the same effective radical flux for the peroxide containing the cumyl, which at 190 °C generates a phenyl radical that is more reactive to abstraction than the methyl radicals generated from the decomposition of the *t*-butoxy group [43]. Bremner and Rudin observed similar effects in polyethylene crosslinking, with two peroxides capable of decomposing to give cumyloxy radicals (dicumyl peroxide and α,α'-bis-*t*-butylperoxy diisopropylbenzene), giving significantly more crosslinking than a peroxide that could not (Luperox 130, 2,5-dimethyl-2,5-di(*t*-butylperoxy) 3-hexyne) [44]. (However, these observations could equally well be explained by the reduced k_d of Luperox 130 and initiation in all cases by methyl radicals formed by β-scission.)

Parent reported that in the grafting of the monomers maleic anhydride or vinyl triethylsiloxane (both of which homopolymerize poorly) to ethylene-vinyl acetate copolymers, the influence of the initiator could be adequately explained in terms of peroxide decomposition rates alone [29].

Further possible effects of the initiator arise in multiphase systems. The *t*-butyl peroxylaurate/ferrous sulfate redox couple generates hemispherical blobs of polybutadiene on the surface of polydimethylsiloxane latex particles, while the potassium persulfate/ferrous sulfate redox couple generates well-formed core-shell polymers [45]. This is less likely to indicate differences in grafting efficiency than to indicate that the sulfate-terminated polybutadiene oligomers formed from the latter system are surface active and to enable the spreading of the polybutadiene phase over the water/polymer interface.

Effect of the Backbone

The backbone polymer will affect the kinetics of grafting and crosslinking from two directions: (1) How readily can reactive centers be generated on the backbone by reactions with the initiator? (2) How readily can the reactive centers that were generated react with the monomer (to give grafting) or with other reactive centers (to give crosslinking)? If the radicals formed are highly reactive towards the monomer, they will readily give grafting, but if they are unreactive they are more likely to give crosslinking. The most readily abstractable moieties are those that give rise to the most stable radical species. The most suitable double bonds for addition will have low steric hindrance (a minimum of substitution at the "tail") and high electronic stability of the resultant radical species (radical-stabilizing substituents at the "head"). Backbone radicals may also rearrange by β-scission [46] or abstract atoms from other species present to give a more stable radical species.

Huang and Sundberg found that grafting of a range of monomers to a polybutadiene containing significant amounts of 1,2 addition was primarily by addition to the pendant double bonds, with abstraction of the allylic tertiary hydrogens giving rise to retardation by the generation of radicals inactive in addition [42]. Zhang et al. found that the relative degree of crosslinking increased sharply as the butadiene-styrene content of an elastomer increased, while the grafting efficiency of a vinyl monomer containing a carbonate group was independent of the nature of the elastomer. This implies that allylic radicals are largely involved in termination, and that it is other radical species that are chiefly significant in grafting [47].

For most conceivable mechanisms, the grafting rate will be proportional to the concentration of appropriate functional groups on the backbone. This proportionality is most clear at low concentrations, and rates of reactions are likely to increase more slowly or fall at high concentrations due to the difficulty of diffusion through the matrix.

Effect of the Monomer

The relative reactivity of a monomer toward its own polymer and other radicals generated in the system will be most critical for grafting and crosslinking reactions. Monomers that readily homopolymerize will encourage grafting over crosslinking, and are more likely to generate significant amounts of homopolymer (e.g., butyl acrylate); monomers that are reluctant to homopropagate will give short grafts and may enhance crosslinking (e.g., ethyl maleate) [37, 48].

Huang and Sundberg investigated the grafting of a number of monomers onto poly(*cis*-isoprene) (in dilute solution in order to avoid phase separation and the viscosity-dependent effects on propagation and termination) [32]. They determined a grafting coefficient k_{gr} of ~0.1 $dm^3mol^{-1}s^{-1}$ for styrene, ~1.7 for butyl methacrylate, and ~24.0 for butyl acrylate.

In a miniemulsion system of grafting to alkyd resins, Tsavalas et al. found better grafting of butyl acrylate (BA) than methyl methacrylate (MMA) [41]. Here, all radicals are initially generated in the aqueous phase and it is surface-active $R[M]_z^*$ radicals that initiate grafting. The improved grafting with BA instead of MMA may thus conceivably be due to better addition of a backbone polymer radical to BA instead of MMA, better abstraction by the poly(BA) radical from the backbone to generate polymer radicals, better addition of poly(BA) across double bonds in the alkyd, or a higher mobility of $R[BA]_z^*$ than $R[MMA]_z^*$ within the alkyd matrix because of the lower z value for the more hydrophobic BA.

It is often possible to enhance grafting or crosslinking through addition of a second monomer. For instance, grafting of glycidyl methacrylate (GMA) to LDPE was enhanced by a factor of 3 to 4 when a 1.5:1 styrene:GMA feed was used [43]. This is probably due to an enhancement of k_p for the cross-propagation reactions of a styrene-terminated radical with GMA and the GMA-terminated radical with styrene in comparison to GMA homopolymerization. This trivial mechanism may also be the basis for reports of an enhanced grafting rate in other copolymerizations [49, 50].

A more chemically interesting possibility for the participation of a comonomer would be a contribution from enhanced reactivity of the polymer macroradical toward the comonomer rather than the primary monomer. A rough estimation based on the *Q-e* scheme, approximating the polypropylene radical by polyisobutylene [51], suggests that a propylene backbone radical would be about 3 times as reactive toward styrene as toward GMA.

Monomers that are reactive toward the backbone macroradical, but that generate radicals, which are unreactive towards propagation, have been used to enhance crosslinking. Spencer et al. investigated the grafting of vinyltrimethoxysilane (VTMS) onto PE and found that the monomer was not grafted homogeneously, but was concentrated in "clusters" generated by chain transfer from a VTMS radical to a nearby site liable to H-abstraction, which was then able to add additional VTMS (Scheme 3.7) [52]. Rätzsch et al. found that a large number of nonhomopolymerizable monomers giving short chains onto polypropylene also gave these "nests" or "clusters" of grafting sites (vinyltriethylsiloxane, monovinyl ester of ethylene glycol, vinyl butyl ether, vinyl isobutyl ether) [48]. The overall result of maintaining an unreactive polymer radical in these zones, instead of reactivity being transferred to a more reactive polymer radical, is to enhance crosslinking over grafting [53]. Addition of nonhomopolymerizable monomers can also inhibit degradation mechanisms in polymers such as PP where the backbone is vulnerable to β-scission reactions.

Dibutyl and diethyl maleate are reported to suppress rather than enhance crosslinking under similar conditions [54, 55]. This may be related to the generation of sufficient primary radicals to terminate the maleate radicals on the backbone. In Rätzsch's work, nonhomopolymerizable monomers that did not give clusters of grafted units were either allylic (and hence generated radicals

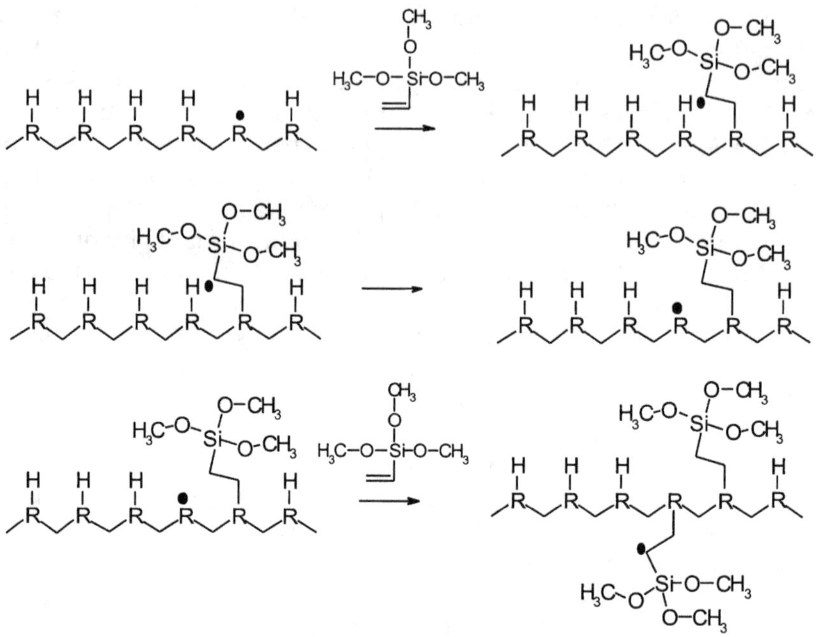

Scheme 3.7. First stages of formation of a "nest" of single-monomer grafts of the nonhomopolymerizable monomer VTMS.

with little propensity to abstract from the backbone) or were unable to swell the polypropylene substrate (e.g., dibutyl maleinate) [47].

In melt processing, peroxide-initiated grafting of ethylene-propylene diene terpolymer with GMA gave more grafting and less homopolymer formation when a reactive comonomer, trimethylolpropane triacrylate, was added [56]. This is probably due to pendant acrylate groups arising from incorporation of the triacrylate in the polymer, providing an alternative mechanism for incorporating poly(GMA) radicals.

Effect of Additives and Solvents

In general, additives employed in radical grafting and crosslinking reactions will act in two ways: as radical traps or as chain-transfer agents. There is a continuum between "comonomers" and "additives": bis(2-furanyl aldazine) and *N,N′*-*p*-phenylenebis(maleimide), for example, have double-bonds susceptible to attack by polymer backbone radicals, but do not propagate or undergo scission [57]. While they are considered "additives," they behave identically to the nonhomopolymerizable "comonomers" discussed above as promoters of crosslinking and inhibitors of chain scission.

Aravindakshan et al. investigated the redox-initiated grafting of methacrylic acid onto starch and found that the presence of triethylamine or mercaptosuccinic acid as chain-transfer agent reduced the amount of effective grafting [58]. This observation is not surprising for a "grafting from" mechanism, as a chain transfer will result in premature termination of grafted side chains and generate radicals that will usually be poor at reinitiating by abstraction or addition. In a "grafting to" mechanism, chain-transfer agents could also play a negative role by terminating chains that would otherwise combine with the substrate.

In the peroxide crosslinking of PE, addition of organic nitroxides can retard the rate of cure by reducing the radical flux, but monofunctional nitroxides also impair the final properties by reducing the crosslink density. Bifunctional nitroxides, however, can enhance formation of intermolecular crosslinks [59]. For the same system (PE crosslinking with peroxide), Anbarasan et al. found the degree of crosslinking achieved in the presence of ester additives was inversely correlated with the degree to which the additives were incorporated into the polymer [60]. It is probable that these additives form radicals by abstraction and terminate radicals on the polymer backbone.

Gaylord observed that the significant enhancement in crosslinking achieved with molten PE in the presence of maleic anhydride vanishes upon addition of any of a large number of sulfur-, nitrogen-, or phosphorus-containing compounds that also inhibit maleic anhydride homopolymerization [61]. This inhibition of grafting was observed even when DMSO, dimethylacetamide, or nonylphenyl phosphite were added at an additive-to-maleic anhydride ratio of 1:10, where insufficient amounts will be present to react with maleic anhydride in some fashion before it attaches to grafting sites [62].

Chauhan et al. investigated the grafting of vinyl imidazole onto cellulose in the presence of various solvents and additives [40]. Water increased the overall rate, as a good source of initiating radicals under irradiation, but gave mostly homopolymerization rather than grafting. Methanol-inhibited grafting and impaired homopolymerization, presumably as a source of abstractable hydrogen for scavenging backbone radicals, and similar effects were seen with zinc chloride. At the same concentration, ammonium persulfate and Mohr's salt both reduced grafting more than homopolymerization, while tetramethylethylenediamine and KPS reduced homopolymerization more than grafting. The key to the influence of these additives is most likely their relative affinity for the cellulose surface. Similar effects were observed in the grafting of acrylamide onto hydroxypropyl cellulose [50]. Under identical conditions acetone gave reasonable grafting (47%), followed by dioxane (40%), benzene (18%), methanol (18%), and water (15%).

PHYSICAL EFFECTS

In the discussion above of chemical effects on grafting and crosslinking kinetics, the state of the system has often been mentioned: whether it is a dilute solution, a viscous melt, or is heterogeneous. This state will have a profound influence on the kinetics. The overall viscosity of the system will govern whether reactions involving the diffusion of large species can be significant, while heterogeneous systems will involve the interplay of chemical and physical effects. It should be noted that many initially homogeneous systems of interest are likely to develop heterogeneity over time. The degree of grafting achieved will generate a compatibilizer that will help to determine how and when phase separation will occur, making grafted systems to an extent for entities that pull themselves up by their own bootstraps.

Higaki et al. investigated the crosslinking of Atom Transfer Radical Polymerization (ATRP) terminated poly(MMA) and nitroxide-terminated polystyrene and found a very significant influence of gelation [63]. As the concentration of the polymer increased, more rapid and complete crosslinking was observed, but beyond the point of gelation there was negligible further increase in crosslinking.

The work of Yang and Rånby [64] provides an illustration of how changes in the physical environment change the kinetics of polymerization. For grafting of acrylic acid with BPO, a 0.89 order dependence on BPO was found for homogeneous solutions, while 0.66 order dependence was found for surface reactions. As the expected theoretical dependence on initiator concentration for "grafting from" reactions should be 0.5 for diradical coupling termination and 1 for monoradical termination, this suggests that both systems are intermediate cases where both mechanisms apply, but the enforced proximity of the polymer radicals in the surface-grafting reaction encourages bimolecular termination.

In the grafting of styrene onto pre-irradiated fluoropolymer films, Guilmeau et al. found the proportion of polystyrene surface coverage to be lower than the final proportion of polystyrene by volume if reactions were allowed to proceed to completion [65] and observed a greater tendency toward heterogeneity with increasing dose rate and sample thickness. These results suggest that in this system styrene preferentially swells the polystyrene phase, and this partitioning dominates the kinetics and the final morphology of the system.

A feature of rapid polymerization in crosslinking systems is that the material may not have time to relax to its equilibrium density; hence, more free volume will be available and k_t and k_p will not fall as rapidly as would be expected in the reactive diffusion regime [66]. Generally, all photoinitiated polymerizations and analogous systems will see a competition between cure speed and cure depth; the more rapidly a crosslinking or grafting polymerization proceeds, the less likely it is to give a homogeneous product.

CONTROLLED RADICAL POLYMERIZATION

A "grafting from" process where the controlled radical polymerization agent (brominated functional group, nitroxide, or RAFT agent) is incorporated on the polymer backbone has frequently been employed to produce grafted polymers [67]. Reduction of the termination rate due to the "living" process can result in much higher grafting yields on surfaces [68].

As a controlled radical polymerization proceeding at a reasonable rate will not reduce the overall concentration of radicals, combination reactions can still be significant. Muehlebach and Rime found that grafting BA and dimethylacrylamide onto poly(BA) by ATRP gave multimodal molecular weight distributions, which indicated that crosslinking could still occur, but the grafting of dimethylaminoethyl methacrylate gave a monomodal molecular weight distribution indicative of grafting only [69]. Most controlled radical polymerizations reduce crosslinking. This may be because the very short lifetimes of free polymer radicals makes them less likely to undergo intermolecular reactions leading to persistent radicals.

Variations in the kinetic behavior of controlled radical polymerizations between bulk polymerizations and surface-grafting reactions have been noted in a number of cases. To achieve living polymerization of MMA on silica gel particles, for example, a much higher concentration of the ATRP catalyst was required to achieve controlled polymerization than for the same reaction in solution [70]. Similar surface congestion was reported in the grafting of styrene on silica by nitroxide-mediated polymerization [71], while the rate of polymerization of 2-(methacryloyloxy) ethyltrimethylammonium chloride to surfaces was inversely correlated with the density of grafting [72]. In the RAFT-mediated grafting of a number of monomers to silica surfaces, styrene and BA polymerized more slowly than in solution, while MMA polymerized more rapidly [73]. These systems all illustrate physical effects that can be expected when controlled radical polymerization methods are applied to heterogeneous systems.

REFERENCES

1. Quirk RP. *Prog Rubber Plastics Technol*, **4(1)**: 31 (1988).
2. Pham BTT, Monteiro MJ, Gilbert RG, Tonge MP. *Macromolecules*, **33**: 2383 (2000).
3. Decker C. *Prog Polym Sci*, **21(4)**: 593 (1996).
4. Dietz JE, Peppas NA. *Polymer*, **38(15)**: 3767 (1997).
5. Edelmann D, Ritter H. *Macromol Rapid Commun*, **15(10)**: 791 (1994).
6. Noureddini H, Timm DC. *Macromolecules*, **25(6)**: 1725 (1992).
7. Kochi JK. *Free Radicals* (New York: Wiley, 1973).
8. Estenoz DA, Leal GP, Lopez YR, Oliva HM, Meira GR. *J Appl Polym Sci*, **62(6)**: 917 (1996).
9. Coote ML. *J Phys Chem A*, **108(17)**: 3865 (2004).
10. Cook WD. *Polymer*, **33(10)**: 2152 (1992).
11. Barner-Kowollik C, Buback M, Charleux B, Coote ML, Drache M, Fukuda T, Goto A, Klumperman B, Lowe AB, McLeary JB, Moad G, Monteiro MJ, Sanderson RD, Tonge MP, Vana P. *J Polym Sci Polym Chem*, **44(20)**: 5809 (2006).

12. Matyjaszewski K. *Macromol Symp*, **182**: 209 (2002).
13. Vana P, Davis TP, Barner-Kowollik C. *Macromol Theory Simul*, **11**: 823 (2002).
14. Tsubokawa N, Iida T, Takayama T. *J Appl Polym Sci*, **75**(**4**): 515 (2000).
15. Bouajila J, Raffin G, Waton H, Sanglar C, Paisse JO, Grenier-Loustalot MF. *Polym Polym Compos*, **10**(**5**): 341 (2002).
16. Kolesnikova NN, Ostroukhova OA, Popov AA. *J Appl Polym Sci*, **104**(**2**): 754 (2007).
17. Pan S-S, Morawetz H. *Macromolecules*, **13**(**5**): 1157 (1980).
18. Cockbain EG, Pendle TD, Turner DT. *J Polym Sci*, **39**: 419 (1959).
19. Gilbert RG. *Emulsion Polymerization: A Mechanistic Approach* (London: Academic, 1995).
20. Lamb D, Anstey JF, Fellows CM, Monteiro JM, Gilbert RG. *Biomacromolecules*, **2**: 518 (2001).
21. Anstey JF, Subramaniam N, Pham BTT, Lu X, Monteiro MJ, Gilbert RG. *Macromol Symp*, **150/151**: 73 (2000).
22. Löser U, Scherzer K, Weber K. *Z Phys Chem (Leipzig)*, **270**(**2**): 237 (1989).
23. El-Wakil AA. *J Appl Polym Sci*, **101**(**2**): 843 (2006).
24. Trommsdorff E, Kohle E, Lagally P. *Makromol Chem*, **1**: 169 (1948).
25. Russell GT, Gilbert RG, Napper DH. *Macromolecules*, **25**(**9**): 2459 (1992).
26. Heuts JPA, Russell GT, Smith GB, van Herk AM. *Macromol Symp*, **248**: 12 (2007).
27. Griffiths MC, Strauch J, Monteiro MJ, Gilbert RG. *Macromolecules*, **31**: 7835 (1998).
28. Chuayjuljit S, Siridamrong P, Pimpan V. *J Appl Polym Sci*, **94**(**4**): 1496 (2004).
29. Parent JS, Tripp M, Dupont J. *Polym Eng Sci*, **43**(**1**): 234 (2003).
30. Fischer JP. *Angew Makromol Chem*, **33**: 35 (1973).
31. Huang NJ, Sundberg DC. *J Polym Sci Polym Chem*, **33**: 2533 (1995).
32. Huang NJ, Sundberg DC. *J Polym Sci Polym Chem*, **33**: 2551 (1995).
33. Huang NJ, Sundberg DC. *J Polym Sci Polym Chem*, **33**: 2571 (1995).
34. Zhou ZF, Huang H, Liu NC. *J Appl Polym Sci*, **85**(**4**): 726 (2002).
35. Zhou ZF, Huang H, Liu NC. *J Polym Sci Polym Chem*, **39**(**4**): 486 (2001).
36. Vivaldo-Lima E, Roa-Luna M, Saldivar-Guerra E. *Macromol Theor Simul*, **14**(**9**): 539 (2005).
37. Pesetskii SS, Jurkowski B, Krivoguz YM, Kelar K. *Polymer*, **42**(**2**): 469 (2000).
38. Pesetskii SS, Jurkowski B, Krivoguz YM, Olkhov YA. *J Appl Polym Sci*, **81**(**14**): 3439 (2001).
39. Aggour YA. *Polym Int*, **53**(**12**): 1930 (2004).
40. Chauhan GS, Singh B, Chauhan S, Dhiman SK, Kumar D. *J Appl Polym Sci*, **100**(**2**): 1522 (2006).
41. Tsavalas JG, Luo Y, Schork FJ. *J App Polym Sci*, **87**(**11**): 1825 (2003).
42. Huang NJ, Sundberg DC. *J Polym Sci Polym Chem*, **33**: 2587 (1995).
43. Torres N, Robin JJ, Boutevin B. *J Appl Polym Sci*, **81**(**3**): 581 (2001).
44. Bremner T, Rudin A. *J Appl Polym Sci*, **49**: 785 (1993).

45. Okaniwa M. *Polymer*, **41**(**2**): 453 (1999).
46. Bettini SHP, Agnelli JAM. *J Appl Polym Sci*, **74**: 247 (1999).
47. Zhang H, Wang J, Cao S, Shan A. *Polym Advan Technol*, **11**(**7**): 334 (2000).
48. Rätzsch M, Bucka H, Ivanchev SS, Mesh AM, Khaikine. *J Appl Polym Sci*, **77**(**4**): 711 (2000).
49. Hoo VM, Whitney RA, Baker WE. *Polymer*, **41**(**11**): 4367 (2000).
50. Chauhan GS, Sharma R, Lal H. *J Appl Polym Sci*, **91**(**1**): 545 (2004).
51. Brandrup J, Immergut EH, Grulke EA, Eds. *Polymer Handbook* (New York: John Wiley & Sons, 1999).
52. Spencer M, Parent JS, Whitney RA. *Polymer*, **44**(**7**): 2015 (2003).
53. Gaylord NG, Mehta M. *J Polym Sci Polym Lett*, **20**: 481 (1982).
54. Aglietto M, Ruggeri G, Luppichini E, D'Alessio A, Benedetti E. *Mater Eng*, **4**(**2**): 253 (1993).
55. Sen A, Mukherjee B, Bhattacharyya AS, De PP, Bhowmick AK. *Angew Makromol Chem*, **191**: 15 (1991).
56. Al-Malaika S, Kong W. *Polym Degrad Stabil*, **90**(**2**): 197 (2005).
57. Coiai S, Passaglia E, Aglietto M, Ciardelli F. *Macromolecules*, **37**(**22**): 8414 (2004).
58. Aravindakshan P, Bhatt A, Kumar VG. *J Appl Polym Sci*, **66**(**2**): 397 (1997).
59. Chaudhary BI, Chopin L, Klier J. *Polym Eng Sci*, **47**(**1**): 50 (2007).
60. Anbarasan R, Babout O, Degueil M, Maillard B. *J Appl Poly Sci*, **97**(**3**): 766 (2005).
61. Gaylord NG, Mehta R. *J Polym Sci A Chem*, **26**: 1189 (1988).
62. Gaylord NH, Mehta R, Kumar V, Tazi M. *J Appl Polym Sci*, **38**: 359 (1989).
63. Higaki Y, Otsuka H, Takahara A. *Macromolecules*, **39**(**6**): 2121 (2006).
64. Yang W, Rånby B. *J Appl Polym Sci*, **62**(**3**): 533 (1996).
65. Guilmeau I, Esnouf S, Betz N, Le Moël A. *Nucl Instrum Meth B*, **131**: 270 (1997).
66. Bowman CN, Peppas NA. *Chem Eng Sci*, **47**: 1411 (1992).
67. Morandi G, Montembault V, Pascual S, Legoupy S, Fontaine L. *Macromolecules*, **39**(**8**): 2732 (2006).
68. Holmberg S, Holmlund P, Wilen C-E, Kallio T, Sundholm G, Sundholm F. *J Polym Sci Polym Chem*, **40**(**4**): 591 (2002).
69. Muehlebach A, Rime F. *J Polym Sci Polym Chem*, **41**(**21**): 3425 (2003).
70. Shen Y, Zhu S, Zeng F, Pelton R. *J Polym Sci Polym Chem*, **39**(**7**): 1051 (2001).
71. Bartholome C, Beyou E, Bourgeat-Lami E, Chaumont P, Zydowicz N. *Macromolecules*, **36**(**21**): 7946 (2003).
72. Cheng N, Azzaroni O, Moya S, Huck WTS. *Macromol Rapid Commun*, **27**(**19**): 1632 (2006).
73. Li C, Benicewicz B. *Polym Prepr (Am Chem Soc, Div Polym Chem)*, **46**(**2**): 274 (2005).

4

ANALYTICAL EVIDENCE

Amit Bhattacharya and Paramita Ray

INTRODUCTION

The modification of polymers can be proved by various analytical tools. The mostly used common techniques are spectroscopic, thermal, study of mechanical properties, determination of molecular weight, morphological, rheological analysis, and contact angle measurement. They are discussed in brief in this chapter.

EVIDENCE FROM SPECTROSCOPIC TOOLS

Spectroscopy is the study of the interaction and interpretation of the interactions of electromagnetic radiations with matter. This is the sophisticated tool to get information about structural make up, the changes, and various paths to changes or modification at the atomic or molecular level of the substance. When photons, the discrete energy packets ($h\upsilon$) of electromagnetic radiations interact with substance, these results in transition in various energy levels as shown in the Table 4.1.

These transitions give a specific impression of the substance as a spectrum. From the spectrum, one can gather information about the structural, constitutional, conformational, and interactional behavior of and about the substance, i.e., the atom or molecule. Ultraviolet-visible, infrared, and nuclear magnetic resonance (NMR) spectroscopy are the most used tools in this regard for

Polymer Grafting and Crosslinking, Edited by Amit Bhattacharya, James W. Rawlins and Paramita Ray

TABLE 4.1. Type of energy transition in each region of the electromagnetic spectrum.

Region of Spectrum	Energy Transition
X rays	Bond breaking
Ultraviolet/visible	Electronic
Infrared	Vibrational
Microwave	Rotational
Radio frequencies	Nuclear spin/Electronic spin

polymer and related materials. These have a very crucial role in the establishment, development, and advancement of the new polymer microstructures as they nurture various aspects of the polymer microstructure such as elucidation of the structure, the structure/properties relationship, modification in the base structure, and related mechanisms of modification or polymerization in both quantitative and qualitative ways.

By UV-Visible Spectroscopy

In this case, electromagnetic radiation results in a transition in the electronic energy levels. As the molecule absorbs energy in UV-visible region, the electrons jump from the highest occupied molecular orbitals (HOMOs) to the lowest unoccupied molecular orbital (LUMO). This energy difference lies between 125 and 650 kJ/mole in most of the molecules. The energy order of various orbitals is as

$$\sigma < \pi < n < \pi^* < \sigma^*.$$

The unoccupied or highest energy orbitals are π^* and σ^*. The transition may occur as shown

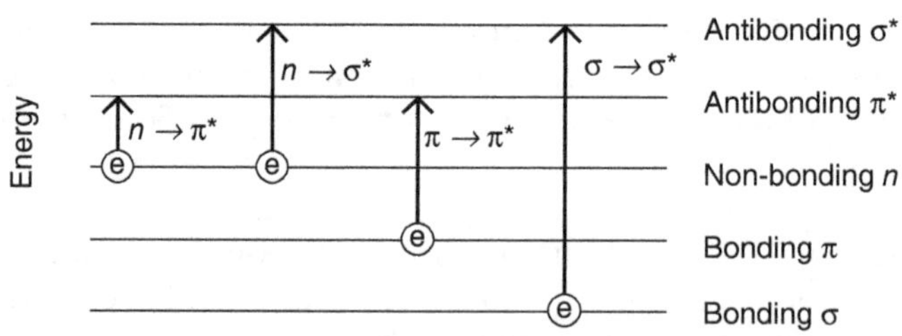

The intensity of the absorption is governed by the Beer-Lambert Law depending on the number of molecules present that are capable of absorbing light and may be formulated as

$$A = -\log(I/I_0) = \varepsilon.c.l$$

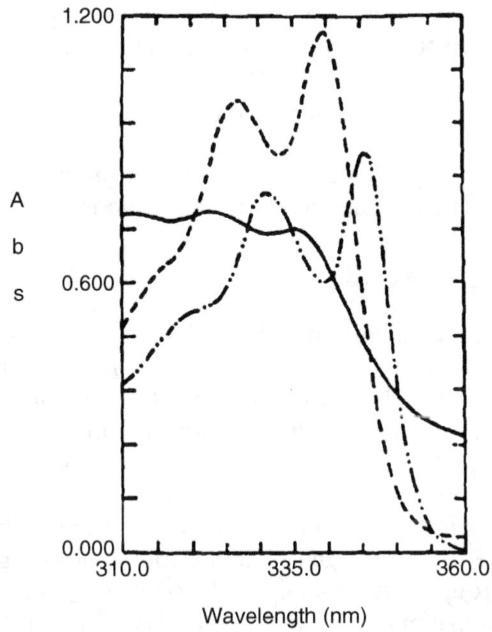

Figure 4.1. Absorption spectra of VCZ (---) and ECZ (—··—) solution in tetrahydrofuran and VCZ grafted on cellulose acetate film (—). Reproduced from *J Polym. Sc Polym Chem* **32**: 2251 (1994).

for a given wavelength, where A = absorbance, I = intensity of the light transmitted, I_0 = intensity of incident light, l = length of the sample cell, c = molar concentration of solute, and ε = molar absorptivity.

When a polymer is subjected to some modification or change, by having the UV-Vis spectra of the base matrix and the treated sample, it can be analyzed for the change if any one of the cases is active to this range.

Figure 4.1 shows the evidence of grafting in the polymer matrix [1]. It is evident from the figure that the absorption peaks of grafted *N*-vinyl carbazole (VCZ) occur almost in the same position as VCZ. However, the spectra should resemble ethyl carbazole as the opening of the double bond of vinyl carbazole results, but due to the rigidity of the molecule (because of grafting), it is not in same position as the ethyl carbazole.

By Infrared Spectroscopy

When infrared radiation (IR) interacts with molecules, it causes transitions in the vibrational levels, due to various stretching and bending movements that occur in the molecules. Thus the IR absorption spectrum of various close-packed absorption bands is obtained, which has specific values for the particular functional group and bonds present in the molecule. The band intensities in the IR spectrum are expressed either as transmittance (T) or absorbance (A), which are related to each other by the relation as

$$A = \log_{10}(1/T).$$

The IR spectra from $4000\,cm^{-1}$ to $667\,cm^{-1}$ is the most revealing for the elucidation of the structural information about the molecule.

By Attenuated Total Reflectance

Attenuated total reflectance is the supportive technique used in this row to measure the absorption spectra of the materials such as flat surfaces or films that are not transparent or hard enough to be placed in the infrared cell. Using this technique, a sample of any film of a few square millimeters and with a thickness of between 20 and 1000 μm can be analyzed. The IR spectrum is a very simple and preliminary informative source. The IR spectrum can be used to trace out the information about the molecular order and even one variant of the degree of the crystalinity in polymers, fibers, and composite films. When some functional group or monomers with some functionalities are incorporated in the base polymers [2, 3], this is the very common technique used to analyze it, as shown in Figure 4.2, where cardo polyetherketone (PEK-c) membranes are photografted with polyacrylic acid (C=O peak at $1725\,cm^{-1}$). The figure also shows that as the grafting increases with the irradiation time, the peak intensity also increases.

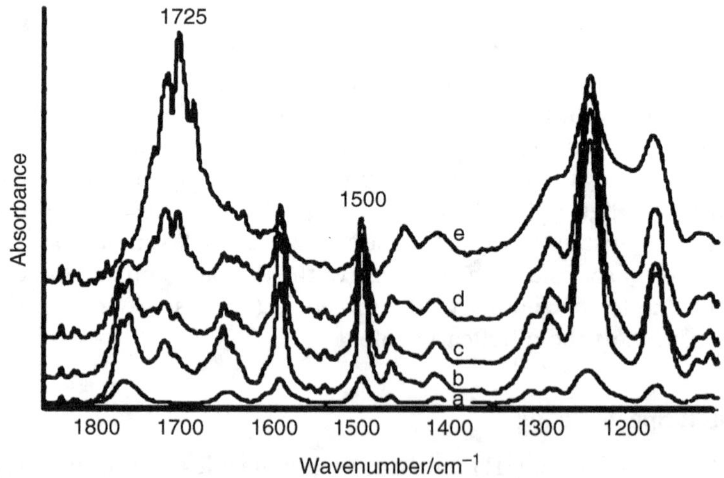

Figure 4.2. FTIR-ATR spectra of a PEK-C membrane modified with grafted PAA for different irradiation times: (a) 0 min, (b) 1 min, (c) 2 min, (d) 3 min, (e) 5 min in AA 10 wt. %. Irradiation intensity: $8.7\,mW/cm^2$. Reproduced from *J Membr Sci*, **255**: 107 (2005).

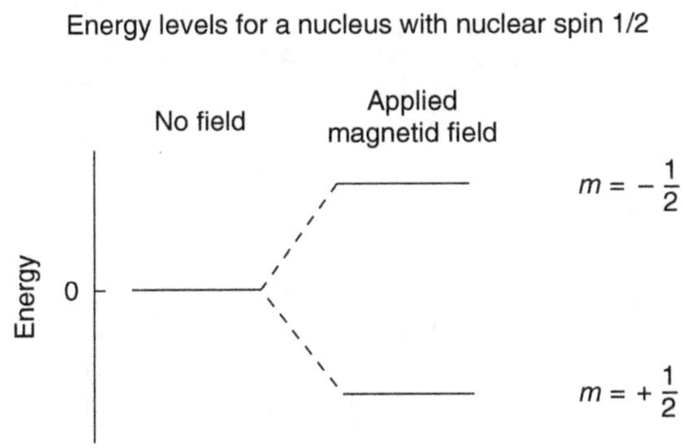

Figure 4.3. Schematic presentation of energy levels with nuclear spin 1/2.

By NMR Spectroscopy

In NMR spectroscopy, radio wave frequency is required from the electromagnetic spectrum. Some of the atomic nuclei possess nuclear spin (I), e.g., ^{1}H, ^{13}C, ^{19}F, ^{31}P, etc., which makes them behave like a bar magnet and have ($2I$+1) allowed spin states. In the absence of the external magnetic field, these states are degenerate. When an external magnetic field is applied, this degeneracy is removed as shown in Figure 4.3.

^{1}H and ^{13}C both have a nuclear spin of 1/2 each, and these split into two states; one remains aligned along the applied magnetic field and another remains aligned against it. The energy difference is given as

$$\Delta E = \gamma (h/2\pi) B_0 = h\upsilon,$$

where γ is magnetogyric ratio, which is a constant for each nucleus, B_0 is the strength of applied magnetic field, h is Planck's constant, and υ is the frequency of the radio waves.

With the help of this spectroscopy, one can analyze the different types of chemical environments around various types of hydrogen in **^{1}H NMR** and carbon in **^{13}C NMR** present in the molecule. In the analysis of the polymer composition (as shown in Figure 4.4 where polyvinyl alcohol is cografted with polyacrylic acid [4]) the distinct peak of polycarboxylic acid appears in the spectrum. NMR is more useful than IR for branch analysis, copolymer sequencing, distribution of copolymer, and tacticity in various polymers can be analyzed more thoroughly in NMR than IR.

By X-Ray Photoelectron Spectroscopy

This technique is used for the analysis of the chemical constituents at the surface of the substance. This is also known as electron spectroscopy for chemical analysis (ESCA). In this technique, under ultrahigh vacuum, X rays are

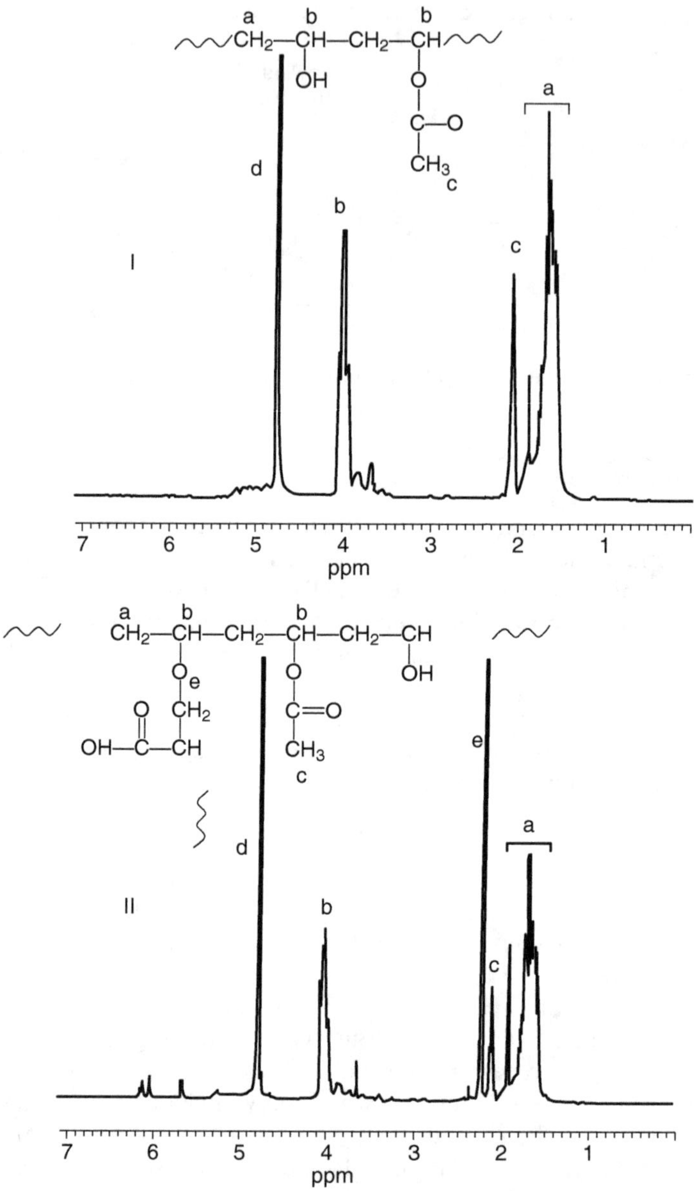

Figure 4.4. (I) 1H NMR spectra of pure PVA and (II) 50% grafted copolymer. Reproduced from *Smart Mater Struct*, **15**: 417 (2006).

made to interact with the top surface that emits the inner shell electrons of the elements present on the surface. The function of binding energy of respective elements can be formulated as

$$E_{binding} = E_{photon} - E_{kinetic} - \varnothing,$$

where $E_{binding}$ is the binding energy of the electron emitted from one electron configuration within the atom, E_{photon} is the energy of the X-ray photons, $E_{kinetic}$ is the kinetic energy of the emitted electron, and Ø is the work function of the spectrometer.

Thus by analyzing the number of electrons emitted by the specific area of the surface from the top 1–10 nm that are specific to the binding energy of particular elements, we can gain a qualitative insight into the various elements present on the surface. The films of poly(ε-caprolactone) (PCL), poly(L-lactide) (PLLA), poly(lactide-*co*-glycilide) (PLGA), and poly(trimethylene carbonate) (PTMC) do not have any nitrogen atoms in their composition but when *N*-vinylpyrrolidone is grafted on these membranes, with X-ray photoelectron spectroscopy (XPS), it can be analyzed qualitatively for the modification by examining the C/O and N/O elemental ratio of the top 1–10 nm surface of the films as shown in Figure 4.5 [3].

Thus with the help of XPS we can determine the top surface chemistry of the polymer films or membranes for modified and unmodified surfaces by analyzing the elemental composition of the various constituents and can also determine the number and thickness of the layers present within the top 10 nm surface. This XPS spectrum (Figure 4.6) shows the –COOH group incorporation on the LDPE film after photografting [5].

BY THERMAL ANALYSIS

Thermal analysis comprises a group of techniques in which the physical properties of a substance is measured as a function of temperature, while the substance is subjected to a controlled temperature program [6, 7]. In thermogravimetric analysis (TGA) the variation of the mass of a substance is measured as a function of temperature. In differential thermal analysis (DTA), the temperature difference that develops between a sample and an inert reference is measured when both are subjected to identical heat treatments. The technique of differential scanning calorimetry (DSC) relies on the difference in energy required to maintain the sample and reference at an identical temperature.

Thermogravimetric Analysis (TGA)

Thermogravimetric analysis (a record of change of the mass with time or temperature) is extremely helpful, for example, when testing polymers like

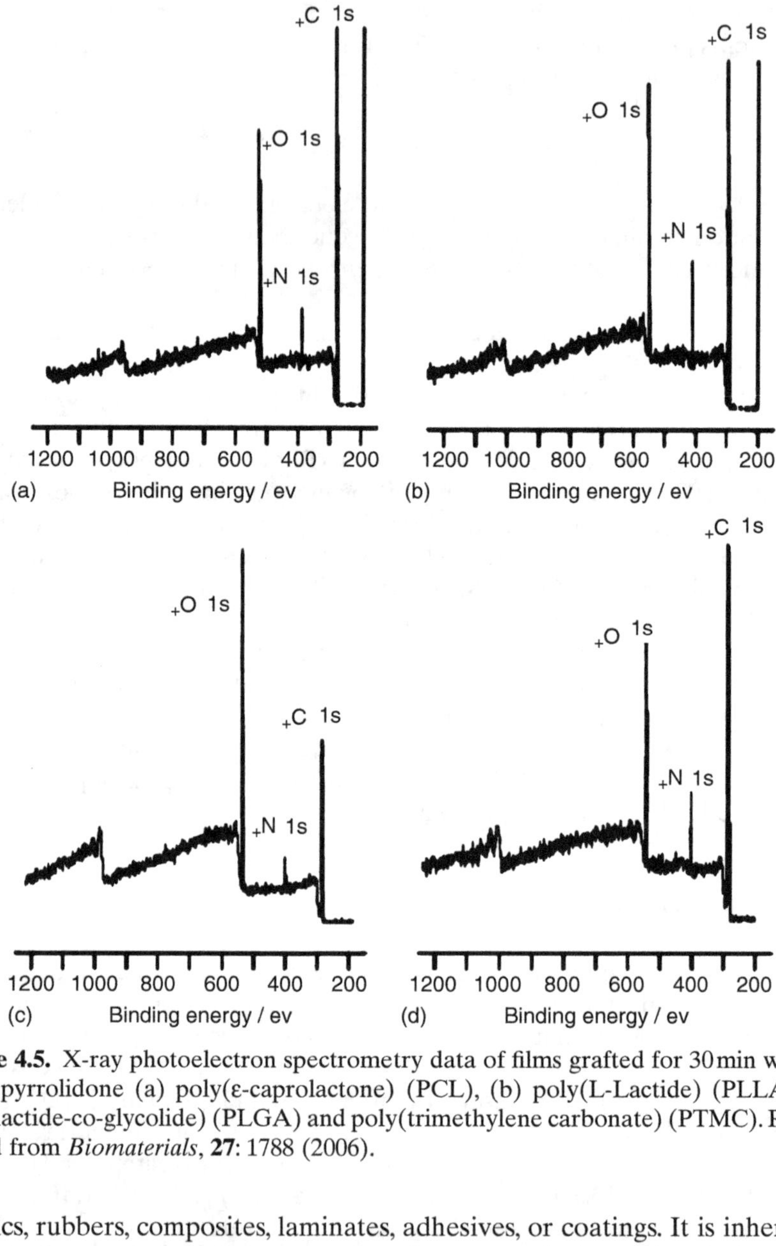

Figure 4.5. X-ray photoelectron spectrometry data of films grafted for 30 min with *N*-vinyl pyrrolidone (a) poly(ε-caprolactone) (PCL), (b) poly(L-Lactide) (PLLA), (c) poly(lactide-co-glycolide) (PLGA) and poly(trimethylene carbonate) (PTMC). Reproduced from *Biomaterials*, **27**: 1788 (2006).

plastics, rubbers, composites, laminates, adhesives, or coatings. It is inherently quantitative, and therefore an extremely powerful thermal technique, but gives no direct chemical information. The main information obtained from TGA is:

- Thermal Stability: related materials can be compared at elevated temperatures under the required atmosphere. The TG curve can help to clarify decomposition temperature and mechanisms.

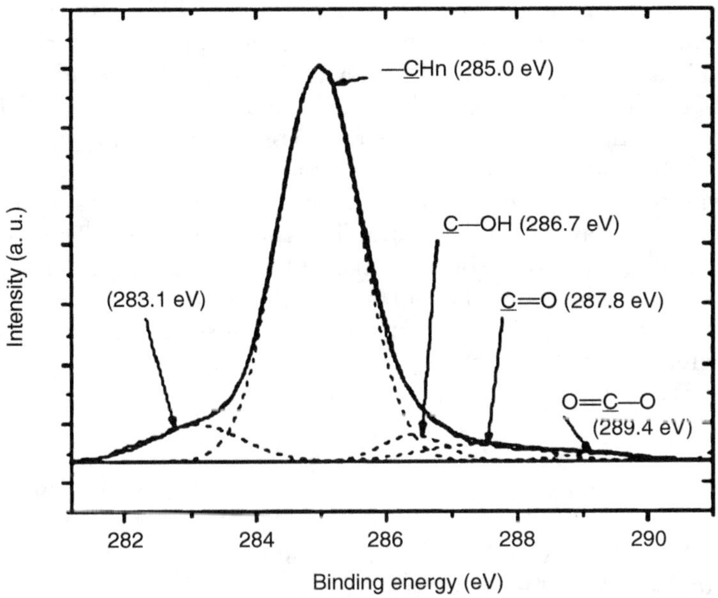

Figure 4.6. XPS spectrum of the modified film. Irradiation time: 60 min; reaction temperature: 70 °C; GD = 0.092 mg/cm^2. Reproduced from *J Appl Polym Sci*, **101**: 2269 (2006).

- Kinetic Studies: a variety of methods exist for analyzing the kinetic features of all types of weight loss or gain, either with a view to predictive studies, or to understand the controlling chemistry.
- Materials Characterization: TGA along with DTA curves can be used to fingerprint polymeric materials for identification or quality control.
- Corrosion Studies: TGA provides an excellent means of studying oxidation or reactions with other reactive gases or vapors. It also can be used to evaluate polymer inflammability.
- Simulation of Industrial Processes: the thermobalance furnace may be thought of as a minireactor, with the ability to mimic the conditions in some types of industrial reactors.
- Compositional Analysis: by careful choice of temperature programming and gaseous environment, many complex materials or mixtures may be analyzed by selectively decomposing or removing their components. This approach is regularly used to analyze filler or ash, oil extender, moisture, and volatile content in polymers.

Differential Scanning Calorimetry (DSC)

This is the thermal analysis technique used to measure the temperatures and heat flows associated with transitions in materials as a function of time and temperature. The measuring principle is to compare the rate of heat flow to the sample and to an inert material (which has no transition in the temperature range of interest) when both are heated or cooled at the same rate. Each is contained in a small holder within an adiabatic enclosure. The temperature of each holder is monitored by a thermocouple, and heat can be supplied electrically to each holder to keep the temperature of the two equal. Changes in the samples are those associated with absorption or evolution of heat that cause a change in the differential heat flow. The difference in energy supplied to the sample against the average temperature is recorded as a peak. The area under the peak is directly proportional to the enthalpy change, and its direction indicates whether the thermal event is endothermic or exothermic.

The greatest impact of this technique has been seen in the study of polymeric materials with crystallinity and melting behavior, glass transition temperature, curing processes, and polymerization. Typical applications for DSC are:

- to determine important transition temperatures like melting temperature (T_m), glass transition temperature (T_g), and to study their nature
- to find out the heat of fusion of a crystalline phase, (ΔH_m), the degree of crystallization, and its kinetics studies
- to determine heat capacity, (C_p), kinetics of crosslinking and degradation reactions, and miscibility in polymer blends and structural relaxation such as enthalpy relaxation during physical changes

Differential Thermal Analysis (DTA)

In this method the specimen polymer and an inert reference material (often powdered alumina) are heated concurrently at a linear rate, each having its own temperature sensing and recording apparatus. The thermal-energy changes (either endothermic or exothermic), which occur in the course of heating, are plotted. This thermogram provides data on the chemical and physical transformations that have occurred, such as melting, sublimation, glass transitions, crystal transitions, and crystallization.

Studies on the thermal properties of grafted and crosslinked polymers are informative in different aspects. It reflects the changes in intrinsic crystallinity, melting behavior, flow properties, nature of degradation (in neutral and oxygen environment), for example, on grafted and crosslinked polymers. The change in thermal behavior depends upon the nature of the base polymer as well as the polymer grafted on it. It also affected by the mode of grafting or crosslinking. The effect of irradiation grafting of polystyrene on three different fluoropolymers like poly(vinylidene fluoride) (PVDF), poly(ethylene tetrafluoro ethylene) copolymer (ETFE), and fluorinated ethylene propylene copolymer (FEP) reflects different thermal behavior. DSC studies reveal that the T_m values of three base polymers remain relatively unaffected with the molar graft level [Figure 4.7(a)] [8]. The shape of the melting peaks were also relatively unchanged by grafting. However, the base polymer crystallinity decreases

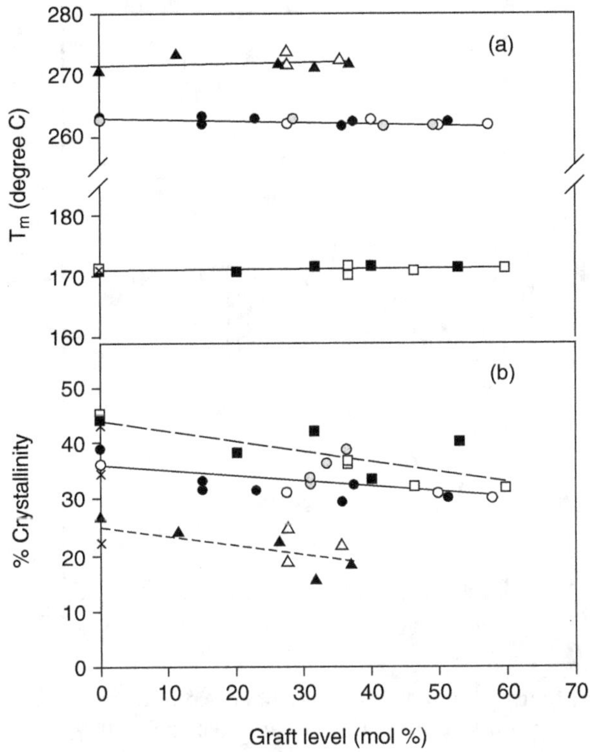

Figure 4.7. (a) T_m's and (b) intrinsic crystallinities of base polymer components: (○) ETFE-g-PS, (□) PVDF-g-PS, and (△) FEP-g-PS films. Prepared through e^-/air (open symbols), ϒ/air (solid symbols), or e^-/N_2 (gray) preirradiation grafting. Reproduced from *J Polym Sci Part B: Polym Phys*, **42**: 2612 (2004).

linearly with the increase in graft level [Figure 4.7(b)]. The decrease in inherent crystallinity is greatest and of about the same magnitude for PVDF and FEP base polymers whereas the slope of the ETFE base polymer is about half the magnitude. Although grafting primarily occurs in the amorphous region of the base polymer, it creates a significant change in the physical dimension of the films. Such large-scale changes in the matrix volume can create reasonable internal stresses that may disrupt the crystalline structure. However, it has been observed that the crystallinity of FEP films remains unaffected if grafting is carried out with a low radiation dose for a long duration [9].

Graft copolymerization of a synthetic polymer on a natural fiber enhances the thermal stability of the fiber [10]. Grafting of polymethyl methacrylate (PMMA) on a lignocellulosic fiber like sisal changes the thermal stability pattern of the fiber (Figure 4.8). PMMA grafting results in an important change in the thermal properties of the sisal fiber. The water evaporation isotherm has been shifted from 90 °C–100 °C to 120 °C indicating improvement

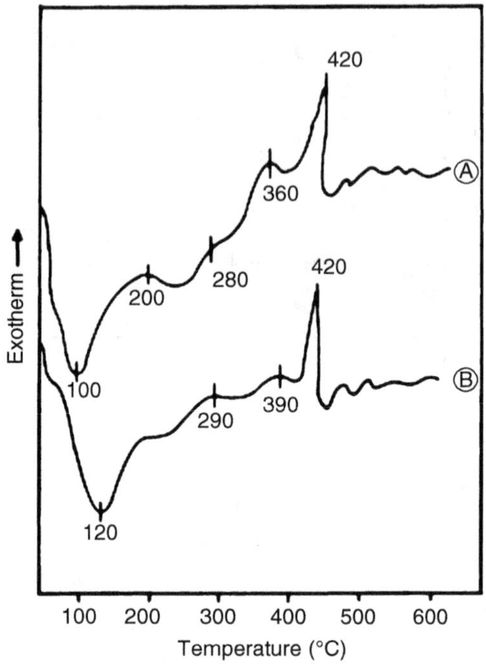

Figure 4.8. DSC thermogram of (A) sisal fiber scoured and defatted, (B) sisal-g-PMMA (16.6%). Reproduced from *Eur Polym J*, **36**: 1597 (2000).

in moisture retention character of the fiber. The hemicellulosic decomposition shoulder has been shifted from 200 °C–230 °C to 290 °C–300 °C indicating overall thermal stability enhancement due to PMMA grafting. The α-cellulosic decomposition peak (at 360 °C) is absent in grafted polymer and a broad peak around 390 °C–400 °C indicates that the α-cellulosic decomposition has combined with the PMMA decomposition exotherm. Thermograms also indicate that the grafting has taken place mostly in the α-cellulosic part, keeping the lignin part free.

T_g is strongly dependent on the mobility of the polymer chains. For modified silica (modified by introducing epoxy functional groups) grafted acrylonitrile-butadiene-styrene nanocomposites T_g increases significantly from 106.6 °C to 117.9 °C with increasing silica content, while the T_g of pure ABS is 99.03 °C (Figure 4.9). Modified silica particles help to achieve a netlike structure in the ABS matrix. The enhancement in T_g is due to the confinement of the ABS chain motion, which is imposed by the covalent bond between ABS and silica particles. The higher the silica content, the larger the restriction of chain motion and the higher the T_g [11].

Crosslinking enhances the thermal stability of polymers. Polyvinyl alcohol is a water-soluble biodegradable polymer. The crosslinking of PVA through its hydroxyl groups opens new application areas such as medical uses, water-

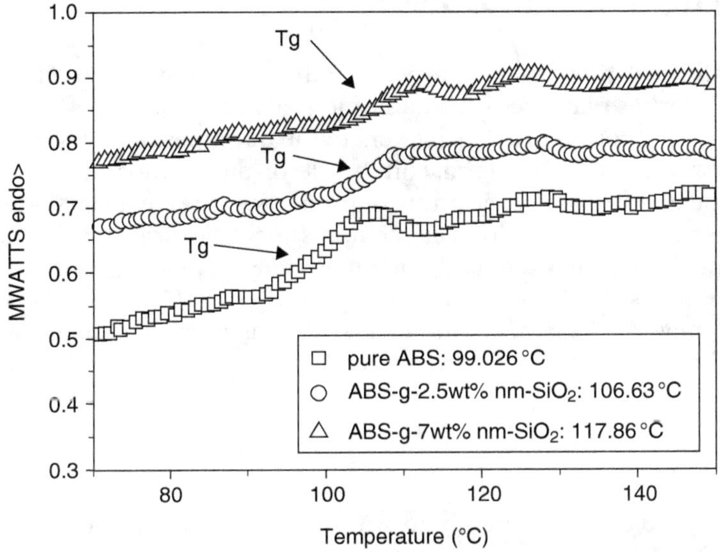

Figure 4.9. DSC thermograms of ABS and grafted nanocomposites. Reproduced from *Eur Polym J*, **36**: 1597 (2000).

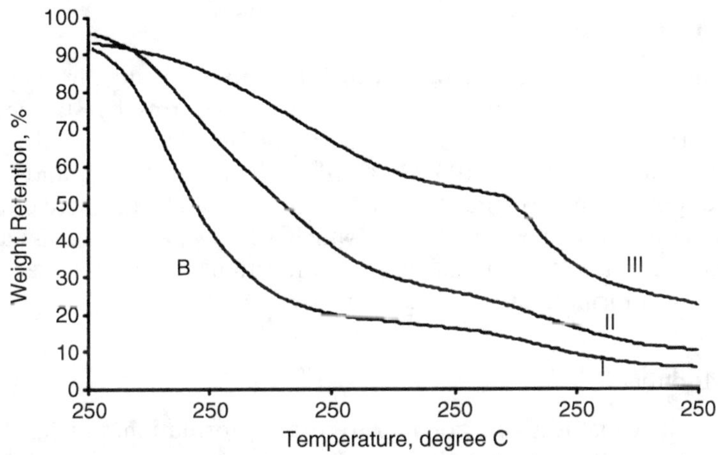

Figure 4.10. Thermogravimetric analysis curves for (I) virgin PVA, (II) heat-treated PVA, (III) maleic-acid-treated PVA. Reproduced from *J Polym Res*, **13**: 161 (2006).

soluble bioactive and photoresist materials, and membranes. Thermogravimetric analysis reveals better heat stability for maleic acid crosslinked PVA compared to virgin PVA only. Intermolecular crosslinking of PVA chains through maleic acid results in a three-dimensional network structure that is mainly responsible for higher heat stability of crosslinked PVA (Figure 4.10) [12].

BY MECHANICAL PROPERTIES

The mechanical properties of a polymer involve its behavior under stress. These properties play decisive role in the practical applications of polymers. Mechanical properties provide answers to the questions of how strong or stiff a polymer is; whether it is hard, soft, brittle or ductile; does it holds up well under repeated stress, etc. The group of properties that are defined as mechanical properties are strength (tensile, compressional, flexural, impact, torsional), elongation (elastic and ultimate), modulus (tensile, rigidity, bulk), toughness, rigidity, hardness, abrasion resistance, and brittleness. Grafting and crosslinking have major effects on different mechanical properties of the modified stock. A brief description of different mechanical properties and their process of measurement are as follows [13, 14]:

Tensile Strength

Strength is the ability of a body to withstand load without breaking. Tensile strength is the limiting stress at which a specimen fractures and is expressed in kgf/cm^2. It is an important property for polymers that are going to be stretched. Fibers, for instance, must have good tensile strength.

Elongation

Elongation is a type of deformation, which is simply a change in shape under stress. In the case of tensile stress, the sample deforms by stretching and becomes longer, which is called elongation. Generally it is expressed as percent elongation, which is $\Delta l/L_0 \times 100$ where Δl is change in length and L_0 is the original length. Elongation may be ultimate or elastic. The ultimate elongation is the strain on a sample when it breaks, which is known as elongation at break. Elastic elongation is the percent of elongation that may be reached without permanently deforming the sample.

Tensile Modulus

Modulus is the capability of a polymer to resist deformation, which is the ratio of stress to strain. It is also called the modulus of elasticity or Young's modulus and is the slope of the stress strain curve (Figure 4.11). In general, fibers have high Young's modulus, elastomers have low values, and plastics lie in between.

Toughness

Toughness is the measure of energy a sample can absorb before it breaks. It is the area under a stress-strain curve. The area under the curve is proportional to the integral of the force over the distance the polymer stretches before breaking.

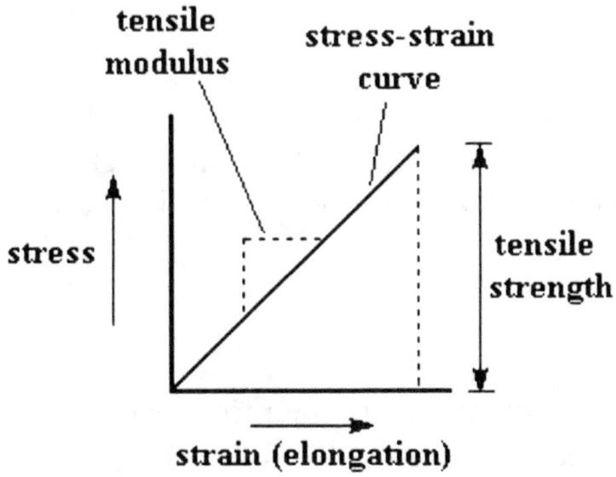

Figure 4.11. Stress-strain curve.

The stress-strain properties are measured by a standard procedure. A dumbbell test piece (following ASTM standards) is cut out with a die. Benchmarks one inch apart are placed on the narrow portion. The test piece is then placed in the tensile testing machine (INSTRON tensile tester) and stretched until it breaks. A stress-strain curve is plotted showing the relationship among different tensile properties.

Tear Strength

Tear strength is the maximum force required to tear a polymer sample. The force acts substantially parallel to the major axis of the test specimen. It is expressed in kg/cm. It is usually determined by crescent tear test. A crescent-shaped piece of definite size (following ASTM standards) is cut out with a die. A small slit is made in the inner curved surface of the test specimen. The specimen is then placed in a tensile machine and loaded to rupture. The load required to separate the specimen is used in calculating the resistance to tear.

Rebound

Rebound is a measure of the resilience of an elastomeric compound. The test is made by allowing a weighted pendulum device to travel through a measured arch and to strike the polymer block. The percentage rebound is read directly from a calibrated scale.

Hardness

Hardness is determined with Shore durometer. A metallic point actuated by means of a spring is pressed against the surface of the stock, and the hardness value is read directly on the scale of the instrument.

Abrasion Resistance

Abrasion is the process of wearing away the surface by friction. There are many devices to measure abrasion resistance. In all cases, an abrasive material is rubbed against the surface of the stock. Either volume loss is measured directly or the work required to produce a definite volume loss is determined.

Permanent Set

The resistance of a crosslinked polymer to permanent deformation when subjected to a compression stress is expressed by permanent set. The extent to which distortion becomes permanent after subjecting a test piece to a known load between plates is known as compression set. It is expressed as a percentage of the original thickness.

Hysteresis

When a crosslinked elastomer is deformed and released, mechanical energy is converted to potential energy and then back. Part of this energy is converted to heat, which cannot be recovered; this is known as hysteresis loss. In the Goodrich flexometer, a cylinder of vulcanized stock is compressed under constant load. A very rapid cyclic deflection is imposed on top of this constant load compression. This leads to a rapid heating up of the sample. The hysteresis is reported in °F as the ΔT or heat rise during the test.

Flexural Modulus

The strength of a material when its top surface is subjected to compression (squeezed together) and the bottom is in tension (stretched) is known as flexural modulus. A simple beam supported at each end and loaded in the middle is used to determine the flexural modulus.

Creep

When a thermoplastic material is subjected to a constant load, the length of the thermoplastic bar will continue to increase. This is called creep. Thermosets are more resistant to creep than thermoplastics. Creep modulus will vary with time; that is, it decreases as the time increases.

Fatigue Strength

Plastics when subjected to cyclic loading are susceptible to brittle crack growth fracture at stress levels well below their tensile and compression strength. The combination of tension and compression is the most severe condition and the strength is known as fatigue strength.

Grafting plays a dominant role in controlling mechanical properties. Invariably the effects depend upon the nature of the base and the grafted polymers and also on the mode and extent of grafting. Grafting of a tough flexible polymer on a rigid plastic enhances the impact strength and reduces hardness and brittleness. Conversely, the grafting of a rigid polymer in a soft flexible matrix increases the modulus of the stock. Grafting of an amorphous polymer on a crystalline may result in a decrease or increase in mechanical properties. If grafting takes place in the amorphous zone there may be an enhancement in the mechanical properties; however, a higher degree of grafting may disrupt the crystalline zones, which may reduce the tensile strength and modulus.

Grafting of polymethyl methacrylate (PMMA) or polybutyl acrylate (PBA) on cellulose fibers results in a reduction in crystallinity of the grafted fiber [15]. Because of the grafting, a decrease in the elastic modulus and tensile strength of the cellulose has also been observed. The effect of PBA grafting on the mechanical properties of cellulose fiber is shown in Figure 4.12. Grafting the amorphous rubberlike polymer PBA affects the amorphous as well as the crystalline zones of the fiber and therefore modulus decreases.

Grafting of methyl methacrylate on cotton yarn enhances the Young's modulus value of the yarn from 148 cN/tex to 240 cN/tex. Elongation at the break decreases from 9.8% to 4.4%. Due to higher glass transition temperature (105 °C) the PMMA chains confer stiffness to the oxidized yarn [16].

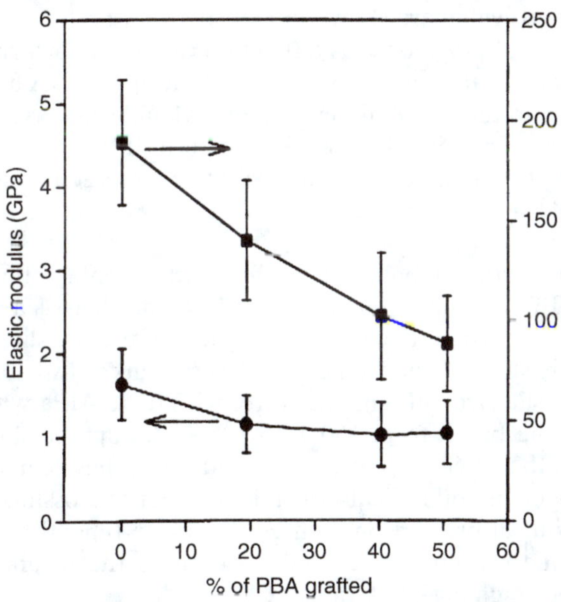

Figure 4.12. Elastic modulus and tensile strength of cellulose fiber grafted with butyl acrylate as functions of the amount of PBA grafted onto cellulose fibers. Reproduced from *Composites: Part A: Appl Sci Manufac*, **30**: 349 (1999).

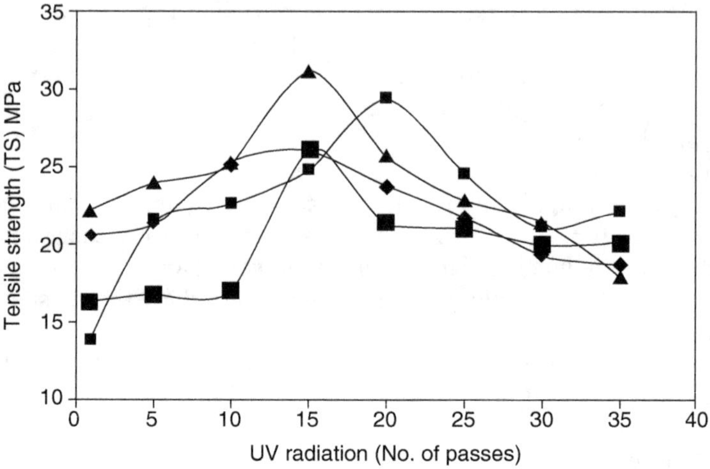

Figure 4.13. Tensile strength of the treated alginate film against number of UV passes (UV doses) with respect to monomer (silane) concentration for 3 min soaking time. (◆) 3% silane, (▪) 5% silane, (▲) 10% silane, (■) 20% silane. Reproduced from *Carbohydrate Polymers*, **72**: 349 (2008).

In general, crosslinking enhances the mechanical properties of polymers. Photocuring of sodium alginate (SA) with 3(trimethoxysilyl) propyl methacrylate silane improves the mechanical properties of SA. The results of the tensile strength (TS) value are shown in Figure 4.13, where values are plotted against the number of UV passes as a function of monomer formulation with the silane monomer treatment. The TS values of the grafted SA films increased, which may be due to the crosslinking of hydroxyl and silane (C–O–Si bond), forming the three-dimensional network structure. It was observed that the tensile strength of the silane-treated film could be increased up to 107% compared to the nontreated one, and elongation at the break increased up to 87% [17].

In the crosslinking of polybutadiene rubber with dipentamethylene thiuram tetrasulfide (DPTT) no polysulfide rings are formed, whereas a great number of –C–C– crosslinks are produced. Enhancement of the crosslinker dose from 2 to 4 phr results in steady enhancement of the crosslink density of the network (Figure 4.14). Tensile strength enhances from 1.2 to 1.4 MPa with a decrease in elongation at the break from 150% to 105%. An appreciable increase in hardness from 44 to 53.5 Shore A was observed [18]. It has been observed that in DPPT curing, persulfenil radicals are able to produce crosslinking via allylic abstraction without disproportionation. At the same time, these radicals can initiate the polymerization of the double bonds in the neighboring chains, producing areas of high crosslink density.

The crosslinking of natural rubber with a binary accelerator system based on 1-phenyl-2,4-dithiobiuret (DBT) and dicyclohexylbenzothiazyl sulfenamide (DCBS) shows an appreciable enhancement in mechanical properties with an

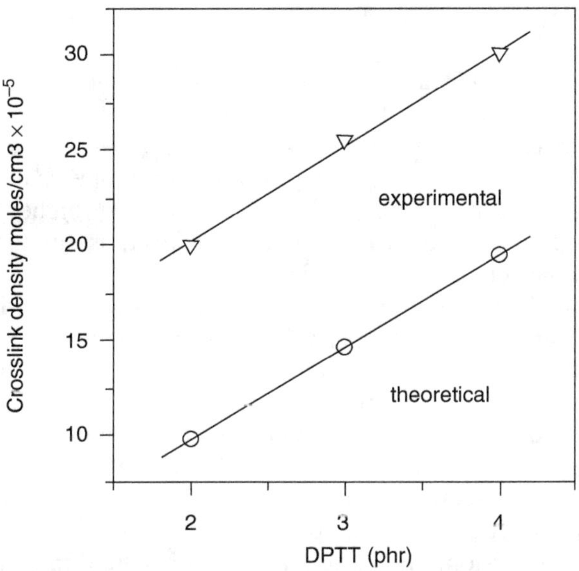

Figure 4.14. Changes in crosslink density for compounds with different concentrations of DPTT. The theoretical line corresponds to entirely monosulfidic structures. Reproduced from *J Appl Polym Sci*, **106**: 3481 (2007).

TABLE 4.2. Mechanical properties of NR cured using DCBS/DBT stock composition of 100 g NR, 5 phr ZnO, 2 phr stearic acid, 0.8 phr DCBS, and 2.5 phr sulfur.

	Batch No.				
	A_5	A_6	A_7	A_8	A_9
DTB concentration phr	0	0.2	0.4	0.6	0.8
Tensile Strength (MPa)	14.22	17.5	29.4	27.7	26.25
Tear strength (N/mm)	18.43	25.89	25.17	26.07	30.47
Elongation at break (%)	930	876	822	850	795
Compression set (%)	78	56	59	78	66
Hardness (Shore A)	31	37	38	40	38
Resilience (%)	23	27	25	26	24
Modulus 300% (MPa)	3.17	4.06	5.01	4.81	4.71

Reproduced from *J Appl Polym Sci*, **87**: 2193 (2003).

increase in the curative dose. Maximum values were achieved at an optimum concentration. The variation of different properties with the crosslinker dose is shown in Table 4.2 [19]. The enhancement in modulus and tensile strength and decrease in elongation at the break are reflections of the increase in crosslink density (decrease in chain flexibility) with the increase in accelerator dose.

BY MORPHOLOGICAL ANALYSIS

Morphological analysis is achieved by different techniques such as scanning electron microscopy (SEM), transmission electron microscopy (TEM) and atomic force microscopy (AFM).

Scanning electron microscopy employs a microscope that uses electrons rather than light to form an image. The combination of higher magnification, larger depth of focus, and greater resolution makes it useful instrument in the analysis of the particular polymeric field.

Transmission electron microscopy is an imaging technique whereby a beam of electrons is transmitted through a specimen, and an image is formed, magnified, and directed to appear either on a fluorescent screen or layer of photographic film or to be detected by a sensor such as CCD camera.

In both techniques, electrons are generated thermoionically. They are then accelerated by an electric field and focused by electrical and magnetic fields onto the sample. The electrons are focused onto or through the sample providing a resolution far better than is possible with light microscopes and with improved depth of vision. The basic feature is presented in Scheme 4.1.

The interaction of the incident electron beam with the specimen results different signals. For scanning electron microscopy, the secondary, back-

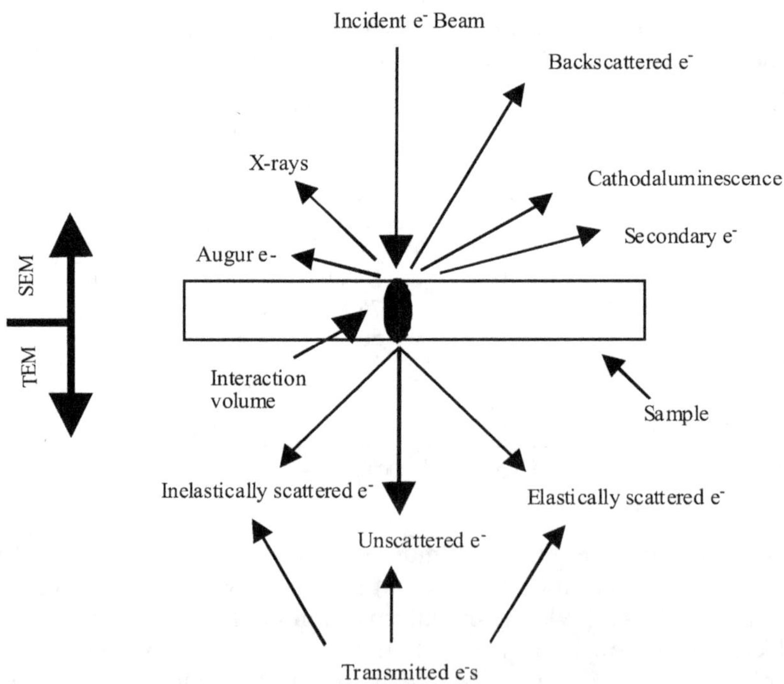

Scheme 4.1. Presentation of the basic features of SEM and TEM.

scattered electrons, as well as X rays, are used for scanning whereas transmitted electrons through the sample are useful for transmission.

Atomic force microscopy (AFM) is, on the other hand, used for imaging, measuring, and manipulating matter at the nanoscale. It measures the topography with a force probe. The probe is a tip on the end of a cantilever that bends in response to the force between the tip and the sample. AFM provides a true three-dimensional surface profile.

The usefulness of AFM is as follows:

(1) It can supply the clear visual picture of virgin polymers as well as their modified ones. It may be two dimensional or three dimensional.
(2) The nature of modification (uniformity) is also observed in the micrometer as well as the nanometer scale.
(3) From the roughness parameters (from AFM), one can get idea of the grafting content on the polymer.

The introduction of different moieties through grafting techniques features different morphologies, e.g., grafting with polyvinyl pyrrolidone on to the polypropylene [20], polyacrylic acid grafting on polyamide (NF-270) membrane [21], or methyl methacylate over the ultrahigh molecular weight of polyethylene [22]. The white layer of polyvinyl pyrrolidone spreads with the grafting (Figure 4.15), with the bumpy structure appearance of methyl methacrylate over the ultra high molecular weight of polyethylene (Figure 4.16). The white layer of polyacrylic acid over the NF-270 membrane is evidenced from the cross-sectional TEM image (Figure 4.17).

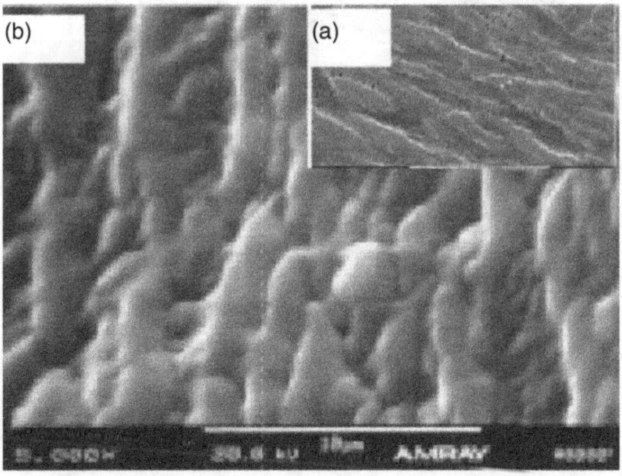

Figure 4.15. SEM images of the membranes inside (a) nascent PPMM without any treatment and (b) PVP-grafted polypropylene with DG = 64.7 wt. %. Reproduced from *Eur Polym J*, **40**: 2077 (2004).

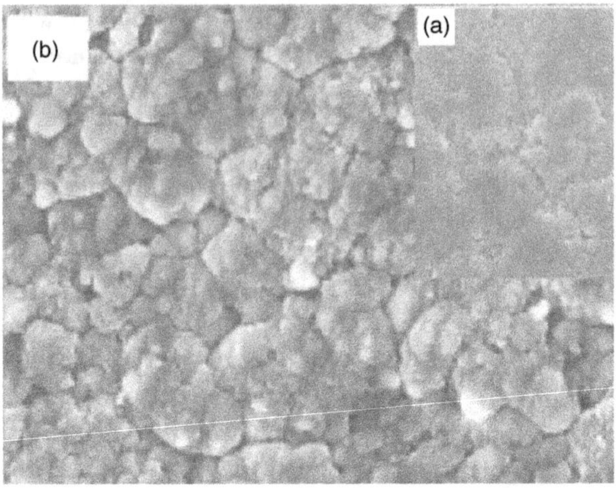

Figure 4.16. SEM photographs of the membranes inside (a) control and (b) 1.04 mg/cm^2 grafting. Reproduced from *J Mat Sci: Mat in Medicine*, **11**: 593 (2000).

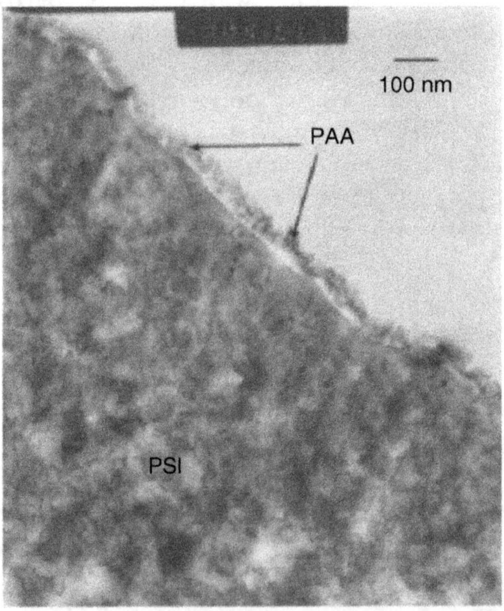

Figure 4.17. TEM cross-sectional images of a heavily modified NF-270 membrane (magnification 60K). Reproduced from *J. Memb Sci*, **209**: 283 (2002).

The AFM images of polyethylene glycol methacrylate (PEGMA) and sulfopropyl methacylate (SPM) grafted BW-30 commercial reverse osmosis (RO) membranes [22] are shown below (Figure 4.18). With the grafting, the roughness parameter is interestingly decreased [22, 23] (of course it depends upon the roughness of virgin membrane).

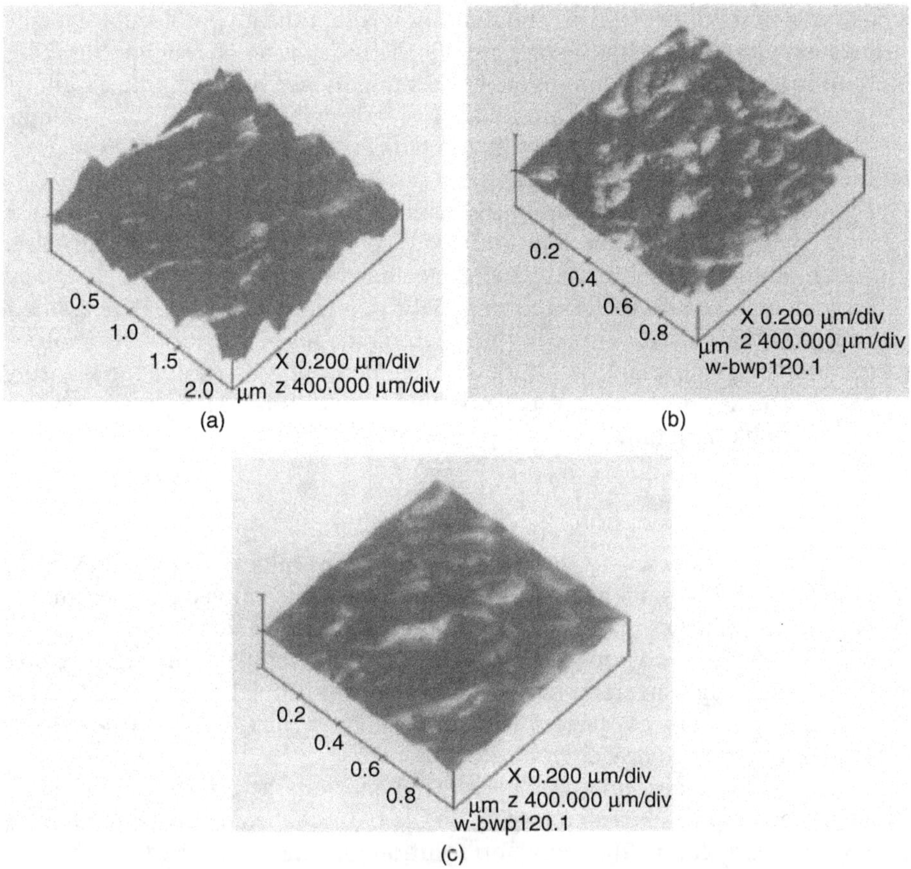

Figure 4.18. AFM images of BW-30 RO membranes (Dow-Filmtec): (a) original, (b) modified with PEGMA, and (c) SPM. The average roughness in nanometers is (a) 93, (b) 30, (c) 28. Reproduced from *J. Memb Sci,* **209**: 283 (2002).

BY MOLECULAR WEIGHT ANALYSIS

The two most fundamental properties of polymers are chemical composition and molecular weight. While the chemical composition can usually be determined relatively easily by a variety of techniques, the determination of molecular weights may not be as straightforward.

Of the primary methods to determine molecular weight, the methods based on osmotic pressure, light scattering, and ultracentrifugation are most often used.

In the osmotic pressure method, the most important point is to select a suitable membrane, through which the macromolecules are passed through. Although this method has been used for the determination of molecular weight in a number of systems, one limitation is that the low-molecular-weight species may undergo permeation through the membrane. The light-scattering method is based on the inhomogeneity of the polymer solution. The extent of light scattering may be correlated with the molecular weight of the polymer species, and hence one can determine the molecular weight of the desired graft by the extent of light scattering [24].

The ultracentrifugation technique is based on the sedimental rate of polymer particles, when the polymer solution is subjected to strong centrifugal force. The sedimental rate is related to the molecular weight.

Secondary methods of molecular weight determination may be classified into three categories, e.g., viscosity measurement, gel permeation chromatography, and end group analysis. Of these three methods, viscosity as the measuring parameter has long been used in polymer studies [25]. It is based on the Mark-Houwink equation

$$[\eta] = KM^{\alpha},$$

where η is intrinsic viscosity, and K and α are constants, which are, however, functions of both solvent and polymer type. The molecular weight is derived from relationship of the viscosity average molecular weight, e.g., $\overline{M_v}$. This $\overline{M_v}$ is found to lie between $\overline{M_n}$ (the number average molecular weight) and $\overline{M_w}$ (the weight average molecular weight), but much closer to $\overline{M_w}$. This method appears to be the fastest, simplest, and most inexpensive technique to determine the molecular weight of the polymer.

Gel permeation chromatography is a polymer characterization method in which molecules are separated largely on the basis of their size. This technique consists of percolating the polymer solution through a packed column of porous particles where the stationary phase constitutes the solvent inside the gel pore, and the mobile phase is the solvent of the percolated solution. The molecules that are too large to enter the gel pores permeate through the column whereas the smaller particles will enter the pores and will be delayed; hence the species that will come out first through the column will be larger molecules followed progressively by the smaller ones. Polymer molecular weight distribution (MWD) represents an intrinsic property that provides direct correlation with many end-use physical properties and universal criterion for polymer characterization [26–28]. Figure 4.19 presents the difference between the molecular weight distributions of the graft copolymer and the parent one. The molar mass of the grafted product is shifted toward a lower retention time. This confirms an increase in the molecular weight of the graft copolymers compared to the ungrafted material. A multiangle light-scattering detector is also used today to determine molecular weight distribution. The technique is based on the influence of laser light on the macromolecules. Lasers will induce the polarizibility in the macromolecule; thus scattering of the light will be effective when monitoring the polymer macromolecules [29, 30].

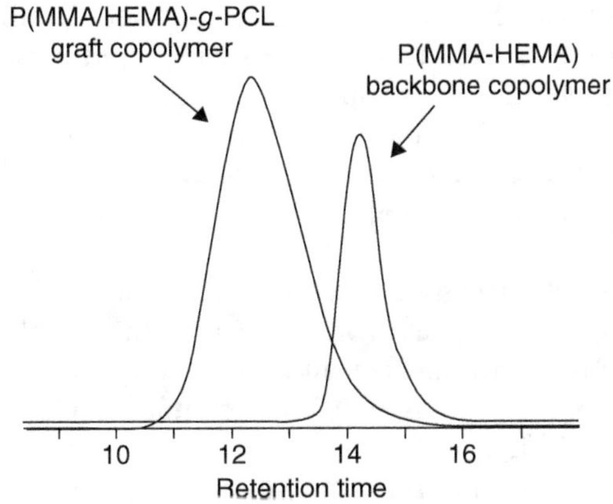

Figure 4.19. Comparison of the GPC traces for the poly(MMA-co-HEMA) copolymer and the final poly(MMA-co-HEMA)-g-PCL grafted copolymer. Reproduced from *Macromolecules* **39**: 9080 (2006).

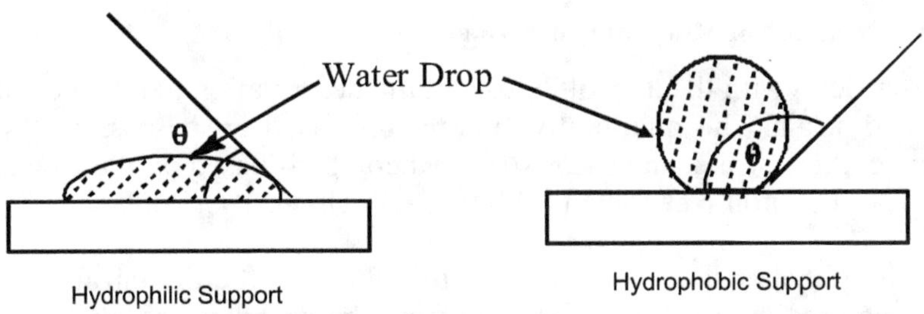

Figure 4.20. Contact angle featuring the nature of support.

BY MEASUREMENT OF PHYSICAL PROPERTIES

The grafting/crosslinking influences different physical properties, for example, the contact angle or rheological properties.

Measurement of the Contact Angle

The measurement of the contact angle of the matrices can determine the nature of the polymers. The contact angle is the quantitative measure of the wetting of a solid by a particular liquid. The contact angle is defined as the angle at which a liquid/vapor interface meets the solid surface (Figure 4.20). The common techniques are the sessile drop, the pendant drop, or the Wilhelmy technique.

The sessile drop method uses an optical subsystem to capture the profile of a pure liquid on a solid substrate. It provides an idea of the nature of the polymeric substrate where the liquid drop is stationed. On the other hand, the pendant drop method is used to measure surface and interfacial tensions of liquids. Here the geometry of a drop is analyzed.

The Wilhelmy technique is based on the wetting force on the solid by a particular liquid. It is calculated from average advancing and receding contact angles on solids of uniform geometry. Both sides of the solid (film) must have the same properties.

If the liquid is very strongly attracted to the polymeric solid surface (for example, water on a strongly hydrophilic solid), the droplet will completely spread out on the polymeric surface, and the contact angle will be close to 0°. Less strongly hydrophilic solids will have contact angles up to 90°. If the surface is hydrophobic, it will show the contact angle >90°.

The grafted moieties of different functional groups introduce the polymeric surface. Polymers grafted with monomers (e.g., acrylic acid, acrylamide) [31–34] characterize hydrophilic surface whereas monomers (e.g., vinyl trimethoxy silane, fluorinated methacrylates) result in hydrophobic surfaces [35, 36].

By Measurement of Rheological Properties

Rheology deals with the flow of fluids and the deformation of polymers under stress and strain. The quantitative parameters used are viscosity, elasticity, shear rate, shear strain, and shear stress. Rheology is based on the Newtonian as well as the non-Newtonian (dilatant, Bingham, pseudoplastic) flow of liquids.

For the polymer liquid, the shear rate influences the organization of the microstructure. The work done in producing the anisotropic global structure and accompanying flow is of two types: a recoverable energy associated with structure formation, which is identified with the elasticity, and a lost energy dissipated in structural formation and sliding, which is associated with the viscosity. Generally, the anisotropy is increased with the rate of flow and accompanying increase of the shear forces. Measurement of viscosity as a function of shear rate is used to measure complex rheological properties such as viscoelasticity (G', G'') as function of frequency (time) or temperature.

Thus, the rheometer is a kind of mechanical spectrometer that measures the viscoelastic properties of materials beyond just viscosity. There are two distinctively different types of rheometers depending on the geometry of applied stress. Rheometers that deal with shear stress are called shear rheometers (capillary, cone and plate, rotational cylinder), whereas rheometers that apply extensional stress are extensional rheometers (acoustic, pulled string).

As grafting/crosslinking introduces the heterogeneity of the polymers, the rheological properties also change. Figure 4.21 presents the differences in the viscoelastic behavior of agar and agar-g-polyvinylpyrrolidone. Both moduli (G' and G'') decreased for the agar-g-PVP hydrogels and became more fre-

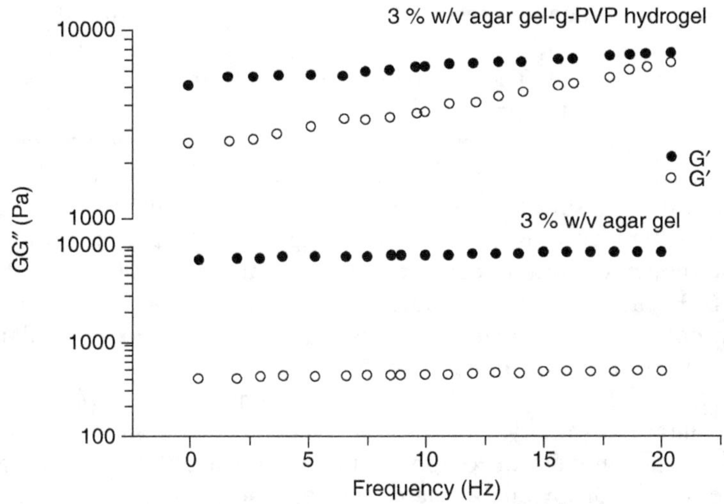

Figure 4.21. Viscoelastic behavior of (a) agar and agar-grafted-PVP. Reproduced from *J Appl Polym Sci*, **102**: 3654 (2006).

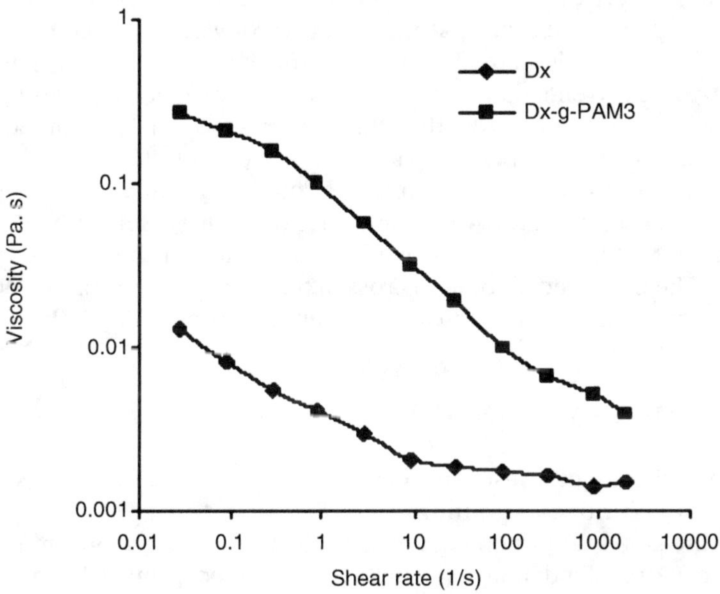

Figure 4.22. Viscosity versus shear rate curves of the aqueous solutions of Dextran and Dextran-g-polyacrylamide. Reproduced from *Carbohydrate Polymers*, **69**: 371 (2007).

quency dependent; i.e., it tended to be more liquidlike under an applied strain of 0.01% [37].

It is observed that by grafting, shear stable and viscofying polymeric materials are obtained. Figure 4.22 illustrates the shear viscosity vs shear rate in an acrylamide grafted dextran system [38]. Though upon grafting the shear viscosity increases; in both the cases, the shear viscosity decreases with the increase in shear rate. The samples show shear thinning non-Newtonian behavior (pseudoplastic) grafting.

For crosslinked polymers, the values of G′ and G″ decrease to zero at low frequencies (otherwise known as the terminal zone). A decrease in the molecular weight tends to shift the terminal zone to higher frequencies. Whereas the addition of chemical crosslinks results in significant changes in the rates at which the dynamic moduli decrease at low frequencies; in the case of elastic materials, G achieves a plateu value.

Melt rheology is a valuable technique to detect changes in molecular weight and structure resulting from secondary reactions that occur during the grafting process. The rheological properties of the graft-modified polypropylene were greatly influenced by the type of initiator, concentration of initiator, concentration of monomer, and use of comonomers. The grafting of glycidyl methacylate (GMA) onto polypropylene in the presence of peroxide initiators resulted in a significant amount of β-chain scission and ultimately reduced the melt strength of the copolymer. The rheological properties of the GMA graft-modified PP samples suggest that some crosslinking of PP occurs during the grafting process, most likely due to epoxy ring opening reactions. On the other hand, with the increase of styrene grafting on polypropylene, the melt strength is enhanced. These trends were also apparent for the copolymers prepared with both GMA and styrene; therefore it is possible to produce highly functionalized PP with good melt strength [39]. The presence of crosslinkers in the system influences the properties (viz., dynamic viscoelasticity) of the hydrogels. The effect of the amount of crosslinker on the dynamic viscoelasticity for hydrogels is shown in Figure 4.23. The values of G′ increased with an increase in the percent addition of methylene bis acrylamide in the pulp acrylamide hydrogels. The increased amount of crosslinker decreases the volume occupied by large pores, and thus decreases the dynamic viscoelasticity [40].

BY X-RAY DIFFRACTION ANALYSIS

The X-ray technique is also an alternative means to detect the structural changes that occur due to grafting.

X rays are generated when high-energy electrons impinge on a metal target (iron, copper, or molybdenum). When an X-ray monochromatic beam impinges on a sample, two processes can be observed, depending on the nature of the sample. Coherent scattering of X rays occurs if the sample is of periodic structure. This process occurs without a change of wavelength and without loss of phase relationship between the incident and scattered rays. The process is called the diffraction X-ray effect and is measured by wide-angle X-ray diffraction. However, incoherent scattering results when the sample is of an unperiodic structure that possesses different electron densities (crystalline and

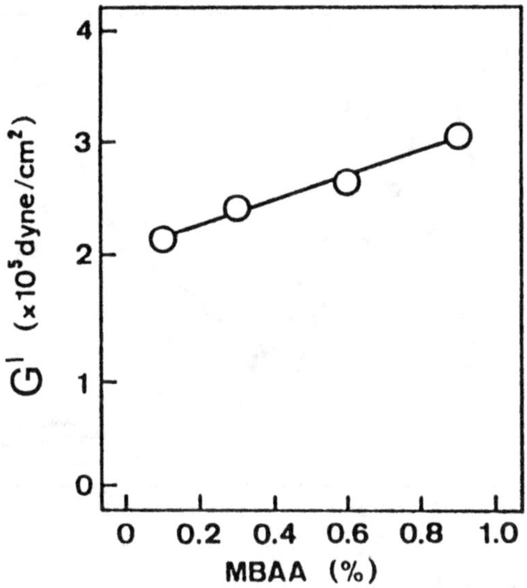

Figure 4.23. Effect of the percent addition of a crosslinker on the value of G″ for 4% hydrogels of pulp-acrylamide hydrogels at 30°C. Reproduced from *J Appl Polym Sci* **45**: 805 (1992).

amorphous regions). This process occurs with a change of wavelength and a change in the phase relationship between the incident and scattered rays. This process is termed the diffuse X-ray diffraction effect and is measured by small angle X-ray scattering.

An X-ray diffraction study of graft copolymers formed by certain monomers on polymers has shown that grafting changes the structure of the parent polymeric backbone [41–43]. Of course, it depends upon the nature of the polymer. Sometimes it does not reflect changes due to grafting or crosslinking [44–46].

Figure 4.24 presents the comparative study of the X-ray diffraction pattern of hydroxyl propyl methyl cellulose and the polyethyl acrylate grafted one [41]. The peaks of the grafted one are comparatively dispersive, and the percentage of crystallinity decreases. Thus, XRD presents evidence of grafting, which enlarges the proportion of the amorphous regions and reduces the percentage of crystallinity.

The variation of crystallinity with the grafting percentage can also be illustrated as in Figure 4.25 [47]. The decrease in crystallinity upon the grafting of styrene and acrylic acid on fluorinated ethylene propylene copolymer (FEP) and the post-grafting treatment (sulfonation) is due to the dilution effect of crystallites because of the incorporation of styrene-co-acrylic acid into the amorphous region of the FEP matrix as well as crystal disruption.

Thus, X-ray diffraction provides valuable information for the differences of polymer perfection as well as their crystallinity.

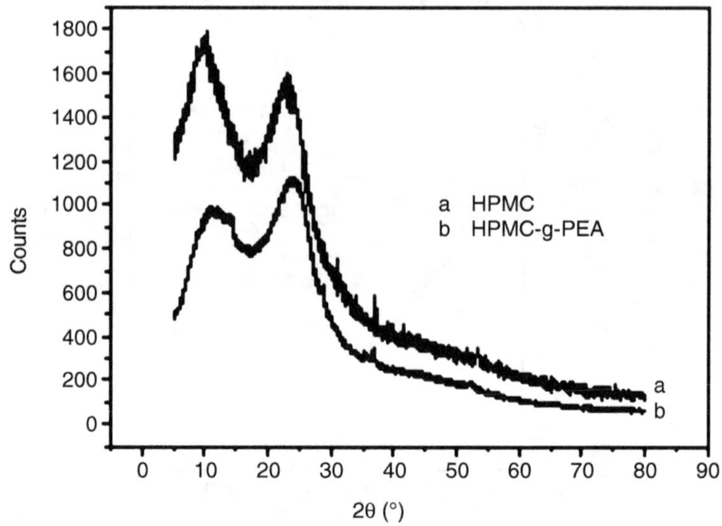

Figure 4.24. X-ray diffraction pattern of (a) hydroxyl propyl methyl cellulose and (b) hydroxyl propyl methyl cellulose-g-polyethyl acrylate. Reproduced from *Carbohydrate Polymers* **68**: 626 (2007).

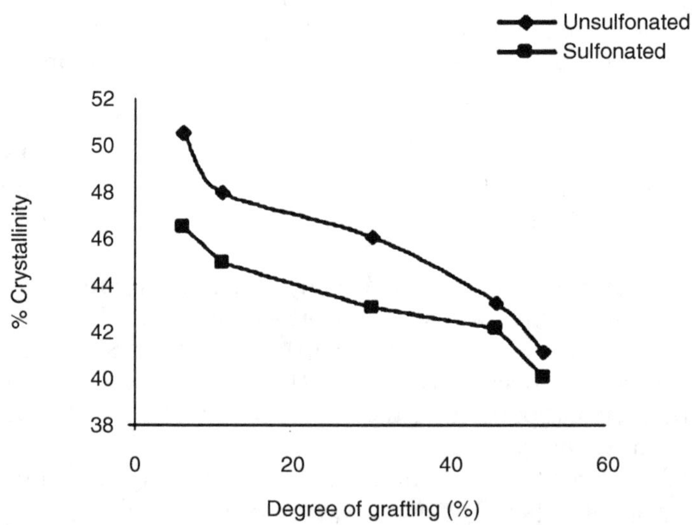

Figure 4.25. Variation of the crystallinity with the degree of grafting. Reproduced from *J Appl Polym Sci* **97**: 1426 (2005).

REFERENCES

1. Basu S, Bhattacharya A, Mondal PC, Bhattacharyya SN. *J Polym Sc Polym Chem*, **32**: 2251 (1994).
2. Qiu C, Xu F, Nguyen QT, Ping Z. *J Membr Sci*, **255**: 107 (2005).
3. Kallrot M, Edlund U, Albertsson A-C. *Biomaterials*, **27**: 1788 (2006).
4. Kurkuri MD, Lee JR, Huang Han J, Lee I. *Smart Mater Struct*, **15**: 417 (2006).
5. Zhang Z, Kong L, Deng J, Yang P, Yang W. *J Appl Polym Sci*, **101**: 2269 (2006).
6. Brown ME. *Introduction to Thermal Analysis: Techniques and Applications* (Dordrecht: Kluwer Academic Publisher/Springer, 2001).
7. Haines PJ, ed. *Principles of Thermal Analysis and Calorimetry* (Cambridge: Royal Society of Chemistry, 2002).
8. Brack HP, Ruegg D, Buhrer H, Slaski M, Alkan S, Schrer GC. *J Polym Sci Part B: Polym Physc*, **42**: 2612 (2004).
9. Gupta B, Scherer GC. *Angew Makromol Chem*, **210**: 151 (1993).
10. Gangopadhyay R, Ghosh P. *Eur Polym J*, **36**: 1597 (2000).
11. Zheng K, Chen L, Li Y, Cui P. *Polym Eng Sci*, **44(6)**: 1077 (2004).
12. Gohil JM, Bhattacharya A, Ray P. *J Polym Res*, **13**: 161 (2006).
13. Gravey BS in *Introduction to Rubber Technology*, M Morton, ed. (New York: Van Nostrand Reinhold, 1973).
14. Crawford RJ. *Plastics Engineering* (Oxford: Butterworth Heinmann, 1998).
15. Ganche-Escamilla G, Cauich-Cupul JI, Mendizabal JE, Puig JE, Vazquez-Torres PJ. *Composites: Part A: Appl Sci Manufac*, **30**: 349 (1999).
16. Princi E, Vicini S, Pedemonte E, Gentile G, Cocca M, Martuscelli E. *Eur Polym J*, **42**: 51 (2006).
17. Mollah MZI, Khan MA, Hoque MA, Aziz A. *Carbohydrate Polymers*, **72**: 349 (2008).
18. Posadas P, Fernandez A, Brasero J, Valentin JL, Marcos A, Rodriguez A, Gonzalez L. *J Appl Polym Sci*, **106**: 3481 (2007).
19. Aprem AS, Joseph K, Laxminarayanan R, Thomas S. *J Appl Polym Sci*, **87**: 2193 (2003).
20. Liu ZM, Xu ZK, Wang JQ, Wu J, Fu JJ. *Eur Polym J*, **40**: 2077 (2004).
21. Kwon OH, Nho YC, Lee YM. *J. Mat. Sci. Mat in Medicine*, **11**: 593 (2000).
22. Freger V, Girlon J, Belfer S. *J. Memb Sci*, **209**: 283 (2002).
23. Yogesh, Paul P, Basu S, Bhattacharya A. *J Appl Polym Sci*, **105**: 609 (2007).
24. Kuhn R. *Die Makromolekulare Chemie*, **185**: 1003 (1984).
25. Krishnamoorthi S, Mal D, Singh RP. *Carbohydrate Polymers*, **69**: 371 (2007).
26. Wang L, Dong W, Xu Y. *Carbohydrate Polymers*, **68**: 626 (2007).
27. Villarroya S, Zhou J, Thurecht KJ, Howdle SM. *Macromolecules*, **39**: 9080 (2006).
28. Pengfei L, Maolin Z, Jilan W. *Rad Phys Chem*, **61**: 149 (2001).
29. Kutsevol N, Guenet JM, Melnik N, Sarazin D, Rochas C. *Polymer*, **47**: 2061 (2006).

30. Nguyen HK, Lemieux P, Vinogradov SV, Gebhart CL, Guerin N, Paradis G, Bronich TK, Alakhov VY, Kabanov AV. *Gene Therapy*, **7**(**2**): 126 (2000).
31. Chen J, Nho YC, Kwon OH, Hoffman AS. *Rad Phys Chem*, **55**: 87 (1999).
32. Hamilton LM, Green A, Edge S, Badyal JPS, Feast WJ, Pacynko WF. *J Appl Polym Sci*, **52**: 413 (1994).
33. Park JY, Acar MH, Akthakul A, Kuhlman W, Mayes AM, *Biomaterials*, **27**: 856 (2006).
34. Ichijima H, Okada T, Uyama Y, Ikada Y. *Die Makromolekulare Chemie*, **192**: 1213 (1991).
35. Abidi N, Hequet E, Tarimala S. *Textile Res J*, **77**: 668 (2006).
36. Hoebergen A, Uyama Y, Okada T, Ikada Y. *J Appl Polym Sci*, **48**: 1825 (1993).
37. Prasad K, Mehta G, Meena R, Siddhanta AK. *J Appl Polym Sci*, **102**: 3654 (2006).
38. Krishnamoorthi S, Mal D, Singh RP. *Carbohydrate Polymers*, **69**: 371 (2007).
39. Wong B, Baker WE. *Polymer*, **38**: 2781 (1997).
40. Yoshinobu M, Morita M, Sakata I. *J Appl Polym Sci*, **45**: 805 (1992).
41. Wang L, Dong W, Xu Y. *Carbohydrate Polymers*, **68**: 626 (2007).
42. Dong Z, Liu Z, Han B, He J, Jiang T, Yang G. *J Mater Chem*, **12**: 3565 (2002).
43. Bhumkar DR, Pokharkar VB. *AAPS Pharm SciTech*, **7**: E1 (2006).
44. Carreon MP, Aliev R, Ocampo R, Burillo G. *Polymer Bulletin*, **44**: 331 (2000).
45. Wang L, Dong W, Xu Y. *Carbohydrate Polymers*, **68**: 626 (2007).
46. Parikh DV. *Textile Research J*, **77**: 612 (2007).
47. Phadnis S, Patri M, Chakraborty BC, Singh PK, Deb PC. *J Appl Polym Sci*, **97**: 1426 (2005).

5

BROADER SPECTRUM: EXAMPLES

INDERJEET KAUR AND PARAMITA RAY

INTRODUCTION

Graft and crosslinked polymers are specialty polymers that can lead to desirable properties that are not inherent to the parent backbone. As grafting is generally related to side chains of the backbone polymer, it does not encompass the far interior, and it may be envisaged that this will cause little perturbation of the molecular properties. If the penetration is greater, major changes are observed. Crosslinked polymers exhibit significant bulk property development. Some properties may therefore be linked with the surface only, whereas others may be grouped together due to bulk changes. In the following sections, the impact of grafting and crosslinking on different properties are discussed.

AFFINITY TO SOLVENTS

The attachment of hydrophilic monomers (e.g., acrylic acid, acrylamide, acrylonitrile, etc.) has been found to lead to an increase in wettability, adhesion, dyeing, and rate of release of oil stains by a detergent solution from the grafted surface. On the other hand, if the monomer is hydrophobic, the result is decreased wetting. Of course, it depends upon the nature of wetting liquids. If grafting is not restricted to the surface alone, but encompasses the bulk of the backbone polymer, then properties such as sorption, flame resistance, or crease resistance, etc. are affected.

Apart from the behavior of imbibing a large amount of water, hydrophilic monomer grafted polymers are biodegradable and may be sensitive to stimuli

---

Polymer Grafting and Crosslinking, Edited by Amit Bhattacharya, James W. Rawlins and Paramita Ray

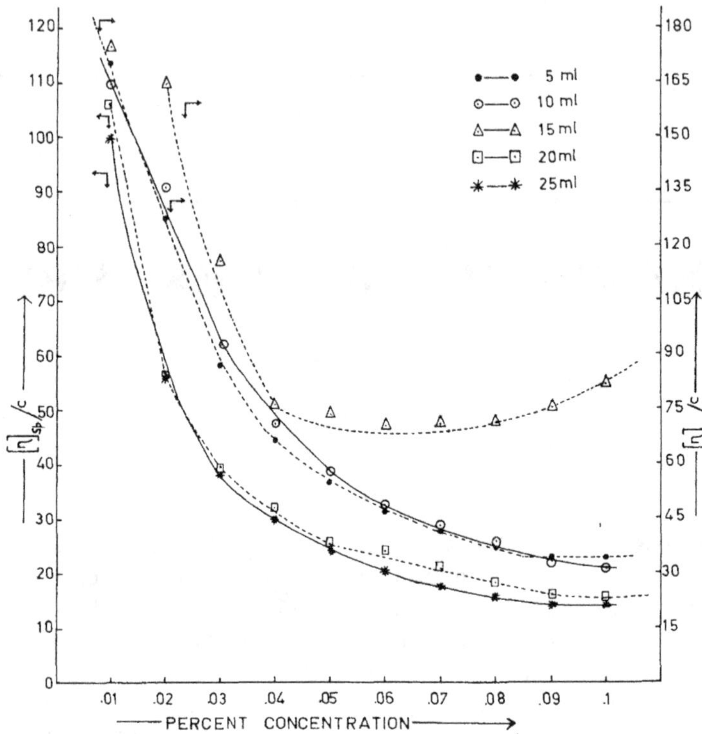

Figure 5.1. Effect of the amount of water on $[\eta]_{sp}$ /C of starch-grafted PAAM composite. Reproduced from *J Appl Polym Sci*, **47**: 1165 (1993).

(pH and temperature). As such they are also described as SMART POLYMERS. Water uptakes by the hydrogels (grafting by acrylamide (AAm), 2-acrylamido-2-methyl propane sulfonic acid, NIPAAm, etc.) relate to significant volume transitions, when small changes are made in their environment [1]. Modification of polymers in this direction lead to the synthesis of "hydrogels,", which are discussed in later sections.

Polyelectrolyte behavior is conveyed by grafting hydrophilic monomers on to the polymer [2–4]. This is determined from the viscosity measurements (Figure 5.1) for the starch-*g*-poly (AAm). These types of copolymers have been evaluated as agents to increase the viscosity of aqueous fluids and may find use in the EOR process [5].

The presence of polar pendant groups in the grafted samples helps the polymer to swell in different solvents, both polar and nonpolar, though it depends upon the nature of substrates. The percentage of swelling of both polyamide (PA)-g-poly (methyl acrylate) (PMA) and PA-g-poly (methyl methacrylate) (PMMA) increases with the grafting in different solvents as in DMSO>DMF>pyridine>water [6]. The swelling behavior is attributed to the interaction of the polar solvents with moderately polar amide groups of PA

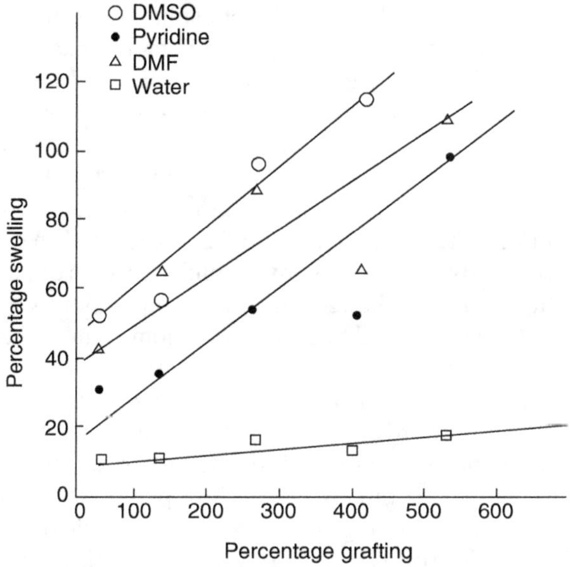

Figure 5.2. Effect of grafting on swelling of PA-g-poly [MA-co-AAM] in various solvents. Reproduced from *Polymers & Polym Composites*, **6(7)**: 465 (1998).

and pendant ester moieties of the grafted PMA and PMMA chains. At higher grafting levels beyond 400%, the percentage of swelling of grafted films shows a steep rise in all solvents except water. This is attributed to the fact that the increase in the number of grafted chains with an increasing percentage of grafting facilitates the interactions with the solvents. Both PMA and PMMA are themselves insoluble in water [6]. Similar behavior is also achieved for the polyamide film on the grafting of binary mixture of MA and acrylamide [7] (Figure 5.2) where water shows minimum swelling in comparison with dipolar aprotic solvent DMSO.

Similarly, functionalization of polyethylene by the grafting of polar monomers such as methacrylonitrile (MAN) and a binary mixture of MAN with MA has shown the maximum swelling (12.53%) in benzene and minimum (4.86%) in water [8]. Tefzel grafted with a mixture of MA and styrene shows low swelling in water, but in benzene and DMF the swelling is appreciable [9].

The flocculation properties of hydroxyethyl cellulose (HEC)-g-poly (AAm) on the outlet effluents of the electroplating, leather, and spinning industries amount as a function of the graft copolymer. It is seen that the polymer behaves as an effective flocculent for all these industrial effluents [10].

Very high water uptake (35,000%) and a high uptake of Fe^{2+}, Cu^{2+}, and Cr^{6+} ions on pectin-cl-poly (AAm PSA) (pectin crosslinked with 2-acrylamido-2-methyl-1-propanesulfonic acid) network [11] is achieved. The sorption property of the metal ions is developed on interpenetrating networks (IPNs) of cellulose derivatives (e.g., hydroxyl propyl, cyanoethyl, hydroxy ethyl, and hydrazinodeoxy cellulose and cellulose phosphates) grafted with 2-acrylamido-2-methyl propane sulfonic acid [12] and methylacrylamide (MAAm) [13] crosslinked with *N,N*-methylene-bis-acrylamide (NNMBA). These hydrogels have a high retention capacity, and the partitioning of metal ions is appreciably high between the hydrogels and the liquid phase. Grafting glycidyl methacrylate (GMA) with comonomer acrylic acid (AAc), and acryl amide (AAm) and acrylonitrile (AN) with the cellulose extracted from pine needles, develops the sorption of ions Fe^{2+}, Cu^{2+} and Cr^{6+} ions [14]. Ion exchange capacity is also influenced by grafting of hydrophilic acid groups [15] as water uptake is a function of ion exchange capacity or equivalent weight. However, very high water uptake causes a decrease in the membrane selectivity together with a considerable loss in the mechanical strength [16].

Networks based on HPC (hydroxypropyl cellulose) and *N*-iso propyl acrylamide crosslinked with NNMBA show thermosensitivity [17]. Post-grafting reaction develops various special properties for the resultant polymers: treatment of ethylacrylate-*g*-crosslinked poly (vinyl pyrrolidone) with potassium hydroxamate shows metal binding properties [18], and partial alkaline hydrolysis generates ionogenic groups, which develop flocculating properties. Saponified Starch-*g*-PAN copolymers are used as thickening agents as water absorbents and are known to form highly swollen insoluble gel [19].

The crosslinking of a polymer greatly influences its affinity toward solvents. Crosslinks impart rigidity in the long chain macromolecules, and all synthetic networks swell when they are exposed to low-molecular-weight solvents. The degree of swelling is a function of several factors of which the length of the network chain is the prime one [20]. In fact, swelling depends on the molecular weight between the crosslinks [21]. Conversely, the network structure of a crosslinked polymer may be studied by means of its swelling characteristics in solvents. The polymer-solvent interaction parameter between a crosslinked network and its solvent has been found to vary with the volume fraction of polymer in the swelled mass, which again changes with the crosslink density of the cured compound [22]. For styrene butadiene rubber (SBR) crosslinked with different doses of sulfur/accelerator, it is possible to find out the relative distribution of each type of crosslinks (polysulfides, disulfides, and monosulfides) from a swelling test, considering the variation of the polymer-solvent interaction parameter with the volume fraction of polymer in the swelled mass [23]. Although the swelling of a crosslinked polymer by a suitable solvent is predominantly controlled by crosslink density, the type of crosslink also play a dominant role. When SBR is crosslinked using four different vulcanization systems such as conventional (CV high sulfur:accelerator ratio), efficient (EV low sulfur:accelerator ratio), peroxide, and a combination of peroxide and sulfur, there is a remarkable difference in their swelling properties. The maximum solvent uptake follows the order EV>mixed>CV>peroxide (Figure 5.3). The highest solvent uptake properties are showed by SBR having mono- and disulfidic linkages, whereas those having C-C crosslinks showed the lowest solvent uptake [24].

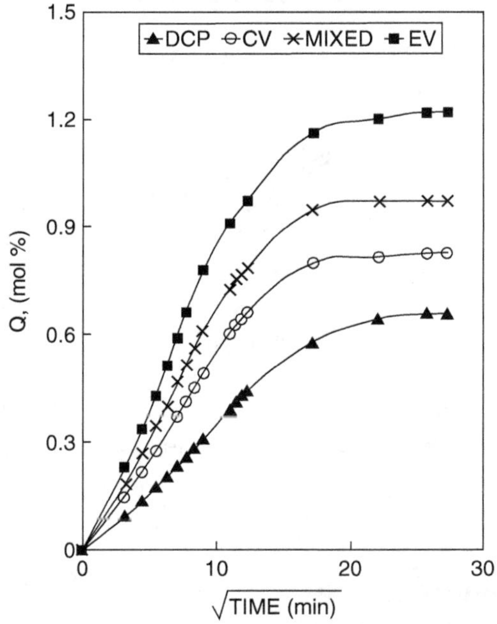

Figure 5.3. Effects of the types of crosslinks on the mole percent hexane uptake of SBR membranes. Reproduced from *J Membr Sci*, **163**: 1 (1999).

Studies on the swelling properties of interpenetrating polymer networks (IPNs) are important to understand the influence of the component segments on IPN structure. In general, the higher the presence of the soft segment in an IPN, the greater the selling ratio (the ratio of the volume of sample after swelling to the volume of sample before swelling). A study [25] on a simultaneous polyurethane (PU) vinyl ester (VE) full IPN showed that the swelling ratio of the integrated network increased with increase in the blocked PU content, which is the soft segment of the network blend. The increase in the filler (kaolin) contents in such IPNs (PU:VE 15:85) results in a decrease in the swelling ratio as filler content increases the hard segment of the IPN. The swelling ratio also decreases with the increase in the initiator content of vinyl ester. The higher the initiator content, the greater the degree of VE polymerization (crosslinking), and the greater the hard segment content existing in the IPN. Comparison of swelling behavior of chitosan gel and a semi-IPN of chitosan and polyethylene oxide (crosslinked with glyoxal) indicates a higher capacity to swell for the semi-IPN at different pH levels like 1.2 and 7.2. Semi-IPN contains 6% more bound water than the reference gel. The presence of hydrophilic PEO chains induces an osmotic effect in the network during swelling. From swelling studies, it was concluded that the semi-IPN had a promising potential because of its higher pH-dependent swelling properties, which could allow a pH-controlled release in oral administration of a drug [26].

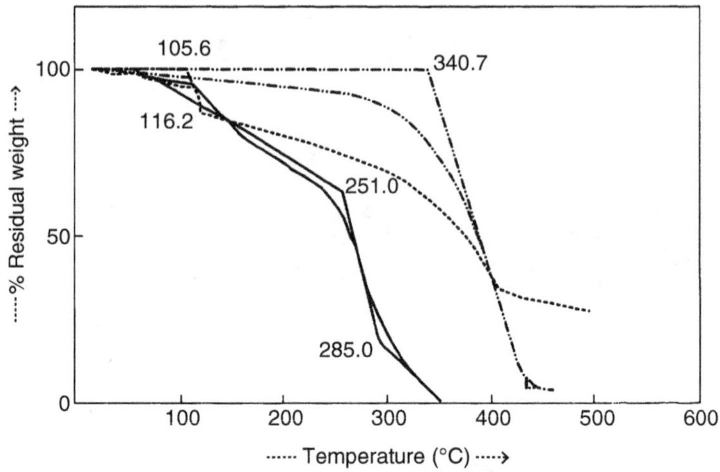

Figure 5.4. Primary thermograms of (-..-) IPP, (-) PAAM, and (---) IPP-g-PAAM. Reproduced from *J Appl Polym Sci*, **91(4)**: 2364 (2004).

THERMAL PROPERTIES

It is necessary to characterize the end-use products in terms of thermal properties. In general, thermal analysis co-relates the characterization of the properties of a material as a function of temperature. Graft and crosslinked polymers show different thermal behavior with respect to the parent polymer backbone.

The grafting of a monomer in any crystalline polymer takes place mainly in the amorphous zone of the parent backbone, but an increase in the percentage of grafting also affects the crystalline regions; hence, thermal stability of the parent polymer decreases, e.g., grafting of vinyl monomers to the PE [27], isotactic PP [28], Teflon-FEP film [29], Tefzel film [30], etc. The decrease in thermal stability of isotactic PP-g- polyacrylamide is reflected in Figure 5.4.

Grafting may result in the formation of stable network to the polymerlike formation of stable imides from pendant amide groups in cotton-grafted methacrylamide [31], which enhances the heat stability.

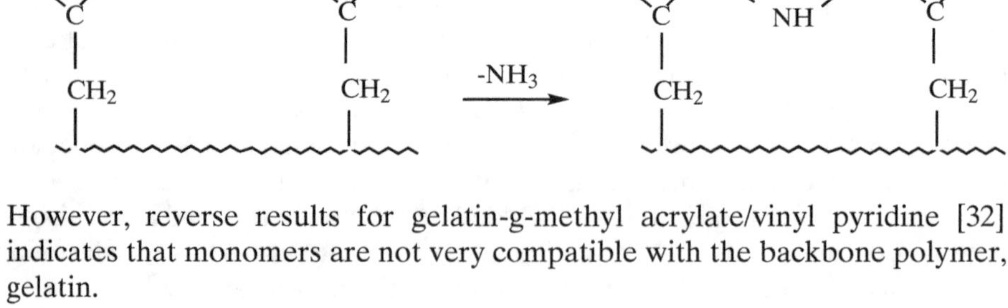

However, reverse results for gelatin-g-methyl acrylate/vinyl pyridine [32] indicates that monomers are not very compatible with the backbone polymer, gelatin.

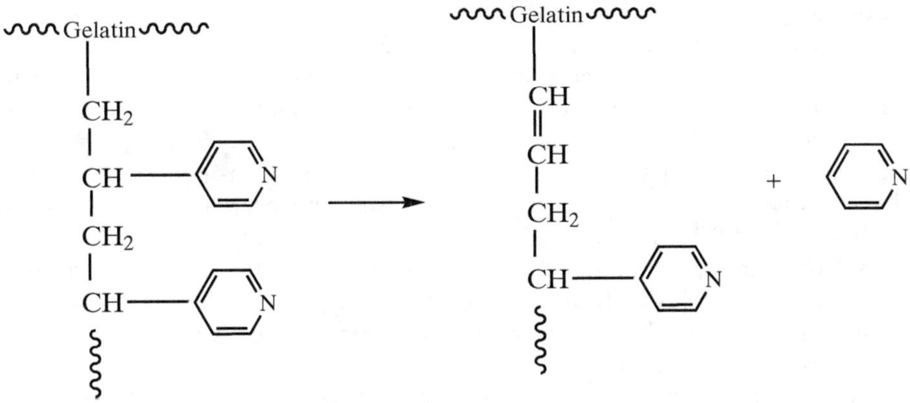

Hence, post-grafting reactions result in different thermal behaviors. Phosphorylation of cotton-g-MAAm, quarternization of cotton-g-4-vinyl pyridine (4-VP) [33] influence the thermal resistance of the parent polymer. Post-grafting reactions of gray cotton and MAAm grafted cotton fabric with PCl_3 and diethyl phosphate (DEP) decrease the initial and final decomposition temperature except when treated with PCl_3/Py. It shows a much higher final decomposition temperature (FDT) (451.06°C) (Figure 5.5). Phosphorylation

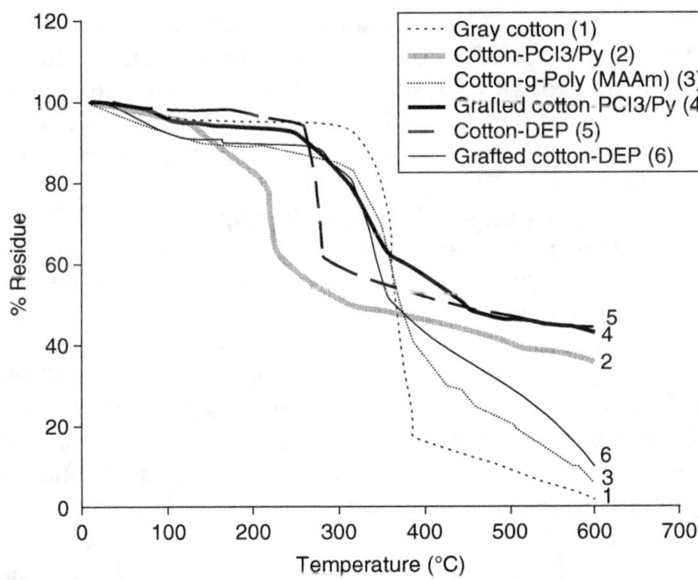

Figure 5.5. Primary thermogram of (1) gray cotton, (2) cotton-PCl3/Py, (3) cotton-g-poly (MAAm), (4) grafted cotton-PCl3/Py, (5) cotton-DEP, and (6) grafted cotton-DEP. Reproduced from *Ind J Fibre Text Res*, **32**: 312 (2007).

followed by grafting improves the thermal resistance of the fabric to the extent that it exhibits flame- resistance properties [34].

Transformation in polymers under thermal treatment involving physical transformations corresponds to glass transition (T_g) and melting temperature (T_m), crystallization, and crystal-crystal transitions. Chemical modification therefore has an impact on these properties. Glass transition temperature changes with crystallinity. In PE this decreases with decreasing crystallinity, but in PET and PP a maximum glass transition occurs at intermediate crystallinity. Copolymerization alters T_g and T_m. It varies with the degree of grafting [35]. The former rises with the increase in the degree of grafting due to the increase in the amorphous content by monomer grafting, whereas the latter decreases under the dilution of the crystalline structure.

The crosslinking nature depends on different techniques. The metallocene ethylene-octene copolymer is prepared through two different techniques, i.e., peroxide and silane-water crosslinking [36], and it is seen that the silane-crosslinked polymers retain the elastomeric characteristics of the pure polymer and show remarkably higher extensibility, better thermal stability, and energy storage capacity. Thermal stability also depends on the nature of the crosslinked moiety as well as on the environment. Polyurethane-polysiloxane thermal stability is better than that of polyether-based polyurethane and polyester-based polyurethane. The inert atmosphere does not influence the decomposition of polyurethanes below 350°C. However, the decomposition nature of the polyurethane-polysiloxane crosslinked polymer network is low in oxygen compared to a nitrogen medium between 350°C and 550°C [37].

Crosslinking enhances the thermal stability of polymers. Crosslinked high-density polyethylene (HDPE) may be used as a thermal energy storage material. HDPE crosslinked with 10 Mrad of electron-beam radiation has 48% crosslink content and shows thermal stability at 150°C [38]. The crosslinked HDPE consists of orthorhombic crystalline and an amorphous phase; crosslinking occurs mainly in the amorphous phase. Thermal degradation was not found to occur in the crosslinked HDPE kept at vacuum at 150°C for 1000 h. The charge/discharge performance showed that the crosslinked HDPE has thermal stability during the heating/cooling cycle and was not more susceptible to the thermal oxidative degradation than the uncrosslinked HDPE. Polylactic acid (PLA) has poor heat stability above its glass transition temperature (60°C). Crosslinking of PLA by irradiation in the presence of triallyl isocyanurate (TAIC) shows durable heat stability even above T_m [39]. The crosslinked PLA shows very low crystallinity because of the wide formation of a molecular chain network that inhibits the molecular motion for crystallization. If TAIC crosslinks can be introduced into annealed (at 90°C) PLA film, it shows heat stability at temperatures from T_g to T_m.

Formation of interpenetrating polymer networks from two immiscible resins like epoxy resin, silicone, and polyamidoamine hardener using a crosslinking agent (butyl titanate) and a catalyst (dibutyl tin laurate) results in heat resistant IPNs suitable for high-performance coating applications [40]. The silicone coating withstands up to 533°K, whereas epoxy polyamidoamine coating performed well up to 423°K but the IPNs remain stable up to 478°K in terms of color and gloss retention. The gloss is changed at 533°K, although no cracking or chalking is noticed on the surface. This suggests that such IPNs are stable up to 533°K. This enhancement in the heat-resistant properties of the epoxy resin is due to an extensive entanglement of the silicone resin with the epoxy polymer. Similar enhancement in thermal stability is observed in natural rubber/polystyrene semi- and full IPNs. Such IPNs (natural rubber crosslinked with dicumyl peroxide and polystyrene crosslinked with divinyl benzene) do not show any appreciable weight loss in the temperature range of 200°C–300°C, whereas homopolymers start to degrade at this temperature range [41]. Again, the degradation behavior varies for semi- and full IPNs, and it is dependent on crosslink density. For full NR-PS IPNs, the weight loss in the temperature range of 300°C–400°C is low compared to that of a semi-IPN as low crosslinking of a semi-IPN makes it easier to rupture bonds. The thermal stability increases with the degree of crosslinking. The physical entanglement due to the networking of two phases in full IPN makes the system tough and thermally stable.

MECHANICAL PROPERTIES

With regard to the properties of grafted polymers much effort has been devoted to their mechanical properties. In general, the results have been disappointing, and only comparatively minor improvements have been achieved by grafting. The reason for the loss of mechanical strength is attributed to the disturbance of the crystalline structure of the polymer, which in turn affects the tensile properties of the polymer.

The mechanical properties of acrylonitrile (AN) grafted cotton fibers and fabrics [42] are presented in Table 5.1. In the case of cotton fibers, the breaking strength is decreased and the elongation is increased upon grafting. Stiffness is reduced in every case by the grafting. With the fabric there is modest decrease in the breaking strength, and elongation at the break. However, tear resistance shows a significant decrease upon grafting.

Mechanical properties are also influenced by post-grafting reactions. Phosphorylation of the cotton-g-poly(MMA) and cotton-g-poly(2-VP) shows poor tensile properties [43] (Table 5.2).

Tensile properties of the pine-cellulosic fiber-adhesive-based particle boards (5 mm thick) are compared to the particle boards prepared from ethylacrylate and methy lacrylate grafted pine-cellulosic fibers (5 mm thick) [44]. The tensile strength, modulus of rupture (MOR), modulus of elasticity (MOE), decrease in the case of particle boards prepared from the grafted pine cellulosic fibers (Table 5.3).

TABLE 5.1. Effect of grafting of acrylonitrile to cotton yarn under various conditions on the mechanical properties.

Sample	Grafting %	Breaking Strength	Elongation at break%	Breaking toughness g/tex	Tear Resistance (g)
Cotton Yarn	0	9.8	15.1	116	573
Cotton Fabric	0	40.9	22.0	—	—
Cotton Yarn	28	8.0	15.5	66	—
(grafted)	37	7.4	24.9	34	
Cotton Fabric	25	39.6	20.5	—	447
(grafted)	26	39.0	21.7	—	413

Reproduced from *Text Res J*, **38**: 710 (1968).

TABLE 5.2. Mechanical properties of gray, grafted, and phosphorylated grafted cotton fabric.

Sample	Tensile Strength(N)	Elongation (%)	Extension (mm)
Gray Cotton	105.8	18.2	18.19
Cotton-g-poly(MMAm)	51.8	8.78	8.78
Cotton-g-poly(2-VP)	78.3	11.22	12.91
Cotton-g-poly(MAA) /PCl_3	4.25	1.36	1.36
Cotton-g-poly(MAAm)/DEP (diethyl phosphite)	20.75	6.18	6.24
Cotton-g-poly(2-VP)/PCl_3	19.25	8.58	8.59

Reproduced from Sharma R, Ph.D. dissertation, 2006, H.P. University, Shimla, India.

TABLE 5.3. Mechanical properties of particle boards prepared from PF resin using pine cellulosic and grafted pine cellulosic fibers.

Sample	MOR (N/mm^2)	MOE (N/mm^2)	Tensile Strength perpendicular to surface (N/mm^2)	Tensile Strength parallel to surface (N/mm^2)
Particle board prepared from pine cellulosic fiber	5.65	590.54	0.021	0.037
MA grafted pine cellulosic fiber	4.80	547.40	0.017	0.039
EA grafted pine cellulosic fiber	4.87	449.41	0.018	0.039

Reproduced from *Sci & Eng of Composite Matls*, **12**: 273 (2005).

Abrasion resistance related to textiles is improved by the grafting of vinyl monomers for the cellulose textiles. These also improve their durable press properties.

Strain-hardening behavior in the elongational viscosity, i.e., the rapid increment of the elongational viscosity with elongational time or strain under a constant shear rate, is highly needed for different polymer processing methods such as film processing, thermoforming, blow molding, foaming, etc. Most polymers without long chain branching hardly show any strain-hardening behavior in the elongational viscosity. It has been observed that incorporation of a small amount of crosslinked polymer in a linear polymer remarkably enhances the strain-hardening behavior in the elongational viscosity of the linear polymer without a major change in the steady state shear viscosity. This happens when the linear polymer and precursor of the crosslinked polymer are compatible. The blending of isotactic polypropylene (IPP) with crosslinked EHDM (ethylene-1-hexene-ethylidene norbornene) greatly enhances the strain-hardening behavior in the elongational viscosity of IPP even though the amount of crosslinked polymer was 0.3 wt. % [45]. However, the crosslinked polymer, which was far beyond or below the sol-gel transition point, had little influence on the elongational viscosity as well as the shear viscosity. The stretching of the chain sections between the crosslink points was responsible for the strain-hardening behavior.

The rheological properties of methacrylic crosslinked polymer microparticle solutions depend on the physical interactions. Crosslinked polymer microparticles (CPM) having different functionalities: epoxy, acrylic, carboxyl, hydroxyl, octadecyl, and methyl show very weak effect of functionality on the intrinsic viscosity in any solvent. Exception is the carboxyl groups, which have significant influence by increasing the viscosity because of hydrogen bonding while octadecyl groups reduce the low shear limiting viscosity.

Crosslinked macromolecules synthesized [46] by the polyaddition reaction of acid polyesters with diepoxides and crosslinking them with anhydrides made it possible to increase the energy of deformation in the transition range by a factor of 10, compared with conventional highly crosslinked epoxy resins. A further increase in toughness could be realized by introducing hydrogen bonds, e.g., urethane groups along the chains of the polymers.

Crosslinking has enormous synergistic effects on the mechanical properties of polymers, particularly of elastomers. A 50% enhancement in modulus could be achieved by crosslinking silicone elastomers with peroxides [47]. An increase in crosslink density leads to an increase in Young's modulus and a decrease in strain at the break. For electron-beam crosslinked fluorinated rubber, the tensile strength increases with radiation dose up to a certain level after which there is a drop while the modulus increases continuously with the radiation dose due to the formation of a network structure. The properties like elongation at break, hysteresis loss, and set decrease on radiation. The increased degree of crosslinking leads to the lowering of the loss tangent both at and above the glass transition temperature [48]. Dynamic vulcanization of natural rubber in the thermoplastic elastomer of natural rubber-polystyrene results in the enhancement in mechanical properties as compared to the uncured samples. The nature of networks formed using sulfur, peroxide, and mixed system (sulfur and peroxide) are different and are reflected in their mechanical properties. The mono-, di-, and polysulfidic linkages in sulfur systems impart high chain flexibility in the polymer network. In DCP systems, only rigid C-C linkages are present, and the mixed system contains all these mono-, di-, and polysulfidic linkages as well as the C-C linkages. The peroxide system shows the highest tensile strength, followed by the mixed and sulfur system, whereas for the sulfur system the impact strength is the highest. Hence, the mechanical properties of crosslinked polymers are dependent not only on the network density but also on the nature of crosslinks [49].

The mode of preparation of interpenetrating polymer networks has a great effect on the thermal and mechanical properties of IPNs [50]. Polyurethane- and polycyanurate-based full IPNs were prepared by both the simultaneous and sequential modes. The *sim*-IPNs showed higher densities, and calorimetric and mechanical glass transition temperature than a *seq*-IPN having the same compositions. In the glassy region the relaxation dynamics of *sim*-IPN probed by dynamic mechanical analysis reveals the existence of distinct segmental motion associated with the individual components. It has been proven that the interpenetration process associated with the simultaneous synthesis markedly reduces the free volume in the system, giving rise to significant differences between the local and cooperative molecular mobility of *sim*- and *seq*-IPNs. The effect of variations of blend composition on the mechanical properties of *sim*-semi-IPN of castor oil-based polyurethane (COPU) and polystyrene (PS) has interesting observations [51]. There is little change in the tensile strength, percent elongation at break, elastic modulus, and tear resistance up to 25% of styrene in IPNS indicating that there is little intermixing between the two polymers, and styrene is dispersed in the PU continuous phase. Beyond this there is a sharp increase in these properties, and that reflects a higher degree of physical entanglement between the COPU network and PS. This results in an effective stress transfer between the two polymer components of the IPN upon deformation, leading to an enhancement in mechanical strength. The physical entanglement is most likely due to the formation of a bicontinuous phase structure at a critical styrene composition. The effect of entanglement and crosslink density in the enhancement of mechanical properties has also been reflected in *sim*-PU/allyl novolac resin IPN [52]. An interesting synergism in mechanical properties has been observed for elastomeric IPNs [53]. Elastomeric IPNs made of castor oil, NCO-terminated polyether/telechelic dihydroxy polybutadiene and vinyl, and a methacrylic monomer together possess a permanent set <15%, a tensile strength over 13 MPa, and ultimate elongation over 240%. Thus it behaves as a reinforced toughened elastomer. Gel content of such IPNs is found to be over 95%.

CRYSTALLINITY

Graft copolymerization of monomers of both polar and functional types has always had a decreasing effect on the crystallinity of the polymers. The incorporation of side chain grafts, which have an amorphous nature, exerts a dilution effect on the grafted polymer by increasing the amorphous content at the expense of the crystallinity [54, 55]. Partial disruption in the crystalline structure also occurs since graft copolymerization is initiated by free radicals, which are well known for their random distribution in nature. Hence, free radicals are formed both in crystalline and amorphous regions of the polymer presenting the possibility that some of the formed side chain grafts will penetrate the crystalline zones, causing crystalline disruption and a further decrease in the degree of crystallinity depending on the grafting parameters [56, 57]. Post-grafting reactions also affect the crystallinity of the polymers. For instance, sulfonation of the polystyrene grafted PE, FEP, PVDF, and PFE causes partial disruption in their crystallities adding to the decrease in the crystallinity caused by the dilution effect. Such crystallinity changes together with the amount, the nature, and distribution of the grafted moieties.

Crosslinking also has adverse effect on crystallinity of polymers. A thorough study on the radiation- induced crosslinking of different grades of polyethylene (LDPE, LLDPE, HDPE, etc.) shows that due to crosslinking on the lamellae surfaces, the radiation-induced defects increase at higher radiation doses and as a result the crystallinity and melting temperature decrease [58]. At a high radiation dose the increase in the surface free energy is much more dominant than the increase in configurational entropy of the chains [59]. The decrease in crystallinity can be attributed to the additional lattice distortion primarily introduced by the crosslinks occurring at the lateral grain boundaries while lattice expansion can be associated with the same crosslinking mechanism that begins at the defects both within the crystal and outside the crystals at the lateral grain boundaries [60]. The effects of electron-beam irradiation on the crystallinity of two partially fluorinated homopolymers like polyvinylidine fluoride (PVDF) and the poly(ethylene tetrafluoro ethylene) copolymer are quite different. The degree of crystallinity in irradiated PVDF film is found to increase up to 200 kGy beyond which it marginally decreases, whereas the degree of crystallinity of irradiated ETFE films initially rises up to 100 kGy and then remarkably drops with the increase in irradiation dose [61] (Figure 5.6). The initial enhancement in the degree of crystallinity indicates the presence of transfer in chains from the amorphous to crystalline regions under the influence of chain scission, which increases the chain mobility upon irradiation. Further increase in the irradiation dose enhances the degradation and crosslinking effects and brings more changes in the crystalline structure [62]. The decrease in crystallinity is minor in PVDF but considerable for ETFE. The decreasing trend above 200 and 100 kGy for PVDF and ETFE, respectively, indicates

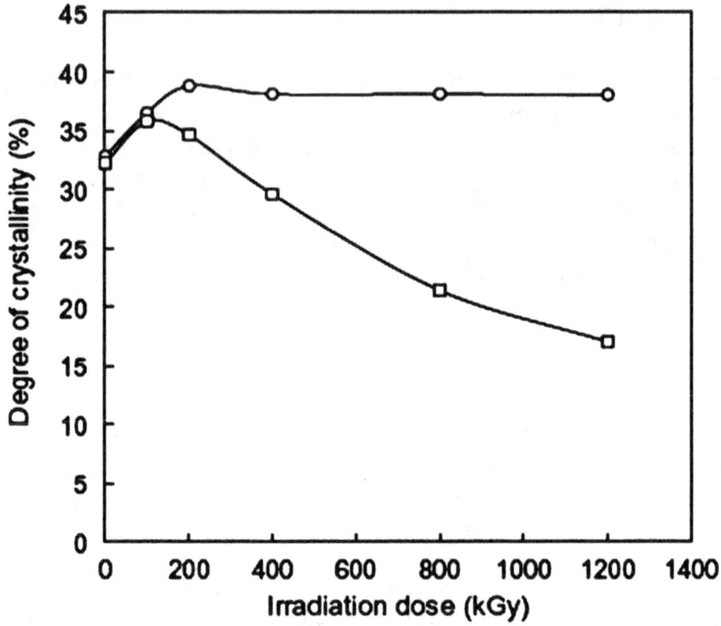

Figure 5.6. Variation of degree of crystallinity of (○) PVDF films and (□) ETFE films with the irradiation dose. Reproduced from *Nucl Instrum Methods Phys Res B*, **201**: 604 (2003).

that crosslinking inhibits the recrystallization in both films, and it results in a gradual decrease in the melting temperature with the irradiation dose. The decrease in melting temperature is a sign of crystal defect. Hence, it may be concluded that beyond the optimum dose of radiation the crystalline damage takes place only in the newly formed crystallites for PVDF, whereas for ETFE the damage takes place in both the new as well as in the original crystals. This may be due to the presence of the radiation-sensitive tetrafluoroethylene comonomer in ETFE.

The grafting has the ability to alter the crystallinity of the polymer and thus alter related properties such as water uptake, ionic conductivity and mechanical stability [63–65]. Modification of the organochemical properties of cellulosic fibers by graft polymerization with selected monomers imparts new chemical properties. The microbiological and light resistances of cellulosic fibers to degradation are increased by grafting [66]. Surface properties of modified fibers are changed to impart soil release [67], dyeability [68], and flame-resistant [69, 70] properties.

Structured polymer layers that have nonlinear optical properties are produced by a process wherein either organic compounds containing ethylenically unsaturated groups are subjected [71] to free-radical copolymerization with photoresponsive molecules (e.g., stilbene, azo, or azomethine compounds containing ethylenically unsaturated groups and donor and acceptor groups) have potential use in production of polymeric waveguides in communications technology.

POLYMER SUPPORTED REAGENT TO FACILITATE REACTION

PS–reagents are defined as functionalized polymers used in stoichiometric amounts in a one-step process to transform low-molecular-weight substrates into products.

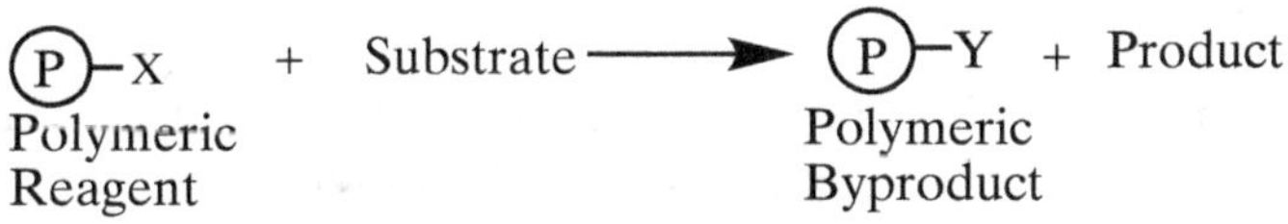

It carries a reactive organic groups bound to a macromolecular support, and such groups may be attached to polymeric carriers by grafting and crosslinking reaction.

There are a number of different polymeric supports that have been used successfully and that are mainly based on polymers or the copolymers derived from vinyl monomers such as poly(vinyl pyridines). A well-known example of a copolymer-based reagent is Merrifield resin, which comprises polystyrenc crosslinking by divinyl benzene [72] used for "solid-phase pepetide synthesis."

This type of polymer-supported reagent is able to carry the reaction, viz., polyaniline-supported molybdenum, poly[1-vinyl (2-pyrrolidonium permanganate)] catalyzed aerobic oxidation of alcohols to aldehydes and ketones [73, 74].

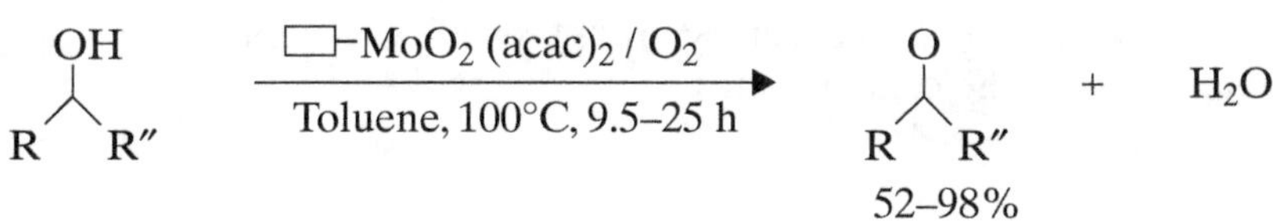

R = alkyl, aryl; R″ = alkyl, aryl, H; □ = polyaniline

The post-grafting reactions of cellulose-g-poly (4-VP) [75] and polyethylene-g-maleic anhydride (PE-g-MAn) [76] are also useful in reduction and oxidation reactions, respectively, as follows:

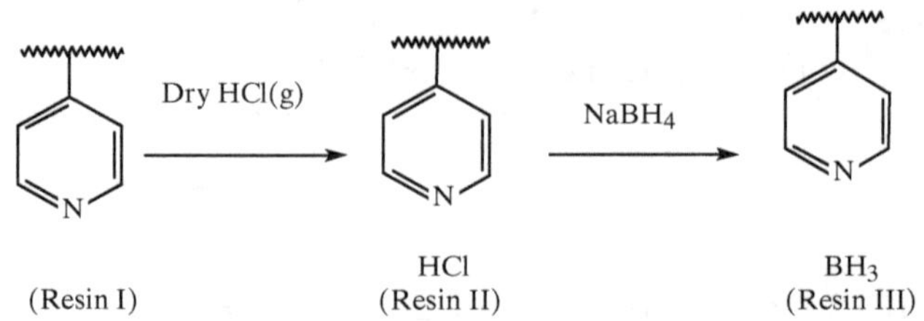

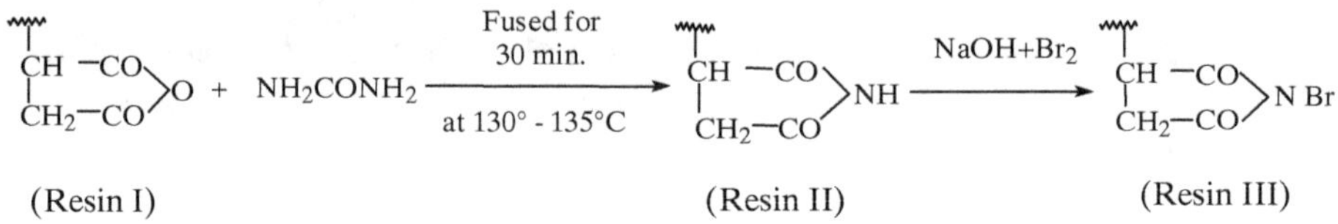

Divinyl benzene crosslinked poly(styrene-co-acrylol chloride) can be used as a versatile intermediate for the preparation of different polymer supported reagents [77]. It can also serve as a polymer protecting group via a nucleophilic process on the acid chloride carbonyl group. Reactions with a nucleophile were fast and proceeded at room temperature. Examples of prepared polymer bound amines, hydrazines, alcohols, thiols, and even nitrile prove the reactivity of supported acid chloride towards nucleophiles. The good mechanical properties of the polymer beads facilitate recycling and multiple uses. Both poly(styrene–co-acrylic acid) and poly(styrene-co-acrylol chloride) efficiently scavenged benzylamine from solutions, the process being faster with acid chloride.

Permanganate functionalized crosslinked polyvinylamine (prepared by Hoffman degradation reaction using hypobromite with divinyl benzene crosslinked polyacrylamide) can be used effectively as an oxidizing agent for the conversion of primary and secondary alcohols to their corresponding carbonyl compounds [78]. The extent of oxidation is found to be strongly dependent on the degree of crosslinking. The best conversion was obtained with the resin with a crosslinked density of 5%. At a lower crosslink density, the polymer becomes more swollen, and the effective catalytic sites are less accessible. Similarly, at higher crosslink density (10%), swelling may be too low to give effective access to the catalytic sites.

REFERENCES

1. Chauhan GS, Kumari A, Sharma R. *Polymers & Polym Compost*, **13(2)**: 151 (2005).
2. Kaur I, Misra BN, Barsola R, Singla K. *J Appl Polym Sci*, **47**: 1165 (1993).
3. Kaur I, Maheshwari S, Gupta A, Misra BN. *J Appl Polym Sci*, **58**: 835 (1995).

4. Kaur I, Misra BN, Chauhan S, Chauhan MS, Gupta A. *J Appl Polym Sci*, **59**: 389 (1996).
5. Cormick CLM, Lin KC. *J Macromol Sci-Chem A*, **16**: 1441 (1981).
6. Kaur I, John V, Gupta A, Misra BN, Mandal UK. *Mater Sci & Eng A*, **267**: 36 (1999).
7. Kaur I, Gupta A, Misra BN, Marina S, Barman PB. *Polymers & Polym Composites*, **6**(**7**): 465 (1998).
8. Kaur I, Gupta A, Misra BN, Chauhan GS. *Desalination*, **110**: 115 (1997).
9. Kaur I, Singh B, Gupta N. *Radiat Phys Chem*, **72**: 489 (2005).
10. Chauhan GS, Mahajan S, Sharma R, Kumari A, Lal H. *Trends in Carbohydrate Chem*, **7**: 137 (2001).
11. Chauhan GS, Kumari A, Sharma R. *J Appl Polym Sci*, **106**: 2158 (2007).
12. Chauhan GS, Guleria LK, Mahajan S. *Desalination*, **141**: 325 (2001).
13. Chauhan GS, Mahajan S. *J Appl Polym Sci*, **86**: 667 (2002).
14. Chauhan GS, Guleria LK, Kumar R. *Cellulose*, **12**: 97 (2005).
15. Nasef MM, Saidi H, Nor HM, Ooi MF. *J Appl Polym Sci*, **78**: 2443 (2000); *J Appl Polym Sci*, **76**: 1 (2000).
16. Kostov G, Turmanova S. *J Appl Polym Sci*, **64**: 1469 (1997).
17. Chauhan GS, Lal H, Mahajan S. *J Appl Polym Sci*, **91**: 479 (2004).
18. Pande CS, Mehta IK, Ambasta BK, Kumari M, Archana. *J Appl Polym Sci*, **77**: 475 (2000).
19. Taylor NW, Bagley EB. *J Appl Polym Sci*, **18**: 2747 (1974).
20. Mark JE, Erman B. *Rubberlike Elasticity: A Molecular Primer* (New York: John Wiley and Sons, 1988), p. 49.
21. Flory PJ, Rehner J. *J Chem Phys*, **11**: 521 (1943).
22. Hergenrother WL, Hilton AS. *Rubber Chem Technol*, **76**: 832 (2003).
23. Marzocca AJ. *Eur Polym J*, **43**: 2682 (2007).
24. George SC, Knorgen M, Thomas S. *J Membr Sci*, **163**: 1 (1999).
25. Chen CH, Chen WJ, Chen MH, Li YM. *Polymer*, **42**: 7961 (2000).
26. Khalid MN, Agnely F, Yagoubi N, Grossiord JL, Couarraze G. *Eur J Pharmaceutical Sci*, **15**: 425 (2002).
27. Kaur I, Misra BN, Chauhan GS, Gupta A. *J Appl Polym Sci*, **69**(**3**): 599 (1998).
28. Kaur I, Singh B, Upasna. *J Appl Polym Sci*, **91**(**4**): 2364 (2004).
29. Kaur I, Misra BN, Kohli A. *Desalination*, **139**: 357 (2001).
30. Kaur I, Kumar R, Singh B, Misra BN, Chauhan GS. *J Appl Polym Sci*, **78**: 1171 (2000).
31. Kaur I, Sharma V, Rajneesh. *Ind J Fibre Text Res*, **32**: 312 (2007).
32. Kaur I, Chauhan GS, Misra BN, Gupta A. *Polym Compos*, **5**(**4**): 291 (1997).
33. Kaur I, Rajneesh, Sharma V. *Def Sci J*, **57**(**2**): 249 (2007).
34. Simionescu CI, Macoveanu MM, Percec S, Cazacu G, Ioanid A, in *Graft Copolymerization of Lignocellulosic Fibers*, DN-S. Hon, ed., ACS Division of Cellulose Paper and Textile Chemistry (New York: American Chemical Society, 1981), p. 57.
35. Gupta B, Scherer GG. *J Appl Polym Sci*, **50**: 2129 (1999), Nasef M. *Polym Degrad Stab*, **68**: 231 (2000).

36. Sirisinha K, Meksawat D. *J Appl Polym Sci*, **93**(**3**): 1179 (2004).
37. Zhu Q, Feng S, Zhang C. *J Appl Polym Sci*, **90**: 310 (2003).
38. Han SO, Dong WL, Han OH. *Polym Degrad Stab*, **63**: 237 (1999).
39. Mitomo H, Kaneda A, Quynh TM, Nagasawa N, Yoshii F. *Polymer*, **46**: 4695 (2005).
40. Dhoke SK, Palraj S, Maruthan K, Selvaraj M. *Prog Org Coat*, **59**: 21 (2007).
41. Mathew AP, Packirisamy S, Thomas S. *Polym Degrad Stab*, **72**: 423 (2001).
42. Blouin FA, Morris NJ, Arthur JC. *Text Res J*, **38**: 710 (1968).
43. Sharma R. "Development of Flame Retardant Cotton and Polyester Fabrics," Ph.D. dissertation, 2006, H.P. University, Shimla, India.
44. Kaur I, Misra BN, Sarkar A, Sharma M, Vibha, Singh B. *Sci & Eng Of Composite Materials*, **12**(**4**): 273 (2005).
45. Yamaguchi Asayuki. *J Polym Sci Part B Polym Phys*, **39**(**2**): 228 (2001).
46. Ludovic V, Jean-Pierre P, Benoît M. *Macromol Mater Eng*, **287**(**1**): 52 (2002).
47. Heiner J, Stenberg B, Persson M. *Polym Test*, **22**: 253 (2003).
48. Banik I, Bhowmick AK. *Radiat Phys Chem*, **54**: 135 (1999).
49. Asaletha R, Kumaran MG, Thomas S. *Eur Polym J*, **35**: 253 (1999).
50. Bartolotta A, Macro GD, Lanza M, Carini G, D'Anselo G, Tripodo G, Fainleib A, Danilenko I, Grytsenko V, Sergeeva L. *Mater Sci Eng A*, **370**: 288 (2004).
51. Sushllla B, Khakhar DV, Nadkarni VM. *Polymer*, **38**(**17**): 4319 (1997).
52. Chiang WY, Chang DM. *Eur Polym J*, **31**(**8**): 709 (1995).
53. Xie HQ, Guo JS. *Eur Polym J*, **38**: 2271 (2002).
54. Gupta B, Hass O, Scherer GG. *J Appl Polym Sci*, **54**: 269 (1994).
55. Nasef M. *Eur Polym J*, **38**: 87 (2002).
56. Hietala S, Holmberg S, Naesman J, Ostrovskii D, Paronen M, Sermaa R, Sundholm F, Torell L, Torkkeli M. *Angew Makromol Chem*, **253**: 151 (1997).
57. Hietala S, Skou E, Sundholm F. *Polymer*, **40**: 5567 (1999).
58. Milicevic D, Trifunovic S, Popovic M, Milic V, Suljovrujic E. *Nucl Instrum Methods Phys Res B*, **260**: 603 (2007).
59. Zoepfl FJ, Markovic V, Silverman J. *J Polym Sci Poly Chem*, **22**: 2017 (1984).
60. Yeh GSY, Chen CJ, Boose DC. *Colloid Polym Sci*, **263**: 109 (1985)
61. Lyons BJ. *Radiat Phys Chem*, **45**: 159 (1995).
62. Nasef MM, Dahlan KZM. *Nucl Instrum Methods Phys Res B*, **201**: 604 (2003).
63. Gupta B, Scherer G. *Chimia*, **48**: 1277 (1994).
64. Gupta B, Buchi F, Scherer G, Chapiro A. *Polym Adv Technol*, **5**: 493 (1994).
65. Hietala S, Skou E, Sundholm F. *Polymer*, **40**: 5567 (1999).
66. Arthur JC, Jr. "Graft Copolymerization onto Polysaccharides" in *Advances In Macromolecular Chemistry*, WM Pasika, ed. (London: Academic Press, 1970), Vol. 2, p. 1.
67. Mares T, Arthur JC, Jr, Harris JA. *Single-Treatment Radiation Process for Imparting Durable Soil-Release Properties to Cotton and Cotton-Polyester Blend Fabrics*, U.S. Patent, 4,063,885, December 20, 1977.
68. Harris JA, Arthur JC, Jr. *Text Res J*, **46**: 219 (1976).

69. Harris JA, Arthur JC, Jr, Goynes WR, Jr. *J Appl Polym Sci*, **23**: 2555 (1979).
70. Harris JA, Arthur JC, Jr, Goynes WR, Jr. *J Appl Polym Sci*, **24**: 201 (1979).
71. Etzbach KH, Frankenthal DE. *Production of Structured Polymer Layers Having Nonlinear Optical Properties*, U.S. Patent 5,393,645, February 28, 1995.
72. Merrifield RB. *J Am Chem Soc*, **85**: 2149 (1963).
73. Velusamy S, Ahamed M, Punniyamurthy T. *Org Lett*, **6**(**26**): 4821 (2004).
74. Zacharias J, Pillai VNR. *Asian J of Chem*, **15**(**1**): 179 (2003).
75. Kaur I, Mahajan RK, Dhiman P. *J Appl Polym Sci*, **108**: 99 (2008).
76. Kaur I, Mahajan RK, Dhiman P. *J Appl Polym Sci*, in press.
77. Krajnc P, Toplak R. *React Funct Polym*, **52**: 11 (2002).
78. Prabhakaran PV, Venkatachalam S, Ninan KN. *Eur Polym J*, **35**: 1743 (1999).

6

IN THE BIOMEDICAL ARENA

Gauri P. Misra,* Eun Seok Gil,* and Tao Lu Lowe

INTRODUCTION

Grafted and crosslinked polymers have been extensively exploited in the biomedical field due to their unique physicochemical advantages and current increasing demands. In this chapter, the design criteria and biomedical applications of four different types of grafted or crosslinked polymers—hydrogels, nanogels, dendrimers, and grafted cyclodextrins—will be discussed in four sections. In each section, the biomedical applications of each type of grafted and crosslinked polymers will be presented by focusing on drug delivery and tissue engineering. Furthermore, other biomedical applications of the polymers, such as biosensing and bioseparation will also be addressed.

In designing grafted and crosslinked polymers for biomedical and pharmaceutical applications, it is essential to design these polymers with desired tunable physicochemical and biomimicking properties. For example, for drug delivery application, the important design criterion is to achieve maximum therapeutic efficacy of encapsulated drugs with minimum toxicity. To meet the above criterion, the following parameters must be taken into consideration in designing grafted or crosslinked polymers for controlled drug delivery: (1) chemical and physical properties of the drugs and the polymers, (2) interactions between the drugs and the polymers, (3) biological environment and delivery sites of the drugs, and (4) routes of delivery, such as oral, ocular, intravenous, intranasal, intravascular, intraperitoneal, intramuscular, and subcutaneous administration. Eventually the designed grafted or crosslinked polymers should protect drugs from degradation, increase the circulation time of drugs in the

*These authors contributed equally.

Polymer Grafting and Crosslinking, Edited by Amit Bhattacharya, James W. Rawlins and Paramita Ray

blood stream, improve the permeability of drugs across biological barriers, and enhance the accumulation and absorption of drugs in disease tissues [1].

Another important biomedical area, where grafted or crosslinked polymers play a key role, is tissue engineering that involves the growth or repair of damaged tissues and organs, such as bones, cartilage, liver, nerves, muscle, skin, myocardium, and small blood vessels. The design criterion for growing new tissues is to provide a suitable biological environment for cell proliferation, migration, and differentiation. To meet this tissue engineering design criterion, grafted or crosslinked polymers should be designed with optimal scaffold characteristics (including chemistry, grafting or crosslinking density, degradation rate, topography, and surface energy) and morphology (including porosity, pore size, and pore connectivity), to provide an adequate cell-extracellular matrix, cell-cell adhesion, and cell-cell communication. In other words, the designed grafted or crosslinked polymers should promote tissue formation mimicking its *in vivo* process with desired chemical, physical, spatial, and biological signals and cues [2–6].

In medical device applications such as biosensors, diagnosis, bioimaging, and bioseparation, grafted and crosslinked polymers as a coating or encapsulation matrix need to be inert in the biological environment while enhancing the stability, sensitivity, and specificity of biomolecules and imaging molecules.

HYDROGELS

Hydrogels are three-dimensional networks of crosslinked polymer chains containing water within the network. Hydrogels are formed either by physical crosslinking, generally referred to as a sol-gel transition, or by covalent bonding (chemical crosslinking) between polymer chains. Physically crosslinked hydrogels are usually formed through H-bonding, hydrophobic and electrostatic interaction, or stereo-complex formation [7–16]. Chemically crosslinked hydrogels can be formulated by radical polymerization of monomers in the presence of crosslinkers, reaction of functional side groups of polymers with themselves or other functional crosslinkers, high-energy irradiation and enzyme reaction [12]. Hydrogels generally swell in water and can contain up to 95 wt. % water within the network. Water content (or swelling) of hydrogels can be tuned by modulating the composition and conformation of polymers: hydrophilic/hydrophobic balance of polymer chains and pendant groups, and degree of crosslinking. Some hydrogels can swell and shrink in response to external stimuli, such as changes in temperature, pH, solvent, electric current, magnetic field, and ultrasound, and are called stimuli-responsive or "smart" hydrogels. Hydrogels also can respond to changes of some biomolecules, such as glucose and single-stranded DNA [17]. Due to their attractive swelling, responsive properties, and resemblance to biological tissues [18], hydrogels have been extensively studied for potential drug delivery, tissue engineering and medical devices applications [5].

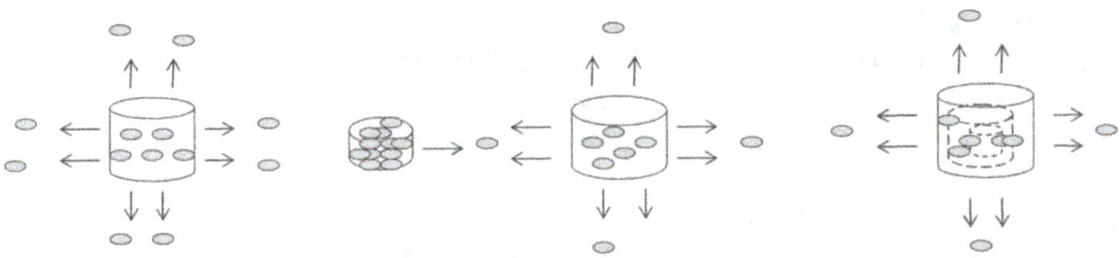

Figure 6.1. Drug release mechanisms: (a) diffusion, (b) swelling, and (c) degradation.

In Drug Delivery

Hydrogels have been explored as oral and implantable drug delivery vehicles for a variety of drugs including proteins, peptides, anti-cancer drugs, and anti-inflammatory agents. In general, there are three important mechanisms controlling the release kinetics of drugs: diffusion of drugs, swelling of hydrogels, and degradation of hydrogels (Figure 6.1). The chemical composition (hydrophilic/hydrophobic balance), ratios of monomers/macromers and solvents, and crosslinking density (porosity) of hydrogels play important roles in controlling the swelling of hydrogels and subsequent diffusion of drugs from hydrogels. Drug release can also be tuned by using degradable hydrogels based on hydrolytically (carbonate, ester, polyphosphate, and phosphazene) or enzymatically cleavable linkage, and changes of environmental stimuli such as temperature, pH, electrical and magnetic fields, and glucose concentration.

One important drug delivery application of hydrogels is insulin delivery. Insulin is a 5.6kDa protein that is clinically used to treat diabetes. Insulin is usually administered by intravenous or subcutaneous injections daily, which is not convenient for patients. Oral insulin delivery encounters degradation due to a proteolytic enzyme in gastrointestinal track, and low absorption by intestinal walls [19]. In order to overcome these barriers, pH-sensitive hydrogels made of poly(methacrylic acid-g-ethylene glycol) [19] or poly(methacrylic acid) crosslinked with poly(ethylene glycol) dimethacrylate [20] were tested as oral insulin delivery vehicles in normal and diabetic rats. The results of Kumar et al. [20] demonstrated that the bioavailability of insulin released from the hydrogel was 4 to 5 times higher than free insulin. The released insulin reduced the blood glucose levels and the effect lasted for 8-10 h. As one-time administration of long-lasting insulin is desired for patient compliance, many types of hydrogels have been designed for sustained insulin release. For example, Atkins et al. [21] fabricated a macroporous hydrophilic hydrogel matrix of crosslinked poly(2-hydroxyethyl methacryate) and the hydrogels released insulin for over one month *in vitro*. Insulin loading and the monomer:-solvent ratio affected the release kinetics. However, only a small fraction (up to 3%) of insulin was released after one month, suggesting that the hydrogels may prolong insulin release beyond that period. Carenza et al. [22] developed thermoresponsive hydrogels based on an acryloyl-L-proline methyl ester crosslinked with trimethylolpropane trimethacrylate, using radiation polymerization. Insulin-loaded hydrogel discs were implanted subcutaneously in diabetic rats, and a 40% reduction in blood glucose levels were observed after implantation, which was sustained for over a month. Future advanced hydrogel development for insulin release should be the pulsatile release of insulin in response to glucose level change. Kataoka et al. synthesized hydrogels composed of poly(*N*-isopropylacrylamide) [poly(NIPAAm)] derivatized with a definite fraction of a phenylboronic acid group as a glucose sensing moiety [23]. Kashyap et al. immobilized a glucose oxidase enzyme, which converts glucose into gluconic acid, on physically crosslinked hydrogels made of chitosan/β-glycerol phosphate disodium [24]. The hydrogels demonstrated insulin release in response to glucose concentration change (Figure 6.2).

Besides insulin, hydrogels have been used for controlled delivery of anti-cancer drugs such as doxorubicin, human interlukin-2 (IL-2, 25.5 kDa protein), and human g-interferon (21 kDa protein). One of the earlier studies involving hydrogels for doxorubicin release was carried out by Ulbrich and colleagues. In their work, doxorubicin was released from degradable hydrogels consisting of copolymers of *N*-(2-hydroxypropyl)-methacrylamide and *N*,O-dimethacryloyl hydroxylamine for up to 70 h *in vitro*. Furthermore, the same group also demonstrated that poly[*N*-(2-hydroxypropyl)methacrylamide)] hydrogels could release doxorubicin in leukaemic mice for 4 days, and the mice could survive for 35 days [25]. Methacrylated dextran and lactate-hydroxyethyl methacrylated dextran hydrogels have been developed for controlled release of the anti-cancer drug IL-2 for up to 100 days [26]. The IL-2 released kinetics that can be tuned by the crosslinking density and the degradation rate of the hydrogels. Thermo- and pH-responsive poly(NIPAAm-co-sodium acrylate-co-n-*N*-alkylacrylamide) hydrogels were reported to control the release of Y-interferon with almost zero-order release kinetics during a period of 130 h. The release kinetics could be modulated by adjusting the hydrophobic component, n-*N*-alkylacrylamide, which controls swelling of the hydrogels.

Hydrogels have also shown capability of encapsulating and releasing other therapeutic agents of different molecular sizes, such as growth factors (human growth hormone, transforming growth factor β1, insulinlike growth factor-1, vascular endothelial growth factor), antibiotics (adriamycin and cephalexin), anti-inflammatory dexamethasone, antibodies (immunoglobulin G, IgG), and high-molecular-weight proteins, [bovine serum albumin (BSA) (66 kDa)]. For examples, Mellott et al. [27] demonstrated that highly crosslinked poly(ethylene glycol) (PEG) hydrogels could continuously release BSA for over 240 days due to their high crosslinking density. The BSA release kinetics also strongly depended on the interaction (possible H-bonding) between the BSA and the PEG chains of the hydrogels. Lowe's group [28] developed poly(NIPAAm-co-Dex-lactate-(hydroxyethyl methacrylic) acid) hydrogels by combining thermoresponsive properties with degradation. BSA was loaded into the hydrogels during the synthesis process in an aqueous medium. Mathematical

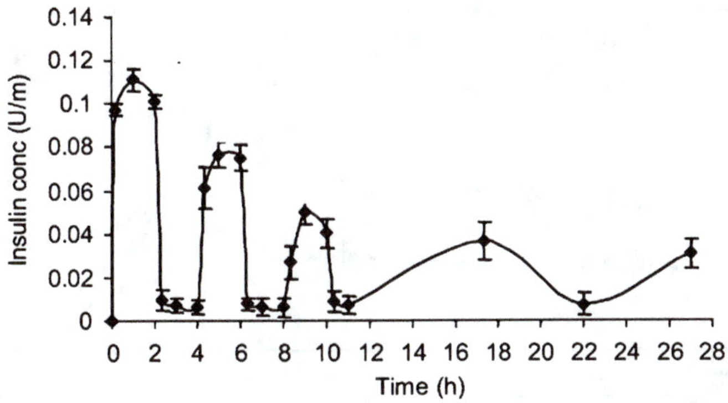

Figure 6.2. Glucose-stimulated insulin release from 1 IU/200 ml insulin-loaded chitosan/β-glycerol phosphate disodium subcutaneous gels in PBS (pH 7.4). The on-off insulin release is observed when the gels are placed in high (3 mg/ml) and low (1 mg/mL) glucose media alternatively. The gels swell at high glucose and release insulin while reverse follows at low glucose. Adapted from *Biomaterials*, **28(11)**: 4051 (2007).

models were developed to describe the swelling kinetics of the hydrogels, and correlation between the degradation and swelling of the hydrogels and the release of BSA from the hydrogels was discussed. Hennink and coworkers used methacrylate-derivatized dextran, hydroxyethyl methacrylate-derivatized dextran, and 2-hydroxyethyl methacrylate-oligolactate-derivatized dextran hydrogels to release IgG at an almost constant rate for 5 to 35 days depending the crosslink density of the hydrogels [29].

Although drug delivery at a constant rate is optimal for many therapeutic agents, a pulsatile drug release is needed for hormone therapy, and for better drug tolerance for patients with chronic diseases. This type of release has been achieved with stimuli-responsive hydrogels. The on-and-off release of drugs is observed in response to temperature change from thermoresponsive, PNIPAAm hydrogels (Figure 6.3) [30, 31]. Pulsed hormone delivery for over 7 days has been achieved by using a poly(NIPAAm-co-methacrylic acid) hydrogel membrane in a prototype device [32]. Similarly, pH-regulated drug release has been obtained by using pH-responsive poly(methacrylic acid-g-ethylene glycol) hydrogels and benzaldehyde acetal crosslinked polyacrylamide [33-34]. As described previously, glucose-induced insulin release has been observed in PNIPAAm hydrogels derivatized with a phenylboronic acid moiety and chitosan/β-glycerol phosphate disodium sol-gels immobilized with glucose oxidase enzyme [23, 24].

In Tissue Engineering

In recent years, a variety of hydrogels have shown potential as scaffolds to engineer cells, tissue, and organs [3–6, 18, 35]. Cartilage repair has been attempted by growing chondrocytes using hydrogels based on polyethylene

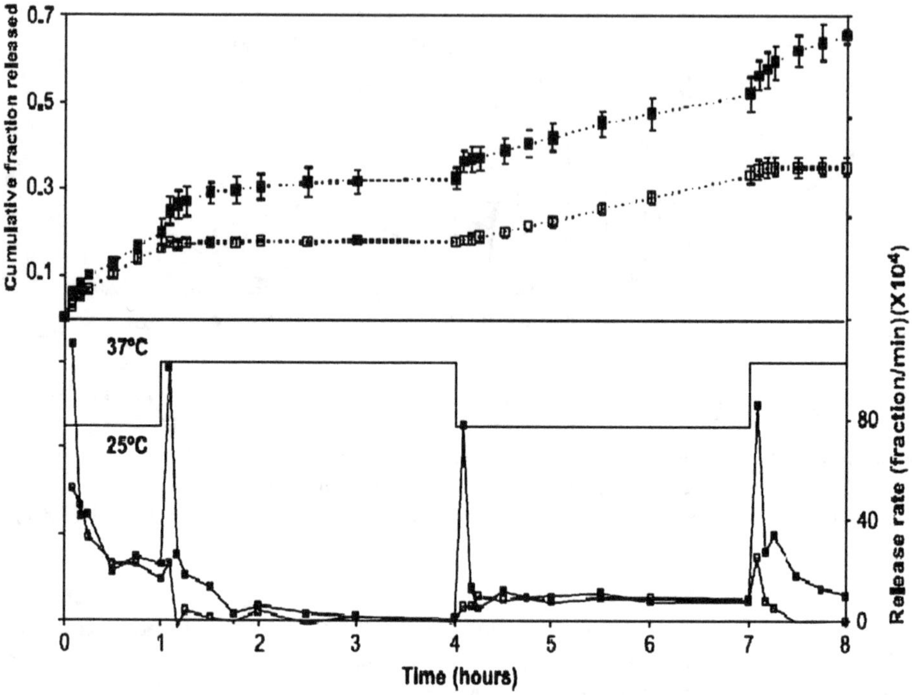

Figure 6.3. Temperature-triggered on-off release of hydrophilic diltiazem HCl (■) and hydrophobic diltiazem base (□) from poly(N-isopropylacrylamide) hydrogels (6mm diameter and 1.5mm height) in PBS (pH 7.4). The hydrogels are transferred between identical baths maintained at either 25 °C or 37 °C. The hydrogels release drugs significantly at 25 °C, but slightly at 37 °C, as shown in the cumulative fraction release curves (top panel) and the corresponding release rate curves (bottom panel). A pulse of diltiazem HCl release is observed when the temperature either increases or decreases stepwise between 25 and 37 °C, due to mechanical squeezing of the hydrogel at 37 °C and high swelling of the hydrogel at 25 °C. Adapted from *J Control Rel*, **98(1)**: 97 (2004).

glycol, chitosan, alginate, collagen, agarose, and hyaluronic acid [36]. When the drug delivery system of degradable oligo[poly(ethylene glycol) fumarate] was built into hydrogel scaffold to deliver insulinlike growth factor-1 and transforming growth factor-β1 for the repair of cartilage defects in rabbits, the cartilage thickness, surface regularity, and chondrocyte clustering were improved compared to untreated defects [37, 38]. Chitosan, PEG, hyaluronic acid, poly(HEMA), and alginate hydrogels have been used for growing osteoblasts responsible for bone regeneration [39–41]. *In vivo* experiments demonstrated promotion of bone formation from subcutaneously implanted degradable PEG hydrogels containing osteoinductive growth factors in rats [42]. Efforts have been made to minimize invasiveness by developing injectable hydrogels for tissue engineering. Anseth et al. demonstrated the potential use of in-situ-forming degradable hydrogels based on PEG and PVA in cartilage tissue engineering by encapsulating and growing chondrocyte in the hydrogels [36]. Hong et al. developed an injectable chitosan hydrogel scaffold for chondrocyte encapsulation [43].

The repair of defects or injury in spinal cords, nerves, and central nervous systems is an important necessity for which hydrogels have been explored as clinically feasible biomaterials. Poly(2-hydroxyethyl methacrylate-co-methyl methacrylate) [44–46], hyaluronan [47], and chitosan-based [48] hydrogels have been used for repairing spinal cord injuries in rats [45]. Thermoresponsive poly-(D-lysine) modified macroporous chitosan/glycerophosphate hydrogels have been used for neurite outgrowth [49]. Degradable PEG hydrogels have been used as potential synthetic cell carriers for neural transplantation [50].

The potential of hydrogels in angiogenesis [47] and regeneration of lung tissue [51], liver [52], and muscle [53–56] has been explored. Hepatic tissue growth is needed to repair liver failure [52]. Underhill et al. [52] successfully encapsulated hepatic cells in PEG hydrogels and maintained cell proliferation and functions. Trudel and Massia used dextran and hyaluronan-based hydrogels for studying the survival, migration, and proliferation of vascular smooth muscle cells *in vitro* [53]. The hydrogels showed good cytocompatibility after 2 days of indirect exposure to the vascular smooth muscle cells. Pullulan-based hydrogels, which can be crosslinked with sodium trimetaphosphate in aqueous solution, also showed good potential for encapsulation and proliferation of smooth muscle cells [55].

Hydrogels have also been used in designing artificial organs such as the pancreas and liver [57–58]. For example, beta-islet cells, which secrete insulin in the pancreas, were encapsulated in poly(organophosphazene) and polyvinyl alcohol [58–60] hydrogels. Beta-islets were viable for one month post implantation in rats, and the implanted PVA/islets lowered and maintained blood glucose levels compared to the blank [60].

In Medical Devices

Hydrogels have improved performance of many medical devices. For example, a detachable silicone balloon is used clinically as an occlusive device for the treatment of congenital diaphragmatic hernias. Dislodgement of the balloon is often a problem. Injectable hydrogel based on poly(ethylene glycol) prevented the balloon from dislodgement by providing adhesion to biologic tissue in lambs [61]. In order to retain the biological activity of enzymes in biosensors, enzymes are immobilized in hydrogels. For example, a glucose oxidase enzyme that converts glucose into gluconic acid and hydrogen peroxide (H_2O_2) can be immobilized in or conjugated with hydrogels films and then used in glucose monitoring devices. Similarly, horseradish peroxidase (HRP) can be immobilized in hydrogel for sensing H_2O_2. Sun et al. entrapped HRP in poly(*N*-isopropylacyamide-co-3-methacryloxy-propyltrimethoxysilane) films for designing an H_2O_2 biosensor [62]. Bioresponsive hydrogels, which are made by conjugating small biomolecules (ligands) can selectively bind protein receptors and antibodies, and can be used for biosensing applications [63]. Recently, Charles et al. [64] used galactoside-based polyacrylate hydrogel for the immobilization of cholera toxin antibodies for the development of a sandwich immumoassay.

Molecular imprinting, where hydrogels are formed in the presence of a print molecule and the molecule is removed after the hydrogel formation, is another interesting area of hydrogel application. When the hydrogel is placed in a solution containing the print molecule, the hydrogel can recognize the molecule and swell or shrink.

Many hydrogels, such as silicone polymer, polyether, polyacrylic derivatives, polyalcohol, poly(vinyl pyrrolidone), and poly(acrylamide), are transparent and possess good mechanical strength. These properties have been exploited for fabricating intraocular lenses [65, 66] and artificial corneas [67–69]. Hydrogels can also be used in coating for drainage tubes and stents [70–72], burn dressing [73–77], and controlling flow in microfluidic devices and microelectromechanical systems (MEMS) [78, 79].

Summary

Hydrogels have promising applications in biomedical areas. Both natural as well as synthetic polymers can be used for specific applications. Natural polymers are biocompatible and usually have high molecular weights, which are suitable for some applications. However, natural polymer-based hydrogels are mechanically weak, so that uncontrolled degradation occurs. On the other hand, synthetic polymers with precise molecular weight can be synthesized to form hydrogels in which crosslink density, hydrophobic and hydrophilic balance, and degradation kinetics can easily be controlled. However, synthetic hydrogels may be potentially toxic, which needs to be minimized by purification, the use of less toxic reagents, in situ gelation, UV photocrosslinking strategies, and bioinspired gelation based on endogenous proteins and enzymes.

NANOGELS

Polymeric nanogels, the submicron-sized gel particles, have attracted great attention during recent years because of their potential applications in biomedical applications [80–87]. Nanogels possess all the properties of macroscopic gels. In fact, nanogels are three-dimensionally crosslinked polymers, which can swell and shrink in solvents, and can become solid spheres when the polymer volume fraction is unity [85]. Both natural and synthetic polymers have been used for the preparation of nanogels. Nanogels can be synthesized by using complexation, emulsion polymerization, pulse radiolysis, and self-assembly techniques [84, 86, 88–93]. Self-assembly of polymer chains gives rise to the physically crosslinked nanogels [88, 94]. Both intra- and interpolymer chain associations can occur due to H-bonding or hydrophobic interactions, which lead to the formation of physical nanogels. Two oppositely charged polymers can also form a nanosized complex. Chemically crosslinked nanogels can be obtained using a crosslinker during emulsion polymerization. Surfactants, such as sodium dodecyl sulfate (SDS) are commonly used to form oil in water microemulsion for the emulsion polymerization of nanogels [95]. Under certain conditions, nanogels can also be synthesized without surfactants [90, 97, 98]. In surfactant-free nanogel synthesis, the monomer or polymer itself can stabilize emulsion. The surfactant-free nanogels possess advantages over the traditional methods because they do not require removal of the potential toxic surfactants. In addition, reverse emulsion polymerization (water in oil) can also be used to synthesize crosslinked nanogels [98]. It is also worthwhile to mention that degradable nanogels can be prepared using polymers with degradable linkages, or degradable crosslinkers [99–102].

As responsive polymers play important roles in drug delivery and tissue engineering, responsive nanogels have drawn increasing attention during past decade [103–106]. Thermo-responsive nanogels have been synthesized by aqueous redox polymerization of acrylic acid (AAc) and NIPAAm or vinylimidazole (VI) and NIPAAm in the presence of a crosslinker, *N,N*-methylenebisacrylamide, and surfactant SDS [104–107]. Nanogels of the size 125–600 nm can be obtained by varying the acrylic acid or vinylimidazole contents. Nanogels that are pH sensitive can also be obtained by the crosslinking of functionalized polymers, such as poly(ethylene oxide) and polyethylenimine [86], or photoinitiated free-radical precipitation polymerization of methacrylic acid and poly(ethylene glycol) monomethyl ether monomethacrylate [108].

Nanogels with layered structures (core-shell) can also be synthesized by using hydrophobic and hydrophilic monomers or polymers as cores or shells. Synthesis techniques, such as UV irradiation [109], sonication, and emulsion polymerization [110, 111] are used to prepare core-shell nanogels. A number of core-shell nanogels with tunable size have been prepared using polystyrene shells and crosslinked poly(methacrylamide) cores by sonication [112], polystyrene-b-poly(2-cinnamoylethyl methcacrylate) by radiation [109], poly(acrylonitrile-co-NIPAAm) by emulsion polymerization [110, 111]. Core-shell nanogels that are pH- and temperature-responsive can also be prepared by the use of a suitable monomer—for example, 2-(dimethylamino) ethyl methacrylate, PNIPAAM/poly(2-vinylpyridine) macromonomers and PNIPAAM/poly(4-vinylpyridine) [113–115].

As mentioned above, various types of nanogels with tunable sizes and stimuli-responsive properties can be designed and synthesized for biomedical applications. However, their physicochemical characterization is critical for any applications. Atomic force microscopy (AFM), dynamic laser light scattering (DLS), transmission electron microscopy (TEM), or cryo-TEM, are commonly used to characterize nanogels. TEM is a powerful visualization tool that gives two-dimensional images of nanogels. AFM provides three-dimensional images of nanogels in a quasi-native state. DLS provides the hydrodynamic size of nanogels in solution. Temperature, concentration, and solvent dependence of the size of nanogels can be studied with DLS. In addition, the molecular masses of nanogels can also be obtained by using dynamic and static light scattering (SLS) simultaneously.

In Drug Delivery

Nanogels, owing to their small size, can interact with and be internalized by the cells. Therapeutic drugs, peptides and proteins, and nucleic acids can be loaded into nanogels during or after the synthesis process. The main goal of nanogel-based drug delivery systems is to cross cellular or blood-tissue barriers, such as cell membrane, intestinal, blood-brain, or blood-retina barriers. For precise control of drug release kinetics and targeting, stimuli-responsive polymers can be used to synthesize nanogels. The size and drug release from these nanogels can be controlled by changing stimuli. For example, a pH-responsive nanogel bearing cationic groups can have small size and hold a drug at a physiological pH. Once inside the cell where the pH is ~6, the nanogel can swell and release a therapeutic agent. Furthermore, the pH-dependent degradation of nanogels can also be controlled. For the loading of drugs at ambient temperature and a sustained release at physiological temperature, thermoresponsive nanogels can be designed by incorporating thermoresponsive groups in the nanogels [89, 114, 116–119].

Various *in vitro* studies have demonstrated the feasibility of nanogels as drug delivery vehicles. For example, Bharali et al. demonstrated the release of fluorescein isothiocyanate (FITC)-dextran (19.3 kDa) from poly(vinyl pyrrolidone) nanogels for up to 30 days [96]. Sahoo et al. synthesized both pH- and thermoresponsive hydrogel nanogels and observed the release of FITC-dextran (19.3 kDa) for up to 80 h [119]. Huang et al. made a network of poly(*N*-isopropyl acrylamide-co-allylamine) and poly(*N*-isopropyl acrylamide-co-acrylic acid) nanogels and studied the temperature-dependent release of dextran [120]. These nanogels were able to release both small and large dextran molecules (MW 3000–500,000 Da). The poly[oligo(ethylene oxide) monomethyl ether methacrylate] (POEOMA)-based nanogel encapsulated doxorubicin, and degradation of the nanogel in the presence of glutathione tripeptide, led to the release of doxorubicin [113]. Growth of HeLa cancer cells was observed due to the release of doxorubicin from the POEOMA-nanogels. Other therapeutics agents, such as gene and antisense can be encapsulated and released from nanogels [121]. For example, nanogels made of crosslinked poly(ethylene oxide) and polyethyleneimine demonstrated enhanced permeability of oligonucleotide across monolayers of bovine brain microvessel endothelial cells, an *in vitro* model of the blood-brain barrier. Thus, cationic nanogels have the potential for vector-mediated oligonucleotide delivery to the brain. Similarly, Lee et al. [122] demonstrated the release of small interfering RNA (siRNA) from crosslinked hyaluronic acid nanogels and an uptake of the nanogels by cultured HCT-116 cells (Figure 6.4).

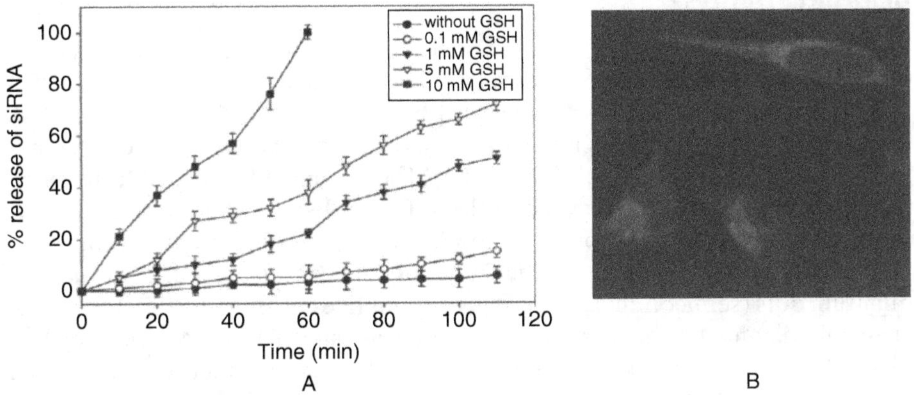

Figure 6.4. (A) Release of interfering RNA (siRNA) from hyaluronic acid (HA) nanogels in the presence of glutathione (GHS). (B) Uptake of rhodamine-labeled HA nanogels by HCT-116 cells after 2 h incubation. Adapted from *J Control Rel*, **119(2)**: 245 (2007).

Besides nanogels of one polymeric component, core-shell nanogels made of poly(acrylonitrile-co-NIPAAm) have also been used as drug delivery vehicles [107], especially for encapsulating hydrophobic drugs inside the core. Release of propranodol from the poly(acrylonitrile-co-NIPAAm) nanogels was demonstrated to last for ~2 h. The release of propranodol lasted for ~6 h with slightly hydrophilic poly(acrylonitrile-co-NIPAAm) nanogels. Although many nanogel systems show remarkable encapsulation and release capabilities, their real potential as drug delivery vehicles can be tested only in *in vivo* conditions. A recent *in vivo* study showed that nanogels made of crosslinked poly(ethylene glycol) and polyethylenimine, and complexed with a oligonucleotide, can deliver oligonucleotide to the brain [123].

In Tissue Engineering

As natural tissue is nanostructured, there has been considerable interest in using nanomaterials as implants for tissue regeneration, such as in bone, cartilage, vascular, bladder, and neuronal regeneration. Because of their high surface area and roughness, nanomaterials provide unique nanoscale surface properties compared to normal surface. It has been pointed out that cell proteins can interact with nanomaterials and lead to improved cell functions [124]. In earlier studies, hydroxyapatite nanomaterial, poly(lactide-co-glycolide) nanoparticles, and a nanocomposite of hydrodroxyapatite and poly(lactide) were used for growth of osteoblasts, chondrocytes, endothelial, and smooth muscle cells. Nowadays it is predicted that polymeric nanogels can also be assembled into three-dimensional scaffolds that can mimic the natural tissue environment suitable for cell regeneration applications.

In Medical Devices

Besides drug delivery and tissue engineering, nanogels can also be used for other biomedical applications, such as biological sensors, biological separations, and single enzyme encapsulation [83, 125]. For example, Nomura et al. studied the role of nanogels made of self-assembly of cholesterol-bearing pullulan in assisting protein refolding [126]. These nanogels were able to catch and release proteins in *in vitro* conditions. The molecular chaperonelike activity of the nanogels have potential applications in refolding chromatography and batchwise renaturation [126]. The nanogels can also be used to deliver quantum dots (semiconductor nanocrystals) for live cell imaging [127]. Sparsely crosslinked nanogels, such as polyacrylamide crosslinked with *N,N′*-methylene bisacrylamide, can be used as fluid for high-throughput microchannel DNA sequencing [128]. An improvement of about 23% in DNA sequencing read lengths was obtained when polyacrylamide nanogel was used compared to linear polyacrylamide [128].

Summary

Nanogels so far have demonstrated intriguing potential for biomedical and pharmaceutical applications. Similar to the hydrogels, physicochemical properties of nanogels including degradation can be tuned for specific needs. Drug loading, stability, and toxicity are some of the critical parameters for designing nanogels. Nanogels can easily be functionalized and conjugated with ligands for targeted drug delivery. Although nanogels have achieved successes in biomedical areas, long-term stability and storage, and size uniformity are some of challenges associated with nanogels.

DENDRIMERS

Dendritic polymers are synthetic, repeatedly grafted species. Based on their symmetry and polydispersity, they can roughly be categorized into low molar mass species including dendrimers and dendrons, and high molar mass species including dendronized, hyperbranched, and brush polymers. The dendrimer name comes from the Greek "δενδρον"/dendron, meaning "tree." Synonymous terms are arborols and cascade molecules. Dendrimers and dendrons are repeatedly grafted, monodisperse, usually highly symmetric, and nanosized molecules. A dendron usually contains a single chemically addressable group that is called focal point. Semiglobular or globular structures of dendrimers provide a smaller hydrodynamic volume and a higher density of functionalities on the surface compared to the corresponding linear polymers. Moreover, despite their convoluted architecture, dendrimers are precisely tunable in their molecular size, shape, numbers, and positions of functional groups unlike traditional linear or branched polymers.

Dendrimers are synthesized by a fully controlled step-by-step approach with well-defined molecular structures and narrow molar mass distribution. A dendrimer is composed of a core, an interior (or branches), and a periphery (or end groups). The generation of dendrimers can be usually defined by numbering focal points (cascade points) when moving from the core to the periphery. The more generations the dendrimers have, the higher the molar mass, branch number, and amount of peripheral end groups they possess.

Since they were first reported in the late 1970s [129] and early 1980s [130, 131], dendrimers have been intensively exploited especially for biomedical applications due to their unique branched structures, precise tuning of composition, functionality, and biocompatibility [132]. The core and shell structure of dendrimers has been utilized for the encapsulation of guest molecules such as catalysts, hydrophobic drugs, or chromophores [133–136]. Moreover, spurred by the advances in dendrimer designs and understandings of the relations between their structures and biological behaviors, dendrimers have been actively investigated in the fields of targeted drug carriers, antiviral drugs, tissue engineering scaffolds, and biomedical sensors [137–140].

In Drug Delivery

Monodisperse and highly branched dendrimers have been suggested as an efficient drug carrier by either encapsulation or post-formed surface modification with a variety of therapeutics such as small molecule drugs, DNA, and oligonucleotides [141–143]. Dendrimers have been functionalized with specific tissue targeting molecules such as folic acid to localize drugs to desired tissue sites [144] or hydrophilic molecules such as PEO to increase their blood circulation times [145]. The dendritic structure has been exploited to give more functionality to linear or grafted structure. In a recent report, this combined dendritic structure was applied to have biodegradability and stimuli-responsive supramolecular property, where hydrophilic poly(L-lysine) (PLL) dendrons were conjugated with the both ends of poly(*N*-isopropylacrylamide) (PNIPAAM) grafted with biodegradable poly(L-lactic acid) (PLLA) (Figure 6.5) [146]. When dendrimers are conjugated on linear polymers and become linear-dendritic copolymers, they can provide great potency in increasing drug payload due to their supramolecular micelle structures [147].

Dendrimers have been actively utilized for passive or active target delivery of antitumor drugs. Dendrimers has been studied as antitumor passive targeted carriers because globular structure has been found to have longer retention times in blood circulation than linear structure [137]. As one example of passive targeting drug delivery carrier, the surface groups of carboxylate-terminated polyamidoamine (PAMAM) dendrimer were conjugated with cisplatin [148]. In a mouse model study, the drug-dendrimer complexes accumulated 50 times more in subcutaneous tumors than the free drug at equivalent doses via a passive targeting mechanism (e.g., the enhanced permeation-and-retention effect). In another report, PEGylated and biodegradable dendrimers were used as antitumor passive targeted carriers of doxorubicin [141]. In this study, biodegradable and half-side PEGylated bow-tie dendrimers were designed to be functionalized with doxorubicin (Figure 6.6).

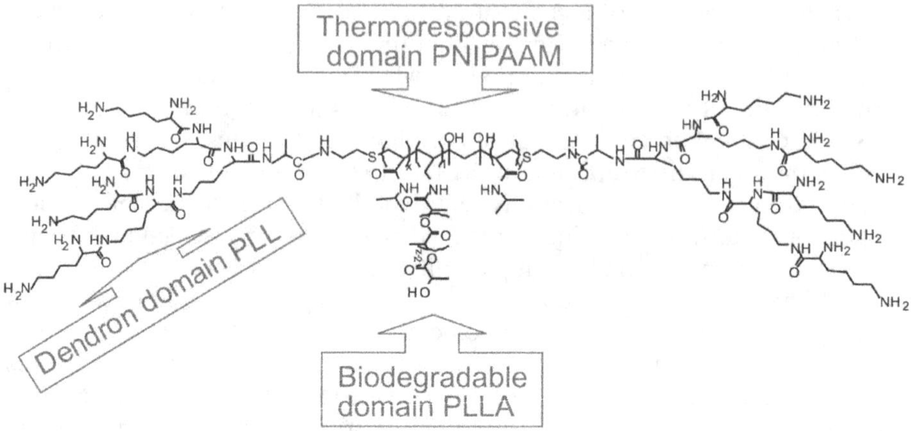

Figure 6.5. Thermoresponsive and biodegradable linear-dendritic nanoparticles composed of PNIPAAM, PLLA, and PLL. Two PLL dendrons impart the dendritic branches and are linked together via PNIPAAM. PLLA is conjugated to the PNIPAAM linker providing the hydrophobic binding pocket. Adapted from *Macromolecules*, **39(23)**: 7805 (2006).

In addition, dendrimers have also been exploited as antitumor targeted carriers via active targeting mechanism. For example, partially acetylated PAMAM dendrimer was conjugated with methotrexate and tumor site targeting ligand folate [149], and the obtained methotrexate-folate-dendrimer could improve the cytotoxic response of the human KB tumor cells 100-fold over the free drug. Gene delivery is one of other potential drug delivery applications of dendrimers. The reason is that the cationic periphery functional groups of dendrimers can form electrostatic complexations with polyanionic DNA, which lead to highly efficient transfection of a broad range of cells and minimal cytotoxicity [150]. However, the charge density and generation of dendrimer need to be optimized for efficient delivery and prevention of enzymatic degradation of DNA, as well as the minimization of the toxicity associated with the positive charge of cationic dendrimers.

In Tissue Engineering

In the past few years, researchers have begun exploring dendrimers as three-dimensional crosslinked scaffolds for tissue regeneration. For example, dendrimers have been studied for the formation of highly crosslinked collagen to support human corneal epithelial cell growth [151, 152], and modified with immunomodulatory and antiangiogenic agents to prevent scar tissue formation [139]. In situ photocrosslinking method is used for fabricating injectable dendrimer scaffolds made of a dendritic linear triblock copolymer containing a poly(ethylene glycol) core and methacrylated poly(glycerol succinic acid)

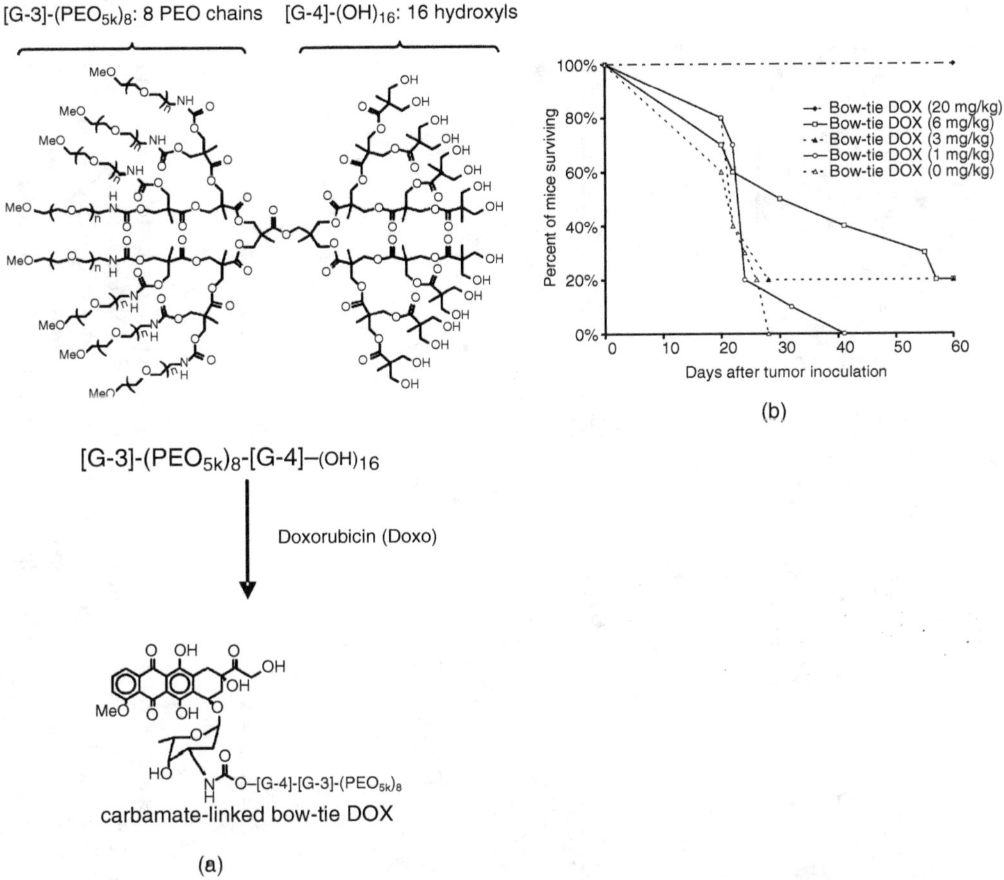

Figure 6.6. Doxorubicin-functionalized, biodegradable, and PEGylated bow-tie dendrimers. (a) Doxorubicin (Doxo) is linked on the surface of the bow-tie dendrimer. (b) Survival versus time for mice bearing C-26 colon carcinomas. The bow-tie dendrimer–Doxo conjugates were effective in a DOX-insensitive solid tumor. Adapted from *P Natl Acad Sci USA*, **103(45)**: 16649 (2006).

dendrimer terminal blocks [153]. The coupling reaction between aldehyde of poly(ethylene glycol) dendrons and cysteine 1,2 aminothiol group of poly(lysine) dendrons was used to form dendrimer hydrogel scaffolds with thiazolidine linkage [154]. By using dendrimer hydrogel scaffolds for cartilage tissue formation, Grinstaff and colleagues found that a compromise between targeted mechanical, diffusion, and biochemical properties is necessary to afford an optimal tissue engineering scaffold for cartilage matrix production *in vitro* or *in vivo* [153]. As illustrated in Figure 6.7, simply increasing the concentration of the macromer of dendrimer hydrogels to improve

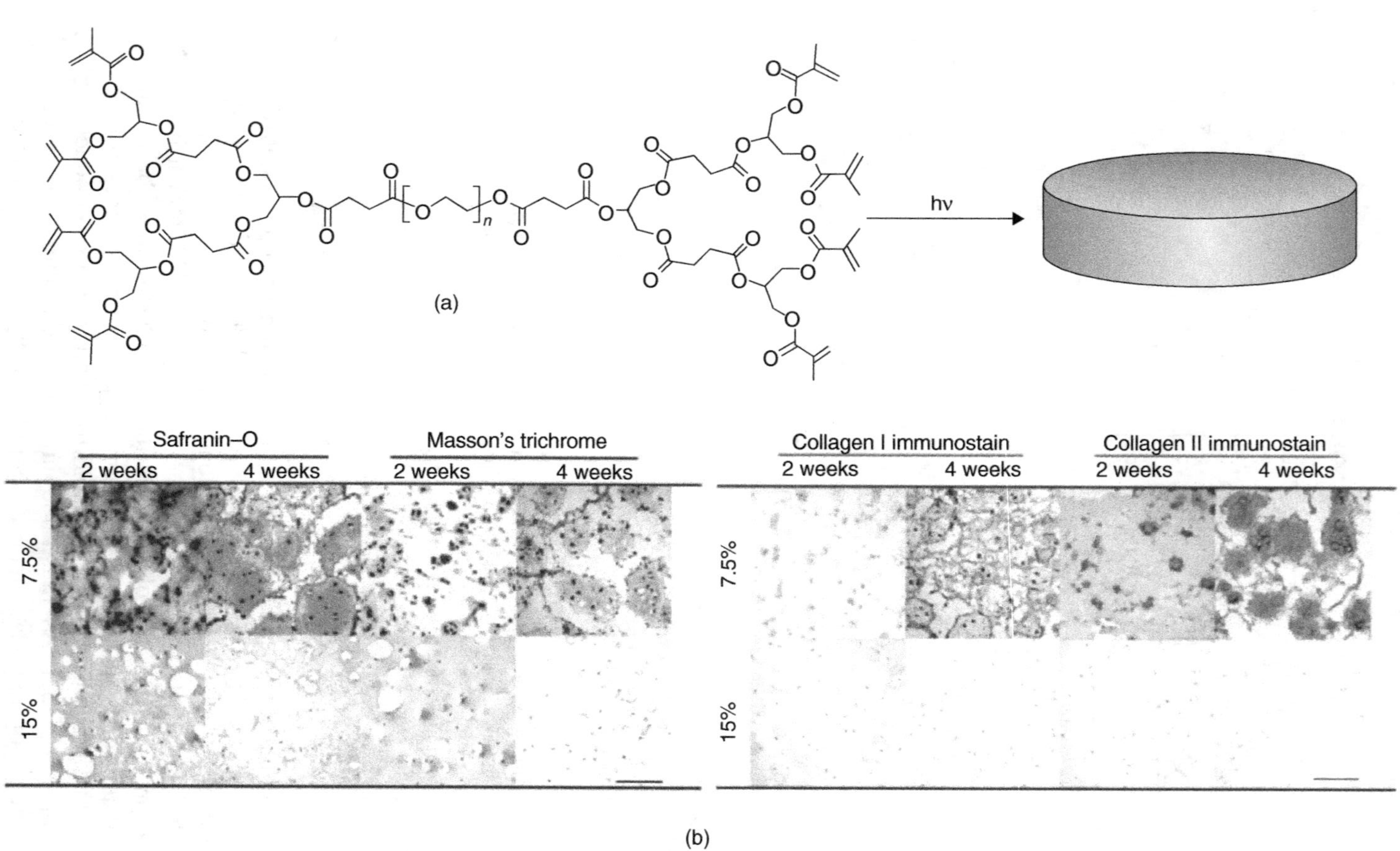

Figure 6.7. (a) Photocrosslinkable dendrimer-based hydrogel scaffolds for cartilage tissue repair. The hydrogel scaffolds were made of a dendritic linear tri-block copolymer containing a poly(ethylene glycol) core and methacrylated poly(glycerol succinic acid) dendrimer terminal blocks. (b) Histological sections of 7.5% and 15% dendrimer concentration hydrogels after 2 and 4 weeks incubation. Left: Red indicates proteoglycans in the Safranin-O stained sections; green indicates collagen in the Masson's Trichrome stained sections. Right: Red indicates type I or II collagen in the immunostained sections; no significant type I collagen was detected at either concentration. The length of the inserted bar is 100 μm. Adapted from *Biomacromolecules*, **7**(**1**): 310 (2006).

mechanical properties was not the best route to a better extracellular matrix deposition for neocartilaginous tissue formation. The reason is that the degradation of the scaffolds is also important for the diffusion of nutrition and metabolic substances in and out of the scaffolds, respectively, to promote healthy cell growth [153].

The research for tissue regeneration with dendrimers is still in preliminary stages. Further research is expected to focus on tailoring the physical, (bio)chemical, and mechanical properties of the dendrimers scaffold through controlling dendrimer generation, degree of branching, crosslinking density, hydrolysable linkage, and end group functionality for repair of tissue defects based on specific performance requirements.

In Medical Devices

Magnetic resonance imaging (MRI) provides a powerful and noninvasive tool for *in vivo* imaging to diagnose many diseases in a variety of tissue types. However, MRI often requires the use of a contrast agent to obtain better images. Gadolinium (Gd), which has been commercially utilized as an MRI contrast agent, has a disadvantage in fast plasma clearance. High molar mass carriers such as dendrimers can extend the lifetimes of small molecular weight Gd in blood vessels. Dendrimers have many functional groups on their periphery that are capable of anchoring many Gd ions. The compounds of dendrimers and contrast agents have been reported to lower the required doses of Gd by efficiently accelerating proton relaxation and to extend the lifetime of Gd in the body [155].

Dendrimers containing metalloporphyrins have been developed for photonic oxygen sensing that provide another noninvasive diagnosing image [140]. Water-soluble dendrimers can encapsulate hydrophobic metalloporphyrins in their cores. The dendrimer sensors bearing metalloporphyrins give an accurate determination of tumor treatment status by sensing the concentration of oxygen in tumors, because the oxygen concentration is inversely related to the phosphorescence lifetime. The noninvasive optical imaging using dendrimer sensors can be obtained by phosphoresce irradiation with visible light or multiple photons of near-infrared light.

Dendrimers have been reported to show intrinsic drug properties themselves in addition to their benefits in encapsulating and carrying therapeutics. The branched architecture of dendrimers can stimulate the removal of prion proteins in infected cells [156]. This result was not observed in linear cationic polymers, indicating that the highly branched multivalent structure of dendrimers should play an important role in this case. In other reports, dendrimers, whose multivalent surfaces were modified with specific ligands, were found to exhibit significant inhibition of infection of host cells by influenza viruses. For example, dendrimers modified with napthyl residues at their periphery inhibited the binding of herpes simplex viruses to cell surfaces [157].

Summary

New designs and applications of dendrimers have been actively introduced in a variety of biomedical fields. The interest in dendrimer-based biomedical application has exponentially increased because the incentive of highly grafted structures has demonstrated their versatility and utility. Dendrimers have been actively explored in biomedical fields including drug delivery, tissue engineering, and biosensors. Despite the potency of dendrimers, their high cost demands that their clinical efficiency should be at least superior to that of other relatively low-cost biomaterials for clinical success. Overall, highly grafted dendrimers give a good example of how grafted polymers can be utilized in biomedical applications.

GRAFTED CYCLODEXTRINS

Cyclodextrins (CDs) are cyclic oligosaccharides composed of D-(+) glucopyranose units attached by α-(1, 4) glucosidic bonds. Generally, three types of natural CDs exist: α-, β-, and γ-CDs with six, seven, or eight glucose units, respectively. CDs with fewer than six glucose units cannot form cyclic structures due to their steric hindrances, and CDs bearing nine or more glucose units are also difficult to obtain because of their difficult isolation and purification processes. CDs have outer hydrophilic surfaces and internal hydrophobic cavities that are capable of hosting guest molecules [158]. CDs endow hydrophobic quest molecules such as small hydrophobic drugs with improved solubility and chemical stability. In addition, CDs form necklacelike supramolecular structures with linear polymers to induce polyroxanes, where cyclic molecules are threaded onto the polymer chain via polymer inclusion complexes (PICs).

The grafted structures of CDs have been intensively studied not only to overcome the drawback of CDs, such as poor water solubility, but also to add functionality. Novel designs of grafted CDs have been introduced and tested in many biomedical applications. The inherent inclusion complexation property and supramolecular formation of parent CDs combined with modified functionality can provide versatile biomedical applications including drug delivery, tissue regeneration, and biosensors, which will be described below.

In Drug Delivery

In the past several decades, CDs and grafted CDs have been utilized successfully as drug delivery systems to improve drug solubility, chemical stability, and bioavailability [158]. The nature of CDs (e.g., inclusion complexation with guest molecules) can be used for efficient drug delivery applications. The inclusion behavior of CDs with drugs shows an equilibrium defined by an equilibrium constant $K_c = K_{1:1} = [\text{Drug-CD}]/[\text{Drug}][\text{CD}]$ [158]. The dissociation of the included drugs from CDs might be expected upon the dilution effect. Also, the included drug in CDs might be displaced by a more affordable quest molecule or might be transferred to a matrix with a higher drug affinity, such as an intestinal cell membrane.

CDs and modified CDs have been exploited for drug delivery vehicles via several administration routes such as oral, parenteral, nasal, ocular, rectal, and dermal [159–165]. Oral drug delivery with CDs and modified CDs has been extensively studied because they can not only improve drug bioavailability with increased drug solubility, drug dissolution rate and extent, and drug permeability across gastrointestinal barriers, but they also reduce drug irritation and bad drug taste [159, 160, 166]. The studies for ocular drug delivery using CDs and their derivatives have shown that CDs and their derivatives such as hydroxypropyl-β-CDs (HP-β-CDs) and sulfobutyl ether-β-CDs (SBE-β-CDs) can enhance the solubility, chemical stability, reduced irritation, and permeability of ocular drugs [167]. In the case of the parenteral drug delivery, high aqueous solubility and minimal cytotoxicity should be considered first. Due to the relatively low solubility and high level of haemolysis of β-CD, β-CD derivatives such as amorphous HP-β-CDs and SBE-β-CDs have been investigated to improve the drawback of CDs for parenteral uses [168]. In a recent report, cationic β-CD polymers containing quaternary ammonium groups were synthesized and investigated for parenteral drug delivery application [169]. The cationic β-CD polymer bearing quaternary ammonium groups not only dramatically increased naproxen solubility, but also showed better hemolytic activities than parent β-CD and neutral β-CD polymers.

Cyclodextrins (CDs) and their derivatives could be one candidate for delivery carriers of therapeutics across biological barriers such as the blood-brain barrier (BBB). To increase the permeability of drugs such as steroids and anticancer drugs across the BBB, the drugs can be covalently conjugated with liphophilc groups (e.g., 1-methyl-1,4-dihydronicotinic acid) through an enzymatically labile linkage as chemical delivery system [170]. Even though the chemical modification of the drug can enhance the permeability of the drug across the BBB, the drawback is that the modified drug has poor water solubility. CD derivatives such as HP-β-CD can increase the water solubility of the modified drug as well as enhance the chemical stability via the HP-β-CD/CDS inclusion complex, when applied for parenteral administration [171]. CD derivatives such as methylated CDs have been reported to enhance the BBB permeability of P-gp substrate type drugs such as doxorubicin by modulating the P-gp efflux function [172].

CD-containing cationic polymers have been developed for effective DNA delivery with low toxicity [158]. CDs can be grafted as pendent groups on the side chains of cationic polymers such as poly(ethylene imine), poly(allylamine) and PAMAM dendrimers [173–175]. These CD-containing cationic polymers have been reported to form polyplexes with anionic DNA by electrostatic interaction and inclusion complexation. These systems can give an efficient gene delivery due to its efficient transfection, low toxicity, and stability against

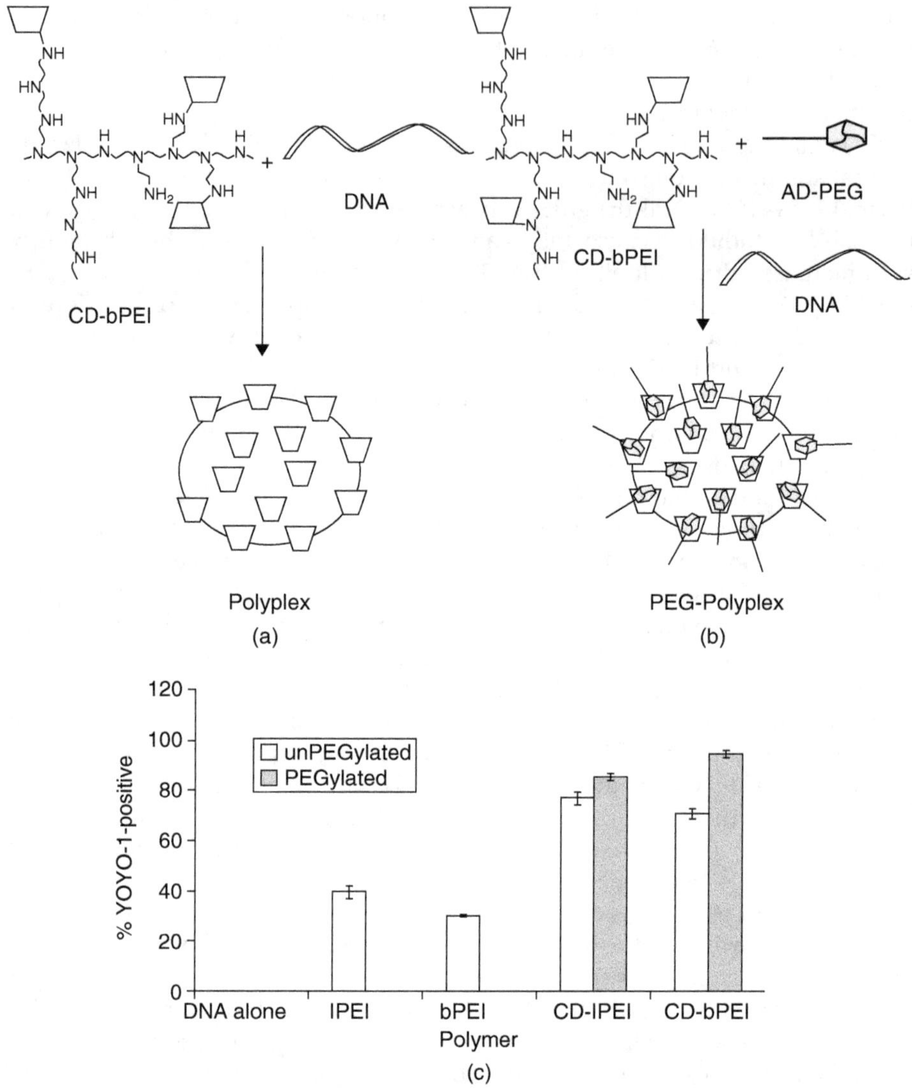

Figure 6.8. Schematics of (a) CD-bPEI (b) and PEGylated CD-bPEI polyplexes. The bar represents 200 nm. (c) Comparison of plasmid uptake efficiencies of polyethylenimine (PEI) and CD-PEI polymers. Polymers were complexed with YOYO-1-labeled plasmids at 10 N/P and exposed to PC3 cells for 15 min. Adapted from *Bioconjugate Chem*, **15**(**4**): 831 (2004).

enzymatic degradation. The CD-containing cationic polymers can be modified by compounds capable of forming inclusion complexes. The stability of CD-containing cationic polymers was enhanced by combining poly(ethylene glycol) conjugated with hydrophobic adamantine that forms the inclusion complex with CD cavities in polyplexes (Figure 6.8) [173, 176]. PEG-combined

CD-containing cationic polymers/DNA polyplexes can prevent the undesired interactions with nonself entities. Moreover, target gene drug delivery using this system was reported to complex transferrin-modified PEG-adamantine with the CD-containing polyplexes [177].

CDs modified with hydrophobic groups that are capable of inducing inclusion binding in CD cavities can form linear polymeric supramolecules. For example, mono[6-*O*-(4-formyl-phenyl) β-CDs was synthesized as solid linear polymeric supramolecules through the binding ability and self-assembly behavior of molecular interpenetration. The smart supramolecular system of modified CDs can provide stimuli-responsive inclusion/release behavior of guest molecules with CD rings in response to environmental stimuli. For example, in one study, the anionic phosphate group, a substrate of alkaline phosphatase, was introduced to β-CDs to release cationic aromatic quest molecules including cationic cancer drugs in response to enzymatic hydrolysis of the phosphatase group [178]. Polymer-conjugated CDs have been reported to show pH- and temperature-responsive gel formation via reversible inclusion complexation of quest molecules in the CD cavity [179]. The system with β-CD-conjugated poly(L-lysine) and 3-trimethylsilylpropionic acid (TPA) showed rapid intermolecular association and dramatic phase transition by pH as well as temperature, which might be caused by dual complexation phenomena, inclusion complexation, and ionic complexation.

In Tissue Engineering

CD-containing biodegradable hydrogels have been investigated for tissue engineering scaffolds in a form of polyrotaxane composed of α-CDs and PEG [142, 180–182]. In these reports, PEG and α-CDs were modified with crosslinkable ester linkages to form hydrophilic PEG hydrogel networks. The hydrolysis rate and erosion profile of the polyrotaxane hydrogels were tunable by modulating the PEG/α-CDs ratios and PEG molar masses [180]. The crosslinked polyrotaxane hydrogels, which show long-term stability and actually tunable hydrolysis, have been applied for polymeric scaffolding in tissue engineering such as cartilage regeneration. In the case of cartilage regeneration, the presence of cholesterol in the PEG- and cyclodextrin-based hydrogels improved the chondrocyte proliferation and glycosaminoglycan (GAG) production and controlled the degradation rate of the hydrogels [183]. PEG-based hydrogels with a hydrolyzable polyrotaxane, combined with hydroxyapatite particles for cell adhesion and survival, were reported as scaffolds for bone tissue engineering (Figure 6.9) [182].

The drug complexation nature of CDs can be applied to a membrane coating for tissue regeneration. For this application, polyvinylidene difluoride (PVDF) membranes was grafted with β-CDs and tested for guided tissue regeneration applicable in periodontology, resulting in the controlled release of antimicrobial agent during the tissue regeneration period [184]. In another report, Dacron (polyethyleneterephtalate, PET) vascular grafts were coated

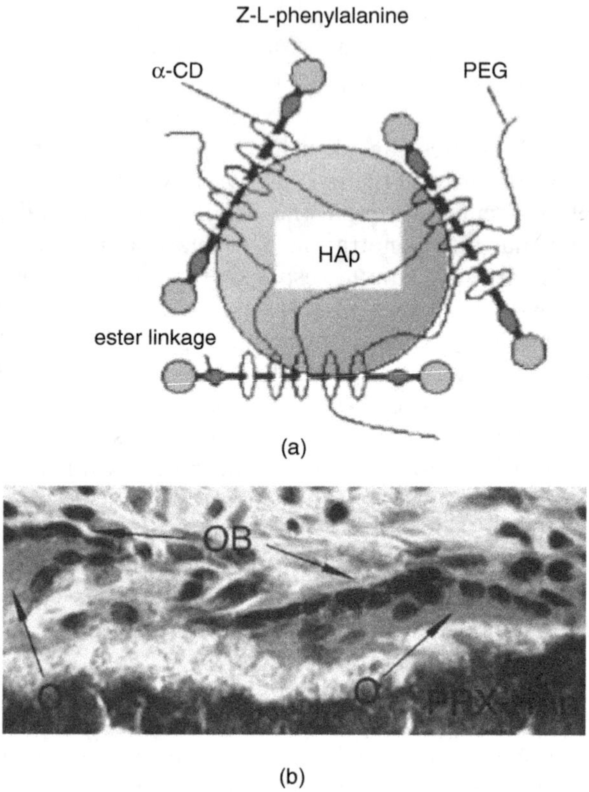

Figure 6.9. (a) Poly(ethylene glycol) hydrogels crosslinked by hydrolyzable polyrotaxane containing hydroxyapatite particles (PRX-HAp) as scaffolds for bone regeneration. (b) Histological image of ossification at 5 weeks after implanting PRX-HAp with rat osteoblasts. Osteoidlike tissues (O) were seen in the region between the PRX-HAp and a queue of osteoblastlike cells (OB). Adapted from *J Biomat Sci-Polym E*, **16(12)**: 1611 (2005).

with CD polymers to a obtain controlled release of Vancomycin from vascular grafts [185]. The CD-polymer-coated vascular graft showed nontoxicity and a linear release profile of Vancomycin over 50 days.

In Medical Devices

CDs and their derivatives have been extensively used for the separation of racemic compounds and drugs [186]. Chemically modified CDs are usually immobilized on the surface of silica in a chiral HPLC column to provide a chiral stationary phase (CSP), because CDs have a different affinity with enantiomeric molecules through inclusion complexation. Chiral HPLC containing CDs provides a useful methodology for separating optically pure isomers. The immobilization of CDs on silica beads have been achieved by several methodologies. The gel form of crosslinked CDs have been chemically immobilized onto silica [187]. Recently, other advanced synthetic techniques have been reported. In one of these reports, CDs containing polymers were prepared as either linear copolymer type or CD-grafted type, and then immobilized onto silica via condensation, forming urea linkage [188, 189]. In the other report, CDs were immobilized on silica beads via spacer arms [190]. The enantiodiscriminatory properties of these CSPs were known to be affected by chain lengths of spacer arms and the derivatizing groups of the CD [191, 192]. When the azido-perfunctionalized CD derivatives were conjugated onto silica beads with urea linkages, it was found that the CD-based CSPs bonded at the C6 position to silica exhibited slightly better chiral recognition ability than the CD-based CSPs bonded at the C2 position.

The supramolecular assembly property of CDs is a powerful tool that can be used to manufacture a high-quality biosensor detecting biological and pharmaceutical molecules. For example, a sensor to detect catecholamine neurotransmitters such as dopamine has been developed by synthesizing β-CD-modified electrodes with mediator and polyphenol oxidase [193]. In this system, β-CD polymers were immobilized on an electrode where tetramethylbenzidine and ferrocene were introduced as mediators to induce switchable inclusion complexes with the immobilized CDs depending on enzyme activity, resulting in voltammetric change. To fabricate enzyme-based biosensors, CDs and their derivatives have been widely utilized for the immobilization of enzymes on metal electrodes and nanoparticles through supramolecular interactions [194]. After thiolated CDs were covalently conjugated on electrode surfaces, biosensible enzymes, conjugated with hydrophobic guest molecules such as adamantan, were employed for supramolecular assembly with the CD cavities. Instead of the thiolated CDs, thiolated CD-containing polymers were used as three-dimensional matrices that are able to provide more sensitive enzyme-based electrochemical biosensors [195]. In this approach, the xanthine oxidase, conjugated with 1-adamantanyl residues, were immobilized on Au electrodes modified with CD-containing polymers via supramolecular assembly, constructing a biosensor that detected xanthine to diagnose several diseases including xanthinuria.

Summary

Grafted CDs can be used for a variety of biomedical applications. The challenge for the design of grafted CDs is to minimize the intrinsic problem of parent CDs (poor water solubility and haemolytic property), to maximize their intrinsic advantage (drug complexation and supramolecular property), and to add new functionality on the parent CDs.

REFERENCES

1. Junginger HE. *Pharm Ind*, **53(11)**: 1056 (1991).
2. Lee KY, Bouhadir KH, Mooney DJ. *Macromolecules*, **33(1)**: 97 (2000).
3. Lee KY, Mooney DJ. *Chem Rev*, **101(7)**: 1869 (2001).
4. Brandl F, Sommer F, Goepferich A. *Biomaterials*, **28(2)**: 134 (2007).
5. Hoffman AS. *Adv Drug Deliver Rev*, **54(1)**: 3 (2002).
6. Griffith LG. *Acta Mater*, **48(1)**: 263 (2000).
7. de Jong SJ, van EerDenbrugh B, van Nostrum CF, Kettenes-van De Bosch JJ, Hennink WE. *J Control Rel*, **71(3)**: 261 (2001).
8. Stenekes RJH, Talsma H, Hennink WE. *Biomaterials*, **22(13)**: 1891 (2001).
9. Tae G, Kornfield JA, Hubbell JA. *Biomaterials*, **26(25)**: 5259 (2005).
10. Jeong B, Kim SW, Bae YH. *Adv Drug Deliv Rev*, **54(1)**: 37 (2002).
11. Berger J, Reist M, Mayer JM, Felt O, Peppas NA, Gurny R. *Eur J Pharm Biopharm*, **57(1)**: 19 (2004).
12. Hennink WE, van Nostrum CF. *Adv Drug Deliver Rev*, **54(1)**: 13 (2002).
13. Bajpai AK, Saini R. *J Mater Sci-Mater*, **17(1)**: 49 (2006).
14. Nam K, Watanabe J, Ishihara K. *Polymer*, **46(13)**: 4704 (2005).
15. Wu XY, Huang SW, Zhang JT, Zhuo RX. *Macromol Biosci*, **4(2)**: 71 (2004).
16. Qu X, Wirsen A, Albertsson AC. *Polymer*, **41(12)**: 4589 (2000).
17. Murakami Y, Maeda M. *Biomacromolecules*, **6(6)**: 2927 (2005).
18. Serra L, Domenech J, Peppas NA. *Biomaterials*, **27(31)**: 5440 (2006).
19. Nakamura K, Murray RJ, Joseph JI, Peppas NA, Morishita M, Lowman AM. *J Control Rel*, **95(3)**: 589 (2004).
20. Kumar A, Lahiri SS, Singh H. *Int J Pharm*, **323(1–2)**: 117 (2006).
21. Atkins TW, Mccallion RL, Tighe BJ. *J Biomed Mater Res*, **29(3)**: 291 (1995).
22. Carenza M, Caliceti P, Veronese FM, Martellini F, Higa OZ, Yoshida M, Katakai R. *Rad Phys Chem*, **57(3–6)**: 471 (2000).
23. Kataoka K, Miyazaki H, Bunya M, Okano T, Sakurai Y. *J Am Chem Soc*, **120(48)**: 12694 (1998).
24. Kashyap N, Viswanad B, Sharma G, Bhardwaj V, Ramarao P, Kumar MNVR. *Biomaterials*, **28(11)**: 2051 (2007).
25. St'astny M, Plocova D, Etrych T, Ulbrich K, Rihova B. *Eur J Cancer*, **38(4)**: 602 (2002).
26. Cadee JA, de Groot CJ, Jiskoot W, den Otter W, Hennink WE. *J Control Rel*, **78(1–3)**: 1 (2002).
27. Mellott MB, Searcy K, Pishko MV. *Biomaterials*, **22(9)**: 929 (2001).

28. Huang X, Lowe TL. *Biomacromolecules*, **6**(**4**): 2131 (2005).
29. VanDijk-Wolthuis WNE, Hoogeboom JAM, vanSteenbergen MJ, Tsang SKY, Hennink WE. *Macromolecules*, **30**(**16**): 4639 (1997).
30. Coughlan DC, Quilty FP, Corrigan OI. *J Control Rel*, **98**(**1**): 97 (2004).
31. Kikuchi A, Okano T. *Adv Drug Deliver Rev*, **54**(**1**): 53 (2002).
32. Misra GP, Siegel RA. *J Control Rel*, **81**(**1–2**): 1 (2002).
33. He HY, Cao X, Lee LJ. *J Control Rel*, **95**(**3**): 391 (2004).
34. Murthy N, Thng YX, Schuck S, Xu MC, Frechet JMJ. *J Am Chem Soc*, **124**(**42**): 12398 (2002).
35. Vinatier C, Guicheux J, Daculsi G, Layrolle P, Weiss P. *Bio-Med Mater Eng*, **16**(**4**): S107 (2006).
36. Anseth KS, Metters AT, Bryant SJ, Martens PJ, Elisseeff JH, Bowman CN. *J Control Rel*, **78**(**1–3**): 199 (2002).
37. Holland TA, Bodde EWH, Cuijpers VMJI, Baggett LS, Tabata Y, Mikos AG, Jansen JA. *Osteoarthr Cartilage*, **15**(**2**): 187 (2007).
38. Holland TA, Tabata Y, Mikos AG. *J Control Rel*, **101**(**1–3**): 111 (2005).
39. Coleman RM, Case ND, Guldberg RE. *Biomaterials*, **28**(**12**): 2077 (2007).
40. Barralet JE, Wang L, Lawson M, Triffitt JT, Cooper PR, Shelton RM. *J Mater Sci-Mater*, **16**(**6**): 515 (2005).
41. Lawson MA, Barralet JE, Wang L, Shelton RM, Triffitt JT. *Tissue Eng*, **10**(**9–10**): 1480 (2004).
42. Burdick JA, Mason MN, Hinman AD, Thorne K, Anseth KS. *J Control Rel*, **83**(**1**): 53 (2002).
43. Hong Y, Song HQ, Gong YH, Mao ZW, Gao CY, Shen JC. *Acta Biomater*, **3**(**1**): 23 (2007).
44. Dalton PD, Flynn L, Shoichet MS. *Biomaterials*, **23**(**18**): 3843 (2002).
45. Nomura H, Katayama Y, Shoichet MS, Tator CH. *Neurosurgery*, **59**(**1**): 183 (2006).
46. Katayama Y, Montenegro R, Freier T, Midha R, Belkas JS, Shoichet MS. *Biomaterials*, **27**(**3**): 505 (2006).
47. Riley CM, Fuegy PW, Firpo MA, Shu XZ, Prestwich GD, Peattie RA. *Biomaterials*, **27**(**35**): 5935 (2006).
48. Freier T, Montenegro R, Koh HS, Shoichet MS. *Biomaterials*, **26**(**22**): 4624 (2005).
49. Crompton KE, Goud JD, Bellamkonda RV, Gengenbach TR, Finkelstein DI, Horne MK, Forsythe JS. *Biomaterials*, **28**(**3**): 441 (2007).
50. Mahoney MJ, Anseth KS. *Biomaterials*, **27**(**10**): 2265 (2006).
51. Cortiella J, Nichols JE, Kojima K, Bonassar LJ, Dargon P, Roy AK, Vacant MP, Niles JA, Vacanti CA. *Tissue Eng*, **12**(**5**): 1213 (2006).
52. Underhill GH, Chen AA, Albrecht DR, Bliatia SN. *Biomaterials*, **28**(**2**): 256 (2007).
53. Trudel J, Massia SP. *Biomaterials*, **23**(**16**): 3299 (2002).

54. Roman I, Vilalta M, Rodriguez J, Matthies AM, Srouji S, Livne E, Hubbell JA, Rubio N, Blanco J. *Biomaterials*, **28**(**17**): 2718 (2007).
55. Autissier A, Letourneur D, Le Visage C. *J Biomed Mater Res A*, **82**A(**2**): 336 (2007).
56. Mann BK, Gobin AS, Tsai AT, Schmedlen RH, West JL. *Biomaterials*, **22**(**22**): 3045 (2001).
57. Hou QP, Bae YH. *Adv Drug Deliver Rev*, **35**(**2–3**): 271 (1999).
58. Park KH, Song SC. *J Biosci Bioeng*, **101**(**3**): 238 (2006).
59. Park KH, Song SC. *J Biomat Sci-Polym E*, **16**(**11**): 1421 (2005).
60. Qi M, Gu Y, Sakata N, Kim D, Shirouzu Y, Yamamoto C, Hiura A, Sumi S, Inoue K. *Biomaterials*, **25**(**27**): 5885 (2004).
61. Chang R, Komura M, Andreoli S, Jennings R, Wilson J, Fauza D. *J Pediatr Surg*, **39**(**4**): 557 (2004).
62. Sun Y-X, Zhang J-T, Huang S-W, Wang S-F. *Sensors and Actuators B*, **124**: 494 (2007).
63. Ulijn RV, Bibi N, Jayawarna V, Thornton PD, Todd SJ, Mart RJ, Smith AM, Gough JE. *Mater Today*, **10**(**4**): 40 (2007).
64. Charles PT, Velez F, Soto CM, Goldman ER, Martin BD, Ray RI, Taitt CR. *Anal Chim Acta*, **578**(**1**): 2 (2006).
65. Yoo MK, Choi YJ, Lee JH, Wee WR, Cho CS. *J Drug Deliv Sci Tech*, **17**(**1**): 81 (2007).
66. Kwon JW, Han YK, Lee WJ, Cho CS, Paik SJ, Cho DI, Lee JH, Wee WR. *J Cataract Refr Surg*, **31**(**3**): 607 (2005).
67. Hicks C, Crawford G, Chirila T, Wiffen S, Vijayasekaran S, Lou X, Fitton J, Maley M, Clayton A, Dalton P, Platten S, Ziegelaar B, Hong Y, Russo A, Constable I. *Prog Retin Eye Res*, **19**(**2**): 149 (2000).
68. Vijayasekaran S, Chirila TV, Robertson TA, Lou X, Fitton JH, Hicks CR, Constable IJ. *J Biomat Sci-Polym E*, **11**(**6**): 599 (2000).
69. Uchino Y, Shimmura S, Miyashita H, Taguchi T, Kobayashi H, Shirnazaki J, Tanaka J, Tsubota K. *J Biomed Mater Res B*, **81**B(**1**): 201 (2007).
70. Pearce RSC, West LR, Rodeheaver GT, Edlich RF. *Am J Surg*, **148**(**5**): 687 (1984).
71. Gorman SP, Tunney MM, Keane PF, van Bladel K, Bley B. *J Biomed Mater Res*, **39**(**4**): 642 (1998).
72. Chew BH, Denstedt JD. *Nat Clin Pract Urol*, **1**(**1**): 44 (2004).
73. Coats TJ, Edwards C, Newton R, Staun E. *Emerg Med J*, **19**(**3**): 224 (2002).
74. Kirker KR, Luo Y, Morris SE, Shelby J, Prestwich GD. *J Burn Care Rehabil*, **25**(**3**): 276 (2004).
75. Martineau L, Shek PN. *Burns*, **32**(**1**): 70 (2006).
76. Osti E. *Arch Surg-Chicago*, **141**(**1**): 39 (2006).
77. Martineau L, Shek PN. *Burns*, **32**(**2**): 172 (2006).
78. Chaterji S, Kwon IK, Park K. *Prog Polym Sci*, **32**: 1083 (2007).
79. Ziaie B, Baldi A, Lei M, Gu YD, Siegel RA. *Adv Drug Deliver Rev*, **56**(**2**): 145 (2004).

80. Nayak S, Lyon LA. *Angew Chem Int*, **44(47)**: 7686 (2005).
81. Chellat F, Merhi Y, Moreau A, Yahia L. *Biomaterials*, **26(35)**: 7260 (2005).
82. Graham NB, Cameron A. *Pure Appl Chem*, **70(6)**: 1271 (1998).
83. Allen TM, Cullis PR. Science, **303(5665)**: 1818 (2004).
84. Ulanski P, Kadlubowski S, Rosiak JM. *Rad Phys Chem*, **63(3–6)**: 533 (2002).
85. Mourey TH, Leona JW, Bennett JR, Bryan TG, Slater LA, Balke ST. *Journal of Chromatography A*, **1146**: 51 (2007).
86. Vinogradov SV, Bronich TK, Kabanov AV. *Adv Drug Deliver Rev*, **54(1)**: 135 (2002).
87. Peppas NA. *Adv Drug Deliver Rev*, **56(11)**: 1529 (2004).
88. Hu J, Yu S, Yao P. *Langmuir*, **23(11)**: 6358 (2007).
89. Hay DNT, Rickert PG, Seifert S, Firestone MA. *J Am Chem Soc*, **126(8)**: 2290 (2004).
90. Zhang GZ, Niu AZ, Peng SF, Jiang M, Tu YF, Li M, Wu C. *Accounts Chem Res*, **34(3)**: 249 (2001).
91. Hu ZB, Xia XH, Marquez M, Weng H, Tang LP. *Macromol Symp*, **227**: 275 (2005).
92. Ulanski P, Janik I, Rosiak JM. *Rad Phys Chem*, **52(1–6)**: 289 (1998).
93. Morimoto N, Nomura SIM, Miyazawa N, Akiyoshi K. *ACS Sym Ser*, **924**: 88 (2006).
94. Nagahama K, Mori Y, Ohya Y, Ouchi T. *Biomacromolecules*, **8**: 2135 (2007).
95. Wu C, Akashi M, Chen MQ. *Macromolecules*, **30(7)**: 2187 (1997).
96. Li M, Zhang YB, Jiang M, Zhu L, Wu C. *Macromolecules*, **31(20)**: 6841 (1998).
97. Li M, Jiang M, Zhu L, Wu C. *Macromolecules*, **30(7)**: 2201 (1997).
98. Bharali DJ, Sahoo SK, Mozumdar S, Maitra A. *J Colloid Interf Sci*, **258(2)**: 415 (2003).
99. Lowe TL, Tenhu H. *Macromolecules*, **31(5)**: 1590 (1998).
100. Oh JK, Siegwart DJ, Lee H-I, Sherwood G, Peteanu L, Hollinger JO, Kataoka K, Matyjaszewski K. *J Am Chem Soc*, **129(18)**: 5939 (2007).
101. Oh JK, Tang CB, Gao HF, Tsarevsky NV, Matyjaszewski K. *J Am Chem Soc*, **128(16)**: 5578 (2006).
102. Van Thienen TG, Lucas B, Flesch FM, van Nostrum CF, Demeester J, De Smedt SC. *Macromolecules*, **38(20)**: 8503 (2005).
103. Zhou SQ, Chu B. *J Phys Chem B*, **102(8)**: 1364 (1998).
104. Varga I, Szalai I, Meszaros R, Gilanyi T. *J Phys Chem B*, **110(41)**: 20297 (2006).
105. Ito S, Ogawa K, Suzuki H, Wang BL, Yoshida R, Kokufuta E. *Langmuir*, **15(12)**: 4289 (1999).
106. Ogawa K, Nakayama A, Kokufuta E. *Langmuir*, **19(8)**: 3178 (2003).
107. Ogawa K, Nakayama A, Kokufuta E. *J Phys Chem B*, **107(32)**: 8223 (2003).
108. Robinson DN, Peppas NA. *Macromolecules*, **35(9)**: 3668 (2002).
109. Guo A, Liu GJ, Tao J. *Macromolecules*, **29(7)**: 2487 (1996).

110. Sahiner N. *Eur Polym J*, **43(5)**: 1709 (2007).

111. Sahiner N, Alb AM, Graves R, Mandal T, McPherson GL, Reed WF, John VT. *Polymer*, **48(3)**: 704 (2007).

112. Liu Y, Wang LX, Pan CY. *Polymer*, **43(25)**: 7063 (2002).

113. Hayashi H, Iijima M, Kataoka K, Nagasaki Y. *Macromolecules*, **37(14)**: 5389 (2004).

114. Kuckling D, Vo CD, Wohlrab SE. *Langmuir*, **18(11)**: 4263 (2002).

115. Li X, Zuo J, Guo YL, Yuan XH. *Macromolecules*, **37(26)**: 10042 (2004).

116. Kazakov S, Kaholek M, Kudasheva D, Teraoka I, Cowman MK, Levon K. *Langmuir*, **19(19)**: 8086 (2003).

117. Qiao XL, Zhang ZJ, Yao S. *J Photoch Photobio A*, **177(2–3)**: 191 (2006).

118. Wu X, Pelton RH, Hamielec AE, Woods DR, Mcphee W. *Colloid Polym Sci*, **272(4)**: 467 (1994).

119. Sahoo SK, De TK, Ghosh PK, Maitra A. *J Colloid Interf Sci*, **206(2)**: 361 (1998).

120. Huang G, Gao J, Hu ZB, John JVS, Ponder BC, Moro D. *J Control Rel*, **94(2–3)**: 303 (2004).

121. McAllister K, Sazani P, Adam M, Cho MJ, Rubinstein M, Samulski RJ, DeSimone JM. *J Am Chem Soc*, **124(51)**: 15198 (2002).

122. Lee H, Mok H, Lee S, Oh YK, Park TG. *J Control Rel*, **119(2)**: 245 (2007).

123. Vinogradov SV, Batrakova EV, Kabanov AV. *Bioconjugate Chem*, **15(1)**: 50 (2004).

124. Liu HA, Webster TJ. *Biomaterials*, **28(2)**: 354 (2007).

125. Yan M, Ge J, Liu Z, Ouyang PK. *J Am Chem Soc*, **128(34)**: 11008 (2006).

126. Nomura Y, Ikeda M, Yamaguchi N, Aoyama Y, Akiyoshi K. *Febs Lett*, **553(3)**: 271 (2003).

127. Hasegawa U, Nomura SIM, Kaul SC, Hirano T, Akiyoshi K. *Biochem Bioph Res Co*, **331(4)**: 917 (2005).

128. Doherty EAS, Kan CW, Barron AE. *Electrophoresis*, **24(24)**: 4170 (2003).

129. Buhleier E, Wehner W, Vogtle F. *Synthesis*, **2**: 155 (1978).

130. Tomalia DA, Baker H, Dewald J, Hall M, Kallos G, Martin S, Roeck J, Ryder J, Smith P. *Polym J*, **17(1)**: 117 (1985).

131. Newkome GR, Yao ZQ, Baker GR, Gupta VK. *J Org Chem*, **50(11)**: 2003 (1985).

132. Lee CC, MacKay JA, Frechet JMJ, Szoka FC. *Nat Biotechnol*, **23(12)**: 1517 (2005).

133. Hecht S, Frechet JMJ. *Angew Chem Int Ed*, **40(1)**: 74 (2001).

134. Liu MJ, Kono K, Frechet JMJ. *J Control Rel*, **65(1–2)**: 121 (2000).

135. Morgan MT, Carnahan MA, Immoos CE, Ribeiro AA, Finkelstein S, Lee SJ, Grinstaff MW. *J Am Chem Soc*, **125(50)**: 15485 (2003).

136. Wakabayashi Y, Tokeshi M, Hibara A, Jiang DL, Aida T, Kitamori T. *J Phys Chem B*, **105(19)**: 4441 (2001).

137. Esfand R, Tomalia DA. *Drug Discov Today*, **6(8)**: 427 (2001).

138. Jiang DL, Aida T. *Prog Polym Sci*, **30**(**3-4**): 403 (2005).

139. Shaunak S, Thomas S, Gianasi E, Godwin A, Jones E, Teo I, Mireskandari K, Luthert P, Duncan R, Patterson S, Khaw P, Brocchini S. *Nat Biotechnol*, **22**(**8**): 977 (2004).

140. Brinas RP, Troxler T, Hochstrasser RM, Vinogradov SA. *J Am Chem Soc*, **127**(**33**): 11851 (2005).

141. Lee CC, Gillies ER, Fox ME, Guillaudeu SJ, Frechet JMJ, Dy EE, Szoka FC. *P Natl Acad Sci USA*, **103**(**45**): 16649 (2006).

142. Lee JH, Lim YB, Choi JS, Lee Y, Kim TI, Kim HJ, Yoon JK, Kim K, Park JS. *Bioconjugate Chem*, **14**(**6**): 1214 (2003).

143. Marano RJ, Wimmer N, Kearns PS, Thomas BG, Toth I, Wilson AS, Brankov M, Rakoczy PE. *J Gene Med*, **5**(**5**): S10 (2003).

144. Chandrasekar D, Sistla R, Ahmad FJ, Khar RK, Diwan PV. *Biomaterials*, **28**(**3**): 504 (2007).

145. Gillies ER, Frechet JMJ. *J Am Chem Soc*, **124**(**47**): 14137 (2002).

146. Kim YS, Gil ES, Lowe TL. *Macromolecules*, **39**(**23**): 7805 (2006).

147. Choi JS, Joo DK, Kim CH, Kim K, Park JS. *J Am Chem Soc*, **122**(**3**): 474 (2000).

148. Malik N, Evagorou EG, Duncan R. *Anti-Cancer Drug*, **10**(**8**): 767 (1999).

149. Quintana A, Raczka E, Piehler L, Lee I, Myc A, Majoros I, Patri AK, Thomas T, Mule J, Baker JR. *Pharm Res*, **19**(**9**): 1310 (2002).

150. Shah DS, Sakthivel T, Toth I, Florence AT, Wilderspin AF. *Int J Pharm*, **208**(**1–2**): 41 (2000).

151. Duan X, Sheardown H. *Biomaterials*, **27**(**26**): 4608 (2006).

152. Duan XD, McLaughlin C, Griffith M, Sheardown H. *Biomaterials*, **28**(**1**): 78 (2007).

153. Sontjens SHM, Nettles DL, Carnahan MA, Setton LA, Grinstaff MW. *Biomacromolecules*, **7**(**1**): 310 (2006).

154. Wathier M, Jung PJ, Camahan MA, Kim T, Grinstaff MW. *J Am Chem Soc*, **126**(**40**): 12744 (2004).

155. Kobayashi H, Kawamoto S, Jo SK, Bryant HL, Brechbiel MW, Star RA. *Bioconjugate Chem*, **14**(**2**): 388 (2003).

156. Supattapone S, Wille H, Uyechi L, Safar J, Tremblay P, Szoka FC, Cohen FE, Prusiner SB, Scott MR. *J Virol*, **75**(**7**): 3453 (2001).

157. Gong YH, Matthews B, Cheung D, Tam T, Gadawski I, Leung D, Holan G, Raff J, Sacks S. *Antivir Res*, **55**(**2**): 319 (2002).

158. Davis ME, Brewster ME. *Nat Rev Drug Discov*, **3**(**12**): 1023 (2004).

159. Kamada M, Hirayama F, Udo K, Yano H, Arima H, Uekama K. *J Control Rel*, **82**(**2–3**): 407 (2002).

160. Yano H, Hirayama F, Kamada M, Arima H, Uekama K. *J Control Rel*, **79**(**1–3**): 103 (2002).

161. Kim JH, Lee SK, Ki MH, Choi WK, Ahn SK, Shin HJ, Hong CI. *Int J Pharm*, **272**(**1–2**): 79 (2004).

162. Loftsson T, Guomundsdottir H, Sigurjonsdottir JF, Sigurosson HH, Sigfusson SD, Masson M, Stefansson E. *Int J Pharm*, **212(1)**: 29 (2001).
163. Jarho P, Jarvinen K, Urtti A, Stella VJ, Jarvinen T. *J Pharm Pharmacol*, **48(3)**: 263 (1996).
164. Kondo A, Irie T, Uekama K. *Biol Pharm Bull*, **19(2)**: 280 (1996).
165. Sridevi S, Diwan PVR. *Eur J Pharm Biopharm*, **54(2)**: 151 (2002).
166. Challa R, Ahuja A, Ali J, Khar RK. *Aaps Pharmscitech*, **6(2)**: E329 (2005).
167. Saarinen-Savolainen P, Jarvinen T, Araki-Sasaki K, Watanabe H, Urtti A. *Pharmaceut Res*, **15(8)**: 1275 (1998).
168. Nagase Y, Hirata M, Arima H, Tajiri S, Nishimoto Y, Hirayama F, Irie T, Uekama K. *J Pharm Sci*, **91(11)**: 2382 (2002).
169. Li JS, Xiao HN, Li JH, Zhong YP. *Int J Pharm*, **278(2)**: 329 (2004).
170. Prokai L, Ouyang XD, Wu WM, Bodor N. *J Am Chem Soc*, **116(6)**: 2643 (1994).
171. Wu WM, Wu J, Bodor N. *Pharmazie*, **57(2)**: 130 (2002).
172. Tilloy S, Monnaert V, Fenart L, Bricout H, Cecchelli R, Monflier E. *Bioorg Med Chem Lett*, **16(8)**: 2154 (2006).
173. Pun SH, Bellocq NC, Liu AJ, Jensen G, Machemer T, Quijano E, Schluep T, Wen SF, Engler H, Heidel J, Davis ME. *Bioconjugate Chem*, **15(4)**: 831 (2004).
174. Ruebner A, Statton GL, James MR. *Macromol Chem Physic*, **201(11)**: 1185 (2000).
175. Kihara F, Arima H, Tsutsumi T, Hirayama F, Uekama K. *Bioconjugate Chem*, **13(6)**: 1211 (2002).
176. Pun SH, Davis ME. *Bioconjugate Chem*, **13(3)**: 630 (2002).
177. Bellocq NC, Pun SH, Jensen GS, Davis ME. *Bioconjugate Chem*, **14(6)**: 1122 (2003).
178. Cho A, Lara KLO, Yatsimirsky AK, Eliseev AV. *Org Lett*, **2(12)**: 1741 (2000).
179. Choi HS, Huh KM, Ooya T, Yui N. *J Am Chem Soc*, **125(21)**: 6350 (2003).
180. Ichi T, Watanabe J, Ooya T, Yui N. *Biomacromolecules*, **2(1)**: 204 (2001).
181. Ichi T, Nitta K, Lee WK, Ooya T, Yui N. *J Biomat Sci-Polym E*, **14(6)**: 567 (2003).
182. Fujimoto M, Isobe M, Yamaguchi S, Amagasa T, Watanabe A, Ooya T, Yui N. *J Biomat Sci-Polym E*, **16(12)**: 1611 (2005).
183. Tachaboonyakiat W, Furubayashi T, Katoh M, Ooya T, Yui N. *J Biomat Sci-Polym E*, **15(11)**: 1389 (2004).
184. Boschin F, Blanchemain N, Bria M, Delcourt-Debruyne E, Morcellet M, Hildebrand HF, Martel B. *J Biomed Mater Res A*, **79A(1)**: 78 (2006).
185. Blanchemain N, Haulon S, Martel B, Traisnel M, Morcellet M, Hildebrand HF. *Eur J Vasc Endovasc Surg*, **29(6)**: 628 (2005).
186. Muderawan IW, Ong TT, Ng SC. *J Sep Sci*, **29(12)**: 1849 (2006).
187. Harada A, Furue M, Nozakura SI. *J Polym Sci Pol Chem*, **16(1)**: 189 (1978).
188. Thuaud N, Sebille B. *J Chromatogr A*, **685(1)**: 15 (1994).
189. Thuaud N, Lelievre G, Deratani A, Sebille B. *Eur Polym J*, **33(7)**: 1015 (1997).
190. Duval R, Leveque H, Prigent Y, Aboul-Enein HY. *Biomed Chromatogr*, **15(3)**: 202 (2001).

191. Tang W, Muderawan IW, Ng SC, Chan HSO. *J Chromatogr A*, **1094**(**1–2**): 187 (2005).
192. Poon YF, Muderawan IW, Ng SC. *J Chromatogr A*, **1101**(**1–2**): 185 (2006).
193. Tu YF, Chen HY. *Biosens Bioelectron*, **17**(**1–2**): 19 (2002).
194. Fragoso A, Caballero J, Almirall E, Villalonga R, Cao R. *Langmuir*, **18**(**13**): 5051 (2002).
195. Villalonga R, Camacho C, Cao R, Hernandez J, Matias JC. *Chem Comm*, **9**: 942 (2007).

7

IN TEXTILES

MAHAMMAD SAFIKUR RAHMAN

INTRODUCTION

Fibers are available from natural sources and are also made by synthetic means. All fibers, natural or synthetic, are high-molecular-weight polymeric macromolecules. Important commercial fibers for making textiles are mentioned in Table 7.1 along with their major strengths and shortcomings. There is a need to modify the fibers/fabrics so that they can be effectively utilized for newer end-use requirements. Some examples of these modifications are imparting dimensional stability to woolen fabric, improving wash-and-wear performance to cotton textiles, and ensuring antimicrobial activity to polypropylene or polyester monofilaments for use as effective surgical sutures. "Grafting" and "crosslinking" are among several techniques used to carry out desired modifications on the fiber either in fiber or fabric form. Synthetic fibers, unlike natural fibers, do not contain reactive functional groups in sufficient numbers, which restricts crosslinking in those fibers. An attempt has been made to highlight applications of these two "macromolecular engineering" techniques (grafting and crosslinking) on some commercial fibers in the following sections.

APPLICATION IN COTTON TEXTILES

Cotton is a unique natural fiber of plant origin and is the most important textile fiber in the world. Cotton fiber produced by the plant is composed

Polymer Grafting and Crosslinking, Edited by Amit Bhattacharya, James W. Rawlins and Paramita Ray

TABLE 7.1. Structure and important properties of commercially available fibers.

Fiber	Structure	Moisture regain, %	Strength	Shortcoming
Cotton	Cellulosic[1]	7–8	Fine denier, absorbs moisture, high comfort	Poor wrinkle resistance, poor biological resistance
Viscose	Cellulosic[1]	12–14	High moisture absorption, soft, high comfort	Poor wet strength, poor wrinkle and biological resistance, sensitive to light
Wool (Protein fiber)	Peptide linkage[2]	14	Naturally crimped, great extensibility, very good warmth	Poor dimensional stability, attacked by moths
Silk (Protein fiber)	Peptide linkage[3]	10	Excellent wear comfort, good luster, exceptional handle	Poor wet resiliency, photoyellowing, attacked by moths
Polyester (Polyethylene terephthalate)	$–O.CO.C_6H_4CO.O.CH_2.CH_2–$	0.4	Excellent dimensional stability, resistant to light and biological agents	Poor moisture absorption, develops static charge, pilling
Nylon (Polyamide) nylon6, Nylon66	$–NH.(CH_2)_5.CO$ $–NH.(CH_2)_6.NHCO.(CH_2)_4.CO–$	4	High tenacity, elasticity, and recovery, good flexing, resistant to abrasion and biological agents	Low moisture absorption, develops static charge, degrades in sunlight

Acrylic	$-CH_2.CH(CN)-$	1–2	Good bulk and extensibility, resistant to light, resistant to chemical, resistant to biological agent	Poor hot-wet strength, low moisture absorption, medium tenacity, pilling
Polypropylene	$-CH_2.CH(CH_3)-$	0	Good mechanical property & knot strength, abrasion resistant, low specific gravity	Sensitive to light, zero moisture absorption, poor creep resistance and poor dyeability

1. Structural unit

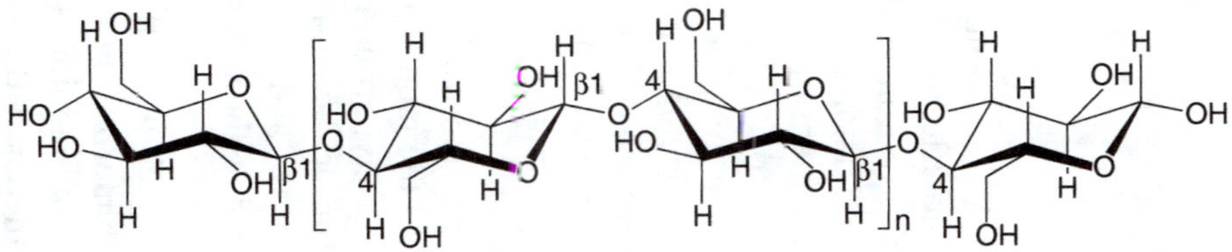

2. –NH–CH(R)-CO–, R: ~21 different types of aminoacid residues having: –OH, –COOH, $-NH_2$, –NH–, –SH, –S–S–, $-C_6H_4OH$ functional groups.
3. Silk fibroin contains the linkage –NH–CH(R)-CO–, R: –H (glycene), $-CH_3$ (alanine), $-CH_2$ OH (serine), $-CH_2.C_6H_4OH$ (tyrosene) are the major amino acids.

almost entirely of cellulose. Cellulose is a linear condensation polymer wherein anhydroglucose units are linked together by 1, 4–ß–glycosidic linkage (structure of cellulose, Table 7.1).

Each anhydroglucose unit of cellulose contains three hydroxyl groups and hence there are many hydroxyl groups and glycosidic oxygen in the cellulose macromolecule. These are responsible for a high degree of hydrogen bonding among the polymeric chains, and results in three-dimensional ordered structures, thus forming crystalline lattices [1]. Portions of the same molecule may run parallel to another chain for a short length, cross other chains, or bend, forming less ordered amorphous regions. When a cotton fabric is wet, the fibers are in contact with water molecules, which get absorbed into the amorphous region of the fiber and disrupt the interchain hydrogen bonding. New hydrogen bonds are formed involving water molecules. When the water molecules are evaporated from the cotton fabric or garment under normal air drying condition, new sets of interchain hydrogen bonds are formed resulting wrinkles on the fabric. Wrinkling is a major shortcoming of cotton textiles. Graft polymerization and particularly crosslinking on cotton fibers have been carried out to overcome this shortcoming and to impart other desirable properties on cotton textiles.

Grafting on Cotton

Cotton is affected by UV radiation. This degrading effect on cotton was noticed over a century ago. Electron spin resonance study of light-irradiated cotton has established the reason of degradation as free-radical generation on cellulose [2, 3]. An understanding of this phenomenon has also opened up possibilities for improvement of the properties of cotton fabric by grafting a variety of monomers onto it [3–5]. As mentioned above, cotton fabrics have poor wrinkle resistance compared to synthetic fabrics and need modification that will lead to improvement in their wrinkle recovery properties. *N*-methylolacrylamide (NMA) has been used as an effective monomer for grafting on cotton. Graft polymerization is initiated by different means such as chemical, thermal, and irradiation techniques [6]. NMA contains two different functional groups namely: a vinyl and a methylol group. The vinyl group takes part in the graft polymerization initiated by cellulose macroradicals formed by irradiation, and the methylol group of NMA undergoes condensation reaction with the hydroxyl group of cellulose at relatively high temperature in the presence of an acidic catalyst. These radiation chemical-finished cotton fabrics have good wash-and-wear properties [7]. NMA and cotton react readily under UV light when the reaction is carried out in a nitrogen atmosphere. Reaction is found to be inhibited by the presence of oxygen. The presence of an inhibitor such as methyl ether of hydroquinone retards but does not prevent the polymerization reaction. The presence of a photosensitizer such as biacetyl (2,3-butanedione) can increase the rate and efficiency of the reaction, but it also increases the stiffness [8].

Glycidyl methacrylate has also been used for modification of cotton. Again, in this case, unsensitized UV-light-initiated polymerization results in formation of (initiating) free radicals on cellulose fiber, thereby minimizing homopolymer formation [9, 10]. However, when several layers of cotton fabrics are immersed in an aqueous solution of glycidyl methacrylate (GMA) and only the surface layer is exposed to UV light, the polymerization reaction is initiated in the surface layer and subsequently in inner layers. Of course, the percentage of graft decreases gradually with depth of the layer, and at the sixth layer, the value of the percentage of graft becomes one-third the value of surface layer [10]. The reaction in the inner layers is believed to have taken place by chain transfer mechanism.

Extensive research work has been carried out at the National Research Centre, Textile Research Division, Cairo, Egypt on chemically induced graft polymerization of cotton fabric with various vinyl monomers in order to modify the surface characteristics [11]. Varieties of redox system as well as oxidizing agents are used for chemical modification. In some cases, cellulose thiocarbonate is prepared first by reacting cotton fabric with carbondisulfide in alkaline medium to provide effective reducing sites on the cellulose backbone. When an oxidizing agent is added to it in presence of a vinyl monomer, polymerization is initiated by the redox mechanism, and a vinyl graft polymer is formed onto cellulose [12, 13].

Graft polymerization of acrylic-based monomers onto a cellulose backbone has been studied as an innovative method for conservation of cellulosic textiles [14, 15]. Grafting acrylate monomers directly from the vapor phase onto cellulose has been found to improve the mechanical properties of degraded textiles without altering the flexibility and protects the surface from formation of superficial coating of other unwanted materials on storage. The grafting of combination of monomers, e.g., ethyl acrylate (EA)/methyl methacrylate (MMA) in a ratio of 75/25 results in a restoration of flexibility ($t_g \sim 10\,°C$). The addition of small quantity of the fluorinated monomer 2,2,2-tifluoromethyl methacrylate (TFEMA) in the above combination, i.e., EA/MMA/TFEMA in the ratio of 73/24.5/2.5 is found to improve water repellency of the grafted substrate [15].

Plasma-induced grafting has also become an attractive technique for surface modification of textiles [16, 17]. The plasma state consists of radicals, photons, and charged particles such as ions and electrons. When these highly active species collide with the surfaces of textile substrates they generate free radicals along with other active sites on the surface [18]. When this active textile substrate comes in contact with a monomer, it initiates a polymerization reaction. Even under conditions of in situ plasma polymerization, the cotton substrate and monomer are simultaneously activated, but homopolymerization is found to predominate only if the monomer is introduced via an inlet too close to the electrode [19]. Cotton fabrics have been found to have higher rate of free-radical generation than other natural and manmade fibers [20, 21]. It has also been assumed that the stability of free radicals generated by the treatment on cellulosic fiber with the plasma is greatly dependent on the fine structure of the fiber substrate [20]. This was evident from the fact that relative free-radical intensity of viscose and mercerized cotton is quite less than cotton when subjected to plasma treatment under identical conditions. This may be due to relatively loose structure of viscose and mercerized cotton compared to cotton, and hence recombination takes place easily in former cases. Generation of a plasma state suitable for textile application is usually done in the presence of a constant supply of different gases such as argon, helium, nitrogen, oxygen, etc. Since the gases are supplied at room temperature, the temperature of the plasma state generated remains below 100 °C and hence is called cold plasma treatment.

Cotton fabric is effectively surface fluorinated through plasma polymerization of hexafluoropropene (C_3F_6) gas [22]. The contact angle measurement and wetability study of the treated fabric has shown that contact angle remains in the range of 140° to 150° and a wettability of 300s, indicating very good hydrophobicity. It is suggested that both plasma-induced polymerization as well as plasma polymerization are taking place simultaneously resulting in an increase in the surface atomic fluorine and reduction in the surface oxygen substantially.

A monomer, 2-hydroxyethyl methacrylate (HEMA) has been graft polymerized quite efficiently on cotton fabric immediately after exposing it to a glow discharge argon plasma generated by a radio frequency of 13.56 MHz [23]. Degassed aqueous solution of HEMA (2.5%) has been used for polymerization, and 85.7% grafting is achieved by this process. It has been observed that a small level of grafting (10% grafting level) increases the breaking strength of the fabric, which of course is leveled at higher percentage of grafting. Similarly, vinyl laurate has been graft polymerized on cotton fabric after generating active radical sites onto it using 2.45 GHz microwave plasma [24, 25]. Among different gas media, argon plasma has been found to create more active sites [24]. Plasma-induced grafting of vinyl laurate resulted in stable cellulose-vinyl laurate copolymers. This plasma-induced hydrophobic finishing on textiles is found to be quite durable, and such treatment practically does not affect the strength property, unlike the traditional resin finishing by wet processing, which usually results in high strength loss of cotton textiles [25].

Crosslinking in Cotton Textiles

Among the various textile fibers and fabrics, the wrinkling property is most pronounced in cotton textiles as discussed above. Efforts have been made since 1920 to make cotton fabric wrinkle resistant or crease resistant [26]. As per estimates in 1978, there are more than 7000 publications and 3000 patents on the subject of the crosslinking of cellulose to impart wrinkle resistance properties [27]. Historically, the developments on cellulosic crosslinking can be presented in six chronological stages: the concept of the finishing of cellulosics with a urea-formaldehyde type of resin (first stage); investigations on chemical and physicochemical aspects of crosslinking using reactive agents (second stage); insight on the mechanism of cellulose crosslinking and its effect on fiber structure and properties (third stage); development of the technology of wrinkle-free fabrics and garments, including post-cure process (fourth stage); concern of freely present and releasable formaldehyde from the finished textiles, and development of low- or no-formaldehyde crosslinking systems (fifth stage); and lastly, ionic crosslinking (sixth stage).

Harro A. Peterson has made a comprehensive presentation of the crosslinking of cellulosic textiles with formaldehyde containing reactants [28, 29]. Natural cotton fiber contains 2000–3000 anhydroglucose rings in the chain joined by glucosidic linkages. Every anhydroglucose unit contains one primary hydroxyl group at the sixth carbon [structure of cellulose, Table 7.1], which is the most reactive hydroxyl group for crosslinking, but other hydroxyl groups also take part in crosslinking reaction [30]. Production of wrinkle-free cotton textiles with low-formaldehyde containing crosslinking agents, namely, dimethylol dihydroxy ethylyne urea (DMDHEU) etherified with a polyhydric alcohol such as diethylene glycol, 2,3 propanediol and amide glyoxal adduct [31–33] in the presence of an acid liberating catalyst is now being practiced globally.

The alternative approach of formaldehyde-free crosslinking using the unique and extremely effective agent 1,2,3,4 butane tetracarboxylic acid (BTCA) in the presence of an acid catalyst (sodium hypophosphite), although technologically preferred [34, 35], has not been accepted by the industry because of the relatively high cost of BTCA [26].

These crosslinking reactions of cellulose, whether with the DMDHEU-based reagent or using BTCA, suffer from strength loss, particularly tear strength (loss of 30%–50% of the original value), reduced abrasion resistance, stiffening, and shade change (for colored textiles).

Under the leadership of Brent Smith and Peter Hauser at the National Textile Center in the United States, considerable study has been carried out on a novel concept of the stabilization of cotton textiles through ionic crosslinking [36]. The concept is to make cellulose anionic and then react it with a polycation, or cationize the cellulose to produce the cationic cellulose first and then react it with a polyanion. In order to impart anionic character to the cellulose, the cellulose is allowed to react with chloroacetic acid (CAA) or chlorosulphonic acid. Similarly, reaction of cellulose with chemicals such as 3-chloro-2 hydroxypropyl trimethylammonium chloride (CHTAC), 1,1-diethyl-3 hydroxy azetidinium chloride or 1,1-dimethyl-3 hydroxy azetidinium chloride converts it to cationic cellulose.

The polycations considered for reacting with anionic cellulose are cationized versions of chitosan, cationized glycerin, calcium chloride, magnesium chloride, cationized ethylene glycol, cationized dextrose, etc. Polyanions considered for reacting with cationic cellulose are BTCA, EDTA (ethylene diamine tetraacetic acid), nitrilotriacetic acid, citric acid, etc.

The two best treatments identified by the group are (1) treating anionic cellulose fabric with cationized glycerin, and (2) treating cationic cellulose fabric with BTCA; the most promising appeared to be the first one. The polyelectrolite chains (e.g., polyanion cellulose) are bound at many multiple sites (e.g., reactive sites of cationized glycerin), resulting in excellent washing durability. This treatment has resulted in improvement in both the dry and wet wrinkle recovery angle (WRA) (as high as 140° for wet WRA and 100° for dry WRA, which is comparable to the standard commercial resin-treated fabric). Ionic crosslinking has been found to result in an increase in tensile strength, which is in contrast to the crosslinking that uses the DMDHEU-based reagent or BTCA.

APPLICATION IN WOOLEN TEXTILES

For centuries, wool has been a popular fiber to make garments that provide warmth. In its simplest form, wool is a natural fibrous protein containing amide groups in the main chain together with other hydrophilic functional groups such as $-NH_2$, –NH–, –SH, –OH, and –COOH (Table 7.1). Wool undergoes undesirable dimensional changes known as felt shrinkage, which is related to the surface of the natural wool fiber covered by overlapping scales. Both grafting and crosslinking have been carried out on wool to impart physical, mechanical, and biochemical properties to woolen fabrics.

Grafting on Wool

Modification of wool has been carried out by the radical grafting of vinyl monomers onto wool involving all the three possible means of generation of radicals: chemical, photochemical, and high-energy irradiation methods [37, 38]. MMA has been found to be an active monomer for grafting on wool. Mishra and coworkers have carried out graft polymerization of MMA onto wool by the chemical-initiation method using manganese acetyl acetonate in aqueous medium [39], butyl hydroperoxide-mineral acid [40], a ferrous ammonium sulfate-hydrogen peroxide system as redox initiator [41], a ceric ion initiator in the presence of nitric acid [42], and chromium acetyl acetonate [43]. A manganese acetyl acetonate-perchloric acid system has also been investigated as an effective system for grafting MMA on wool [44]. The ferrous ammonium sulfate-hydrogen peroxide system is probably the simplest and most effective initiator system for getting high graft yield. This has been found to be effective also for other monomers such as ethyl acrylate and vinyl acetate [41].

Grafting of acrylic acid has been attempted using the ceric ammonium nitrate-nitric acid system for redox initiation [45]. The maximum grafting yield achieved is only 12.24%. A new redox system—potassium monopersulfate/ Iron (II)—has been used for the grafting of acrylamide on wool in aqueous medium; a 40% grafting yield has been obtained by this method [37]. Graft polymerization of 2-methyl 5-vinyl pyridine onto polyester/wool (50:50) blended fabric has also been carried out using benzoyl peroxide as the initiator [46]. The grafted sample has shown improved dyeability toward acid dyes and decreased moisture regain.

MMA has also been graft polymerized onto wool in the presence of high energy irradiation [47–49]. Radiation-induced graft copolymerization onto wool fiber involves the generation of many active free-radical sites on the wool fiber. Electron spin resonance studies on wool fiber irradiated by ^{60}Co γ rays have revealed that most radicals from the cortex are long lived, whereas most radicals formed on the scales were short lived [50]. The study also revealed that water content in wool up to 27.5% does not affect the radical formation.

Incorporation of methanol, a swelling agent for wool fiber, improves the accessibility of the monomer to the grafting site and thus enhances the grafting yield of MMA grafting in presence of γ–ray irradiation [48]. Incorporation of multifunctional acrylates, namely, di(ethyleneglycol) diacrylate and tri(propylene glycol) diacrylate, and a photoinitiator in the monomer enhances grafting. Inclusion of styrene as a comonomer in the MMA solution decreases the homopolymer formation significantly, leading to higher grafting yields. A similar comonomer technique using styrene as the comonomer with the EA monomer has been studied earlier for radiation-induced graft copolymerization onto wool to effectively reduce homopolymer formation [51]. Such enhancement of grafting in the presence of the additives is important, as the required grafting level can be achieved at a reduced dose of radiation. The less the dose of radiation, the less the adverse effect on the properties of wool.

Crosslinking in Wool

Crosslinking in wool has been attempted to improve the dimensional stability of woolen fabric. Diepoxides such as diepoxy butane and diepoxy hexane have been allowed to react with wool aiming to impart resistance to deformation and thus reduce the felting power [52]. Multifunctional epoxides crosslink with various functional groups, namely, amines, alcohols, phenols, thiols, and carboxylic acids present in wool, and thus improve fabric performance with respect to shrink resistance and wrinkle recovery [53–55]. A water-soluble multifunctional epoxide, glycerol polyglycedyl ether, has been applied on wool fabric by a simple pad-dry-cure technique for crosslinking with amino acid residue to obtain better wrinkle recovery [56]. It has been observed that epoxide concentration to the extent of 15% of the weight of fabric is practically the optimum level to achieve the desirable performance in terms of wrinkle recovery and moisture regain properties.

A new crosslinker for wool, 2-iminothiorene hydrochloride (2-IT), is reacted with wool to increase disulfide (–SS–) linkages in the wool. It is believed that wrinkle formation in wool fabric occurs because the number of hydrogen bonds is greater than that of –SS– groups, and hence it is necessary to reverse the order for improvement in wrinkle recovery. Reaction of wool with 2-IT introduces –SH groups into wool fiber, which is finally oxidized to form –SS– linkages. Both wrinkle recovery and setting ability of the modified woolen fabric have been found to be improved [57]. Intermolecular crosslinking has also been carried out by reacting dichloropyrimidine or trichloropyrimidine using 1.8 wt. % of the crosslinker under weakly alkaline conditions (pH 8.5) [58]. Such treatment imparts superior pilling resistance without the deterioration of the fiber properties.

Transglutaminase, a protein crosslinking enzyme, is capable of adding tensile strength to wool protein. Treatment of transglutaminase prevents the damage caused by detergents that contain proteases [59].

APPLICATION IN SILK TEXTILES

Silk as a textile fiber predates written history. High-quality fabrics and luxury items are produced from silk. The fiber is composed of long chain amino acid units joined by peptide linkages with hydrogen bonding between parallel chains. The single polypeptide chains probably pass through both crystalline and amorphous regions, and the amino acid with bulky side chains forms the amorphous region [60]. Silk is an exceptional natural fiber and a top-quality textile material with many desirable properties, such as good mechanical properties, high elastic modulus, moisture regain, and an excellent fabric hand. Unfortunately, it has a fatal weakness—very low wet resiliency. When silk fibers absorb water and swell, the salt linkages between the polymers that give the fiber high dry crease recovery are broken. Thus, silk fabrics have poor wash-and-wear properties. Additionally, silk is susceptible to photoyellowing. Chemical modifications that improve the shortcomings of silk include graft copolymerization and crosslinking reactions.

Grafting on Silk

Chemical modifications of fiber and fabric by carrying out graft polymerization using a variety of grafting monomers, including vinyl monomers, have been performed with varying degrees of success in order to impart desirable properties and overcome the shortcomings (Table 7.2).

Among the different monomers for grafting, styrene and MMA have been widely used in the first stages of industrial applications primarily because of bulkiness and fullness of handle, better wrinkle recovery, and affinity toward disperse dyes [61]. Grafting of vinyl polymers onto silk has been reviewed by Lenka [62].

Grafting reactions of the vinyl and/or styrene monomers are initiated by using a persulfate initiator that immerses the fiber or fabric in an aqueous monomer solution/emulsion containing the initiator [61, 63, 64]. The polymers produced by radical polymerization within the silk fiber either remain within the silk matrix of the fiber core at a reasonable add-on level or partly come

TABLE 7.2. Properties of grafted silk fiber/fabric.

Monomer	Major Effect on property	Reference Cited (Sr. No.)
Styrene	Improved bulkiness and fullness of handle	63
Methyl methacrylate [MMA]	Improved wrinkle recovery	65, 67
Ethoxyethyl methacrylate [ETMA]	Enhanced dyeability (acid dyes)	68
2-Hydroxyethyl methacrylate [HEMA]	Higher thermal stability	61
N(n-butoxymethyl) methacrylamide [BMMA]	Reduced photoyellowing	64

out from the fiber core and remain adhered on the fiber surface at a higher add-on level [64]. Scanning electron micrographs of grafted and modified silk fibers with varying degree of grafting are shown in Figure 7.1.

Although the term "graft" has been used in all such publications, no in-depth study or suggestion on the mechanism of graft polymerization, i.e., generation of a radical site on the fiber molecule, is available.

X-ray study of the grafted silk fibers clearly indicates that the crystalline structure with the oriented β structure remains unaltered by the grafting reaction. The position and intensity of the main X-ray diffraction peak at 20.5° corresponding to the crystalline spacing of 4.39°A does not change regardless of the grafting with monomers of different structures or sizes, e.g., HEMA [61], styrene/BMA [63], MMA [65, 67], or HEMA/MMA [66]. These findings suggest that the polymerization takes place at the relatively loose amorphous region of the fiber matrix. Because of the undisturbed crystalline structure, the breaking load for the grafted silk fibers mostly remain unchanged irrespective of add-on or nature of the monomers used.

Grafting with a monomer containing ether linkage bulky pendant group like ETMA facilitate dyeing of silk [68]. Grafting of vinyl monomers such as MMA and acrylamide has also been carried out on silk fiber initiated by using a semiconductor-based photocatalyst. Cadmium sulfide has been used as the semiconductor for initiation of photoinduced grafting by the free-radical mechanism, and up to 48% grafting in the case of MMA and 26% grafting in the case of acrylamide have been obtained by this technique [69].

Crosslinking in Silk

There are two objectives when examining why silk fibers should undergo crosslinking: to modify the fiber in order to impart crease resistance and washability, and for the study of structure of the fiber. It has been observed that when the dimension of the reactive bifunctional agent is less than 1 nm, reaction with silk is largely monofunctional, but with the 1 nm dimension, the

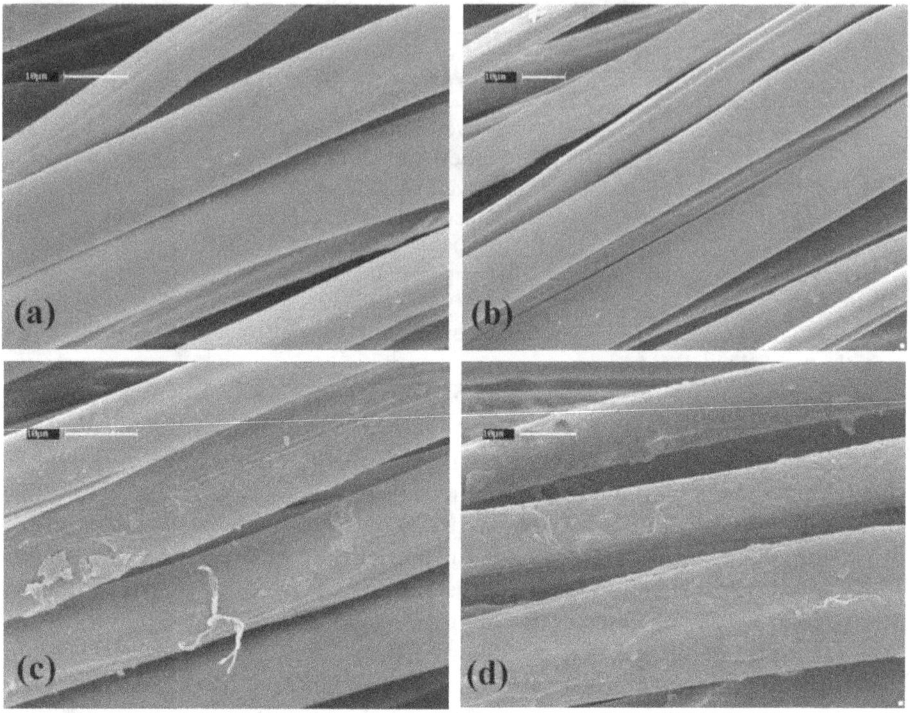

Figure 7.1. SEM pictures of silk fibers grafted or modified by the microwave technique: (a) Epoxide, 1.8% w.g.; (b) MAA, 3% w.g.; (c) IPMA, 14% w.g.; (d) HEMA, 24% w.g. Reproduced from *AUTEX Res. J*, **5(1)**: 40 (2005), http//www.autexrj.org/no1-2005/0131.pdf. Accessed November 11, 2007.

reaction is 100% bifunctional, indicating the distance between the chains in the amorphous region of the fiber [70].

A variety of crosslinking agents from formaldehyde-based crosslinkers [71] and di- or polycarboxilic acid compounds [72–74] similar to the applications on cotton textiles, to epoxides [75–83] and polyurethanes [84] are used for the crosslinking, utilizing the potential functional groups—amine, hydroxyl, and carboxylic acid—of the silk. Even reactive dye molecules having two reactive groups or one bifunctional group undergo crosslinking during dyeing [85]. The structures of a few crosslinkers are shown in Figure 7.2.

Among the crosslinking agents, epoxides have been proved to be markedly effective in improving the properties of silk. Tanaka and Shiozaki have reported the modification of silk with epoxide and concluded from amino acid analysis that epoxide reacts with amino, imino, and hydroxyl groups of silk fibroin [75]. Studies of the structural characteristics caused by different types of epoxide crosslinking indicate that reaction takes place with selective functional groups on the fibroin and also suggest that the reaction takes place in the amorphous region like grafting as discussed earlier [76, 77]. In all these studies, organo-

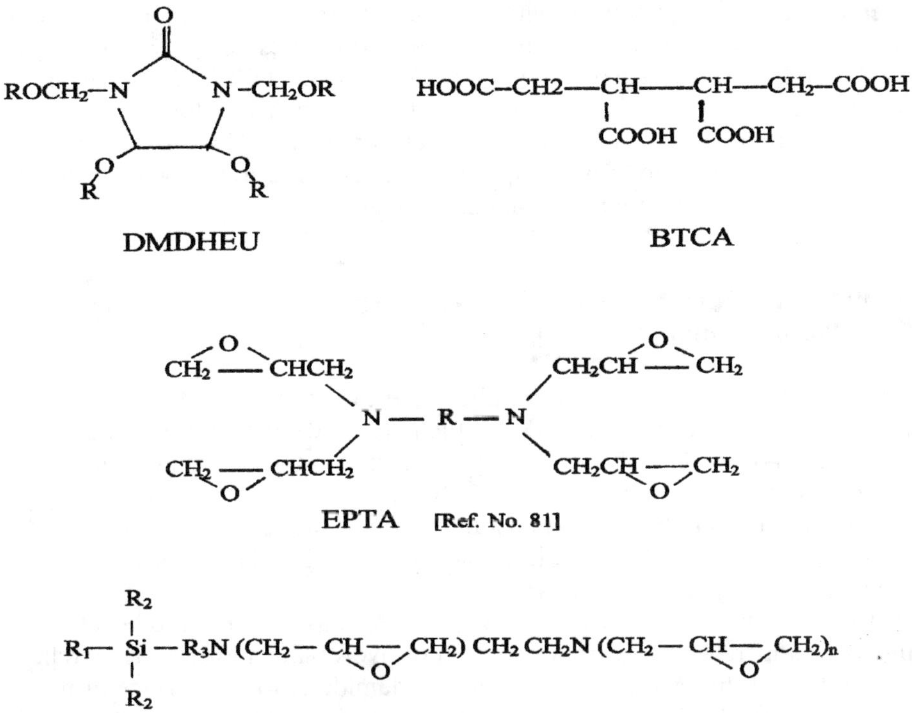

NaSO$_3$CONH (CH)$_6$ NHCOO(CH$_2$O)$_n$ CONH (CH)$_6$ NHCOSO$_3$Na

WPU 200 [Ref. No. 84]

Figure 7.2. Structure of a few crosslinkers for textiles.

chloro solvent was used as the reaction medium. Investigators have further studied the difficulty of this solvent method and have gradually come out with user-friendly and commercially adaptable pad-dry-steam or pad-dry-bake processes without using an organochloro solvent [78, 79]. It has been reported that ethylene glycol diglycidyl ether has been used as a crosslinker and applied onto tussah silk fabric by the pad-batch method using isopropanol (1:1 vol./vol. of epoxide used) in an aqueous medium [80]. A marked increase in wet crease recovery has been observed by this treatment. Cheng and Kai have reported a tetrafunctional water-soluble epoxy resin (EPTA) for treatment of silk fabric [81, 82]. A two-bath/two-stage treatment of silk fabric with this multifunctional epoxide has been found to improve the wet crease recovery angle from 200° to 280°, and 95% of this value is retained even after 20 cycles of home laundering. An epoxy silicone water-soluble crosslinking agent (EPSIA) has also been developed, treatment of which is found to increase by nearly 40% wet crease recovery without sacrificing any good properties of silk [83]. A water-soluble bifunctional reactive polyurethane product (WPU 200) has been developed to carry out the modification of silk at relatively low temperature conditions [84]. An optimum curing temperature of 100 °C has been found to be sufficient in order to impart improved wet resiliency to the fabric, which is also claimed to be durable in washing.

APPLICATION IN POLYAMIDE (NYLON): Graft Polymerization

One important reason that natural fibers like cotton, wool, and silk retain significant share of the apparel market is the perceived comfort of fabric relative to those made from polyester or nylon. This comfort advantage has been attributed mostly to high capacity of absorption of moisture, particularly in hot and humid conditions, and moisture transport along the fiber surface [86]. Between nylon and polyester, the former is better placed in terms of moisture regain property (Table 7.1).

The most studied route to increased comfort has been the modification of fiber by an increase in the concentration of hydrophilic moieties; researchers have attacked the problem by grafting acrylamide onto the surface of nylon fabric [86–93]. "Hydramid," a graft copolymer of polyamide 6 and polyacrylamide, claimed to have excellent performance and comfort properties similar to cotton, has been commercialized [88, 89]. A process based on grafting acrylamide onto the surface of nylon fabric has been patented and commercialized by the International Yarn Corporation [87].

Varma and colleagues have used ceric ion–sulfuric acid systems for initiating graft polymerization [92, 93]. Photoinduced graft polymerization of acrylamide onto polyamide has been carried out by several investigators. Graft polymerization has been performed under UV radiation conditions using fructose as the photosensitizer [90]. Grafting as high as 90% is reported by this method, and moisture regain value has doubled at that grafting level [91]. Grafting on polyamide has also been carried out under direct UV irradiation in the presence of benzophenon acting as the photosensitizer [89]. In this process, the fiber is pretreated with benzophenon solution of dichloromethane and methanol (80:20) in order to impregnate the sensitizer into the fiber matrix. Grafting efficiency of 80% has been achieved by this method.

Graft polymerization of methacrylic acid (MAA) onto nylon6 has been carried out by initiating the reaction using $K_2S_2O_8$–$Na_2S_2O_3$–Cu^{2+}, a reversible redox system [94]. One distinct advantage to using the Cu^{2+} ion is the control of homopolymer formation, and this effect has also been observed when grafting acrylonitrile (AN) onto nylon [95]. Similarly, modified nylon6 fiber containing grafted polydiallyldimethyl ammonium chloride has been graft polymerized using GMA in the presence of $K_2S_2O_8$–Cu^{2+} as a redox initiating system [96]. The modified nylon6 samples are treated with $K_2S_2O_8$ solution first, washed and dried, and then immersed in aqueous solution of GMA and copper sulfate. Graft yield as high as 200% has been obtained by this method. GMA is considered an important vinyl monomer because of the presence of epoxy groups that have the ability to undergo consecutive modifications leading to the incorporation of basic and acidic groups onto the fiber. Such modified fibers may find application as ion exchange fibers. Methacryloyloxyethyl trimethyl ammonium chloride (METAC) has been identified as another such monomer, which after polymerization can be further used for incorporation of desirable moiety. With this objective, graft polymerization of METAC has been carried out on nylon66 knitted fabric using $Na_2S_2O_8$ as an initiator [97]. Grafting to the extent of 21% has been achieved by this method.

γ-ray irradiation-induced graft polymerization of AN and styrene onto nylon6 has been carried out in order to improve the dyeability property of nylon fiber [98]. In general, the dyeability of polyacrylonitrile grafted fiber is markedly improved and polystyrene grafted fibers show good affinity to basic dyes.

Nylon66 has been grafted with polyacrylic acid with the objective of maximizing surface coverage with many potential active sites on the fiber surface for further chemical reaction with active moiety [99]. Thus, polyacrylic acid grafted nylon66 fiber is further treated with ethylenediamine and protoporphyrin IX sequentially [100]. These modified fibers become light-activated antimicrobial fibers, the activity of which is found to increase with the intensity or time of exposure to the visible light.

APPLICATION IN POLYPROPYLENE: Graft Polymerization

Polypropylene fiber is hydrophobic and undyeable in nature, which limit its acceptability as a textile fiber, although it is cheap and has low density and good mechanical properties. Grafting is a possible way for this fiber to impart desirable properties such as dyeability, hydrophylicity, antistatic properties, soil repellency, and thermal stability [101] and even to make its surface suitable for drug immobilization [102].

The grafting of polypropylene through the use of vinyl monomers is the major modification of polypropylene. Both chemical [101, 103] and irradiation means have been adopted for creation of the active site on polypropylene required for initiation of graft polymerization. However, the majority of work on the grafting of polypropylene reported has used the irradiation technique. γ radiation [104, 105], radiation by electron beam [102, 106], and UV radiation [101, 107] all have been successfully used for the creation of radicals on polypropylene surfaces.

Extensive studies have been carried out on the grafting of polypropylene fiber using MAA and acrylic acid for improvement of its dyeability. As expected, acrylic acid grafted polypropylene fiber has shown good dyeability toward basic dyes [101]. MAA grafted polypropylene fiber has shown a sharp increase in dye uptake for basic dyes [108] as well as for disperse dyes [109]. For basic dyes, an increase in dye uptake may be attributed to the increased interaction between the dye molecule and the carboxylic group present in the grafted fiber. Accordingly, dye uptake is found to increase with an increase in the level of grafting, and a sharp increase in dye uptake has been observed up to the 20% graft level. In case of dyeing with disperse dyes, dye uptake has been found to increase sharply within a 1- to 2-h period. Dye uptake is increased linearly with increasing percentage of graft up to about 20%; beyond that the increase is rapid. This might have happened due to alteration of the ratio of phases between the crystalline and amorphous content of the fiber beyond a certain grafting level. Isotactic polypropylene fiber as such is highly crystalline, but the crystallinity decreases when MAA is grafted onto it. The diameter of the fiber also increases with the increase in grafting level. Because of the presence of carboxylic acid functionality in the fiber due to grafting, the moisture absorption property has also improved. In general, marked improvements in thermal properties have been observed for grafted polypropylene fibers [103, 110–112].

As discussed earlier, grafting takes place mostly in the amorphous region of the fiber. The treatment, which causes further loosening of the amorphous region, helps in the grafting process. It has been observed that pretreatment of polypropylene fiber with a solvent such as toluene or tetrachloroethylene enhances the grafting process considerably [101]. Graft add-on has increased to nearly 20% in pretreated polypropylene from a value of 2.3% in the case of untreated fiber.

Polypropylene fiber has been considered as a suitable base material for use as sutures for the stitching of surgical wounds in the area of medical technology. The stitched area of the injury used to become the area of possible infection during the post-operation period. It is highly desirable that such sutures are antimicrobial. Polypropylene filaments have been made functionalized by graft polymerization so that the functional groups could be subsequently used for immobilization of the appropriate drug. An attempt has been made to graft polyacrylonitrile onto polypropylene and partially hydrolyze it to create carboxylic group [113]. γ-ray irradiation of polypropylene filament has been adopted for creation of an active site on polypropylene. However, such modification is found to impart an adverse effect on mechanical properties of polypropylene: The elongation and knot strength of the modified suture deteriorate significantly. An attempt has also been made to graft acrylic acid directly onto polypropylene fiber using the plasma-induced technique [114]. Polypropylene filament is treated in oxygen plasma first in order to create hydroperoxide groups on it as the active sites. Inhibitors like ferrous sulfate or methanol are required to be added in the monomer to inhibit homopolymer formation.

Similar attempts to make polypropylene functional have been made by graft polymerization of 1-vinylimidazole adopting preirradiation as the means for the creation of a radical site on polypropylene that is required for initiation of graft polymerization [115, 116]. In order to control homopolymerization, additives like methanol and acetone have been added [116]. It has been observed that the tenacity of grafted fiber decreases slightly, but the elongation increases with an increase of the percentage of graft [117]. This grafted suture has been found to be antibacterial with the capability of controlled release of a drug when immobilized with an antibiotic like ciprofloxin.

Polypropylene nonwoven fabrics have also been activated by chemical means to make them suitable for grafting. Polypropylene nonwoven of 77.7 g/sq·m, with an average filament diameter of 2.5 μm, has been made by the melt-blown method and is treated with 20 gm/L ammonium persulfate solution at 80 °C for 30 min. This chemically treated polypropylene nonwoven is first rinsed with water and then treated with 0.3 M acrylic acid solution at 80 °C in order to obtain polyacrylic acid grafted polypropylene, which can act as a base for the deposition of polyaniline hydrochloride layer [118]. Such layered textiles become electrically conductive. It has been observed that acrylic acid grafted polypropylene nonwoven fabric retains 85% of its original strength when active sites are generated by chemical means for the grafting, but when γ rays are used instead of chemical means for generation of active sites, it loses its strength to the extent of 50% or more.

APPLICATION IN POLYESTER: Graft Polymerization

Polyethylene terephthalate (PET), commonly known as polyester, was first produced in 1941 and is now the most important single synthetic fiber, not only for apparel but also for other technical applications. PET fibers are highly crystalline, hydrophobic in nature, and hence suffer from very low moisture regain and static buildup. One obvious way to improve these undesirable properties is to graft vinyl monomers onto the PET fibers. PET is a hard substrate and does not contain any chemically reactive functional groups. Hence, low graft yields have been obtained in comparison to other fibers such as cotton or silk. One approach to overcome this limitation is preswelling the fiber in a suitable solvent. This technique is used successfully in the dyeing of polyester to facilitate the penetration of dye molecule into the fiber. One example of this solvent assisted graft polymerization of PET is grafting with n-vinyl imidazole in dimethylformamide using azo-bis-isobutyronitrile as the initiator [119].

Plasma-induced graft polymerization of acrylic acid, a hydrophilic monomer on polyester fabric surface has been found to change the fabric properties significantly. In order to create active sites onto the polyester surface, the fabric is first treated with argon plasma in a glow discharge apparatus to generate active sites. Then the fabric is immediately immersed in degassed acrylic acid solution for polymerization [120]. Significant increases in wetability and dyeability of the otherwise hydrophobic fabric have been observed. The results, when compared with the values obtained by in situ polymerization of polyester in acrylic acid vapor phase plasma, are found to be better for the dyeability and soil repellency. Higher dyeability in the case of graft polymerization may be due to the presence of more carboxylic acid functionality, which may not be the case for in situ polymerization, probably because of simultaneous crosslinking reactions involving carboxylic groups.

Polyester textile substrates are also modified by grafting with acrylic acid [121–123], and combination of n-vinyl-2-pyrolidone (NVP) and MMA [124] in order to create a base with carboxylic acid functionality on polyester substrates to further treat them for use in different biomedical applications. γ-ray-induced acrylic acid graft polymerized polyester fiber is further grafted with chitosan, a naturally based antimicrobial biopolymer containing the amine group in its backbone, via esterification. Afterward, this modified polyester is immobilized with condroitin sulfate [121] in one case and collagen [122] in another to exhibit both antibacterial activity as well as improved cell proliferation for fibroplast activity. Similarly, a polyester textile surface has been kept exposed to oxygen plasma first to produce peroxides on the surface for graft polymerization of acrylic acid [123]. This modified surface is then treated with chitosan and quarternized chitosan to impart antibacterial properties. These treated fabrics are durable to laundering. Radiation-induced graft copolymerized polyester knitted fabric has shown to attract collagen to provide bioreceptive surfaces [124].

APPLICATION IN ACRYLIC: Graft Polymerization

"Acrylic" refers to synthetic polymeric fibers that contain at least 85% acrylonitrile comonomers. Other common neutral monomers are methyl acrylate and vinyl acetate, apart from monomers for specific properties [125]. The fabrics and carpets made from acrylic fibers are easy to care for and esthetically pleasing.

Modifications of acrylic fibers through grafting have been carried out with an aim to improve their properties such as water absorption, fire retardancy, antistatic features, and dyeability. A method for the treatment of acrylic fabric for improved water absorption and antistatic properties is described in the patent literature of Kisaichi et al. [126]. According to the literature, a polymerization initiator is first incorporated into the fiber matrix by treating the fabric in a 0.1% aqueous solution of ceric ammonium sulfate at 90 °C. Polymerization takes place when the washed and dried fabric is immersed in a 1% aqueous solution of 1,3,5-triacryloylhexahydro-1,3,5-triazine at a pH of 3 and at 95 °C. Such treatment is claimed to be durable to washing. Similarly, the grafting of a soyabean protein on acrylic fiber has also been found to increase moisture absorption properties without affecting its mechanical properties [127]. Radiation-induced graft polymerization of vinyl pyrrolidone onto acrylic fiber has been carried out to improve the dyeability of the fiber [128]. Plasma-induced graft polymerization of a phosphorus-containing monomer has been found to impart wash-durable fire-resistance properties to acrylic fabric [129]. Monomers used for this purpose are diethyl(acryloyloxyethyl) phosphate, di-ethyl(acryloyloxymethyl) phosphonate, diethyl-2-(methacryloyloxyethyl) phosphate, and dimethyl(acryloyloxymethyl) phosphonate.

APPLICATION IN OTHER TEXTILES: Graft Polymerization

Graft polymerization of MAA has been tried on viscose [130]. A potassium permanganate–sodium chlorite redox system was used for the initiation of graft polymerization. Viscose is regenerated cellulose and contains a residual thiocarbonate group. This group helps to deposit MnO_2 onto the fiber, thus facilitating the graft polymerization. Loom state viscose fabric is directly used for the treatment. The fabric is first impregnated with an aqueous solution of potassium permanganate to allow the deposition of MnO_2 on viscose fiber, which is then immersed in an aqueous solution of MAA and sodium chlorite to graft polymethacrylic acid onto viscose. A grating of 40% to 60% has been achieved by this process.

Jute, a lignocellulosic bast fiber, has also been modified by graft polymerization of polyacrylonitrile (PAN) using a Cu^{2+}–IO_4^- combination as the initiator [131]. It has been observed that a 20%–30% PAN grafting imparts a desirable balance of properties such as strength, moisture regain, whiteness index, and light fastness rating. Most importantly, this modification has made jute fiber rot resistant.

Polycondensation of nylon onto glass fiber has been carried out as a surface treatment of the fiber for composite fabrication [132]. The first step of the two-step polymerization is the esterification of the pendant –OH group of the glass fiber surface by adipoyl chloride, which then undergoes polycondensation with diamino-1, 6-hexane, and acid chloride in order to obtain nylon66 grafted glass fiber. Similarly, ultrahigh-molecular-weight polyethylene (UHMWPE) has been grafted with MMA to enhance its reinforcing effect in making acrylic bone cement [133]. The fibers are treated first with argon plasma for 5 min followed by UV irradiation in MMA-chloroform solution for 5 h to obtain the grafted fiber. Comparative study of the mechanical properties of surgical bone cement made by using UHMWPE and grafted UHMWPE as reinforcing agent have shown marked improvement in tensile strength, tensile modulus, compressive strength, bending strength, and bending stiffness in case of the composite made with the grafted fiber. This improved reinforcing effect may be attributed to the improvement in interfacial bonding between the grafted fiber and the acrylic bone cement matrix.

FUTURE BREAKTHROUGHS

The inherent self-smoothing characteristics in wool and synthetic fibers created the need for wrinkle-resistant cotton textiles. Despite the development of wrinkle-resistant cotton, it is not without certain shortcomings, the major one being the loss of mechanical properties. Research work is continuing now using novel concepts such as "ionic crosslinking," which strive for high levels of wrinkle resistance without strength loss. Plasma-induced graft polymerization appears to be a sound technique for fabric modification.

The term "textile" is now being used in a much broader perspective and "technical textiles" cover all aspects of fiber use other than for apparel. Technical textiles have opened up the possibility of using newer fibers on one hand, and the modification of existing fibers to meet the demanding requirements on the other. Graft polymerization as the modification technique is being tested, for example, in the modification of commonly used fibers such as polyester, nylon66, and polypropylene monofilaments for effective application as sutures. As the effective reinforcing component in composites, this same technique has been tried for the modification of fibers like UHMWPE and glass fibers for new and high-end applications.

CONCLUDING REMARKS

Grafting surely is a scientific breakthrough that involves making inert substrates lively by generating active reaction sites for the subsequent building of polymer onto polymeric substrates. However, its commercial utilization in textiles has remained limited, the "coating" being the major soft option that offers ease of application. However, coating covers up the entire substrate and imparts stiffness in light of both substrate-to-polymer adhesion and polymer-to-polymer cohesion. Moreover, the polymer is held on the substrate and may reduce air permeability.

The grafts, on the other hand, are part of the substrate and hence are relatively more durable. The substrate remains relatively more pliable. The more important feature of grafting is the retention of desirable substrate properties while at the same time fortifying the substrate through a well-designed graft. From a research point of view, grafting may permit great flexibility in terms of building two or more different homopolymeric grafts onto the same substrate, possibly by sequential grafting, which would result in a true multifunctional textile. However, the uniformity and reproducibility of the treatment, relatively elaborate application method, process control, and suppression of side reaction (homopolymerization) have remained major concerns. Research is underway to overcome some of these shortcomings. Further efforts are needed to take advantage of this molecular engineering tool for modification of textiles at the shop floor.

In the area of crosslinking, the newer demands and technological developments have been hand in glove. The relatively recent demands have been low/zero formaldehyde in crosslinking agents as well as in the finished textiles, minimum/no loss of mechanical properties, a smooth appearance rating of 3.5+, and so on (for cotton textiles). These requirements have largely been met, but further refinements and new developments are in progress.

ACKNOWLEDGEMENTS

Dr. J. J. Shroff, Textile Consultant
Dr. A. K. Sharma, Director, ATIRA

REFERENCES

1. Leon S, Wakelyn PJ, in *Handbook of Fiber Science and Technology Vol. IV. Fiber Chemistry*, M Lewin and EM Pearce, eds. (New York: Marcel Dekker Inc., 1985), p. 809.
2. Sakurada I, Okada T, Kaji K. *J Polym Sci Part C*, **37**: 1 (1972).
3. Hon N. *J Polym Sci Polym Chem Edn*, **14**: 2497 (1976).
4. Guthrie JT. *Polymer*, **16**: 134 (1975).
5. Goynes WR, Harris JA. *J Polym Sci Part C*, **37**: 277 (1972).
6. Arthur Jr. JC. *Adv Macromol Chem*, **2**: 1 (1970).
7. Dziedziela WM, Rosiak J. *J Polym Sci Polym Symp*, **47**: 273 (1974).
8. Reinhardt RM, Arthur Jr. JC. *J Appl Polym Sci*, **24**: 147 (1979).
9. Reine H, Hinojosa JC, Arthur Jr. JC. *J Appl Polym Sci*, **17**: 3337 (1973).
10. Harris JA, Arthur Jr. JC, Carra JH. *J Appl Polym Sci*, **22**: 905 (1978).
11. Hebeish A, Guthrie JT. *The Chemistry and Technology of Cellulosic Copolymers* (New York: Springer-Verlag, 1981).
12. El-Rafie MH, Khalil EM, Abdel-Hafiz SA, Hebeish A. *Acta Polym*, **36**: 688 (1985).
13. Hebeish A, El-Rafie MH, Zahran MK. *J Appl Polym Sci*, **50**: 2099 (1993).
14. Butler CE, Millington CA, Clements DWG, In *Historic Textile and Paper Materials II*, EF Hansen, WS Ginell, and SH Zeronian, eds. (Oxford: Oxford University Press, 1989).
15. Princi E, Vicini S, Pedemonte E, Arrighi V, McEwen IJ. *J Appl Polym Sci*, **103**: 90 (2007).
16. Sishoo R, ed. *Plasma Technologies for Textiles* (Cambridge: Woodhead Publishing Ltd., 2007).
17. Hegemann D. *Indian J Fiber Text Res*, **31**: 99 (2006).
18. Andreozzi L, Castelvetro V, Ciardelli G, Corsi L, Faetti M, Fatarella E, Zulli F. *J Colloid Interface Sci*, **289**: 455 (2005).
19. Benerito RR, Ward TL, Soignet DM, Hinojosa O. *Text Res J*, **51**: 224 (1981).

20. Wakida T, Tao X, Yuen C, Yeung K. *Text Res J*, **59**: 49 (1989).
21. Chen JJ. *J Appl Polym Sci*, **42**: 2035 (1991).
22. McCord MG, Hwang YJ, Qiu Y, Hughes LK, Bourham MA. *J Appl Polym Sci*, **88**: 2038 (2003).
23. Zubaidi H, Hirotsu T. *J Appl Polym Sci*, **61**: 1579 (1996).
24. Abidi N, Hequet E. *J Appl Polym Sci*, **93**: 145 (2004).
25. Abidi N, Hequet E. *J Appl Polym Sci*, **98**: 896 (2005).
26. Vigo TL. *Textile Processing and Properties* (Amsterdam: Elsevier Science B.V., 1994) p. 222.
27. Tesoro GC, in *Handbook of Fiber Science and Technology: Vol. II, Functional Finishes: Part A*, M Lewin and SB Sello, eds. (New York: Marcel Dekker, 1983), p. 1.
28. Petersen HA, in *Handbook of Fiber Science and Technology: Vol. II, Functional Finishes: Part A*, M Lewin and SB Sello, eds. (New York: Marcel Dekker, 1983), p. 47.
29. Petersen HA. *Rev Prog Color Rel Topics*, **17**: 7 (1987).
30. Labana SS, in *Encylopedia of Polymer Science and Engineering, 2nd ed., Vol 4*, HF Mark, NM Bikales, CG Overberger, and G Manges, eds. (New York: John Wiley & Sons, 1986), p. 386.
31. Petersen HA, Pai PS. *Text Res J*, **51**: 282 (1981).
32. Sello SB. *J Soc Dyer Color*, **101**: 99 (1985).
33. BASF Product Literature, http://www.basf.com/pc_textiles/pdfs/FixapretCL.pdf. Accessed October 29, 2007.
34. Welch CM. *Rev Prog Color Rel Topics*, **22**: 32 (1992).
35. Xu W, Li Y. *Text Res J*, **70**: 957 (2000).
36. National Textile Centre Annual Report, November 2006, http://www.ntcresearch.org/projectapp/index.cfm?project=C04-NS01. Accessed October 27, 2007.
37. Giri G, Sahoo PK, Samal RK. *J Appl Polym Sci*, **40**: 471 (1990).
38. Samal RK, Giri G. *J Macromol Sci—Rev Macromol Chem Phys C*, **32**: 55 (1992).
39. Sood DS, Singha AS, Mishra BN. *J Macromol Sci Part A*, **20**: 237 (1983).
40. Sharma RK, Mishra BN. *J Macromol Sci Part A*, **20**: 225 (1983).
41. Mishra BN, Singha AS, Sharma RK. *J Appl Polym Sci*, **27**: 1321 (1982).
42. Mishra BN, Chandel PS. *J Polym Sci Polym Chem Edn*, **15**: 1545 (1977).
43. Sood DS, Rawat BR, Mishra BN. *J Appl Polym Sci*, **30**: 135 (1985).
44. Samal S, Sahu G, Nayak PL. *J Macromol Sci Part A*, **20**: 153 (1983).
45. Mishra BN, Kaur I, Sharma RK. *Polym Bull*, **4**: 635 (1981).
46. Abdel-Fatheh SH, Shalaby SE, Allam EA, Hebeish A. *J Appl Polym Sci*, **21**: 3355 (1977).
47. Varma DS, Sudhir RK. *J Appl Polym Sci*, **23**: 393 (1979).
48. Australian Institute of Nuclear Science and Engineering, Progress Report: Project No. 98/122 (1998), http://www.ansto.gov.au/ainse/ainse1.html. Accessed November 11, 2007.
49. El Naggar AM, Marie MM, El Gendy EH, El Miligy AA. *Radiat Phys Chem*, **47**: 623 (1996).

50. Wang H, Liu R, Tu T, Xie L, Sheng K, Chen Y, Tang X. *J Radiat Res*, **45**: 77 (2004).
51. Garnett JL, Kenyon RS. *Polymer Lett Edn*, **15**: 421 (1977).
52. Fearnley C, Speakman JB. *Nature*, **166**: 743 (1950).
53. Ito H, Muraoka Y, Umehara R, Shibata Y, Miyamoto T. *Text Res J*, **64**: 440 (1994).
54. Muraoka Y, Ito H, Umehara R, Shibata Y, Miyamoto T. *Text Res J*, **64**: 514 (1994).
55. Umehara R, Shibata Y, Ito H, Sakamoto M, Miyamoto T. *Text Res J*, **61**: 89 (1991).
56. Kang TJ, Moon SJ. *Text Res J*, **68**: 719 (1998).
57. Kuzuhara A, Hori T. *Text Res J*, **72**: 526 (2002).
58. Shuku K, Kashihara H., Anti-Pilling Treating Method for Protein Fiber Material, U.S. Patent 7108724, Filed December 31, 2003.
59. Cortez J, Bonner PLR, Griffin M. *J Biotechnol* **116**: 379 (2005).
60. Charles DL, in *Encylopedia of Polymer Science and Engineering, 2nd ed., Vol 15*, HF Mark, NM Bikales, CG Overberger, and G Manges, eds. (New York: John Wiley & Sons, 1989), p. 309.
61. Tsukada M, Freddi G, Monti P, Bertoluzza A, Shiozaki H. *J Appl Polym Sci*, **49**: 1835 (1993).
62. Lenka S. *J Macromol Sci Rev Macromol Chem Phys C*, **22**: 303 (1982).
63. Tsukada M, Freddi G, Monti P, Bertoluzza A. *J Appl Polym Sci*, **49**: 1565 (1993).
64. Tsukada M, Shiozaki H, Crighton JS. *J Appl Polym Sci*, **48**: 1409 (1993).
65. Tsukada M. *J Appl Polym Sci*, **35**: 965 (1988).
66. Tsukada M. *J Appl Polym Sci*, **35**: 2133 (1988).
67. Kobayashi S, Sugiyama M, Yoshida H. *Rep Tokyo Met Text Res Inst*, **15**: 137 (1979).
68. Tsukada M, Freddi G, Matsumura M, Shiozaki H, Kasai N. *J Appl Polym Sci*, **44**: 799 (1992).
69. Ojah R, Dolui SK. *J Appl Polym Sci*, **105**: 2164 (2007).
70. Robson RM, in *Handbook of Fiber Science and Technology Vol. IV. Fiber Chemistry*, M Lewin and EM Pearce, eds. (New York: Marcel Dekker, 1985), p. 646.
71. Lee WK, Sin KM. *Test Asia*, December 1991: 86.
72. Tsukada M, Shiozaki H. *J Appl Polym Sci*, **37**: 2637 (1991).
73. Yang Y, Li S. *J Text Inst*, **84**: 638 (1993); *Text Chem Color*, **26**: 25 (1994).
74. Leksophee T, Supansomboon S, Sombatsompop N. *J Appl Polym Sci*, **91**: 1000 (2003).
75. Tanaka Y, Shiozaki H. *Makromol Chem*, **129**: 12 (1969).
76. Tsukada M, Nagura M, Ishikawa H, Shiozaki H. *J Appl Polym Sci*, **43**: 643 (1991).
77. Shiozaki H, Tsukada M, Gotoh Y, Kasai N, Freddi G. *J Appl Polym Sci*, **52**: 1037 (1994).

78. Xiaojun P, Jitao W, Jie S. *J Soc Dyers Color*, **109**: 159 (1993).
79. Tsukada M, Shiozaki H, Goto Y, Freddi G. *J Appl Polym Sci*, **50**: 1841 (1993).
80. Freddi G, Shiozaki H, Allara G, Goto Y, Yasui H, Tsukada M. *J Soc Dyers Color*, **112**: 88 (1996).
81. Hu C, Sun K. *J Soc Dyers Color*, **114**: 359 (1998).
82. Hu C, Sun K. *J Soc Dyers Color*, **116**: 204 (2000).
83. Cai Z, Qiu Y. *Text Res J*, **73**: 42 (2003).
84. Hu C, Jin Y, Sun K. *Text Res J*, **72**: 1009 (2002).
85. Ball P, Meyer U, Zollinger H. *Text Res J*, **56**: 447 (1986).
86. Lofquist RA, Saunders PR, Tam TY, Twilley IC. *Text Res J*, **55**: 325 (1985).
87. Dyer ME. Treatment of Polyamides to Improve Antistatic, hygroscopic and Dye Receptive, Patent GB 2073217A, 14 October 1981.
88. Sarmany J. *Magy Textiltech*, **33**: 113 (1980).
89. Bogoeva-Gaceva G, Pimonenko NY, Petrov G. *Text Res J*, **63**: 51 (1993).
90. Mukherjee AK, Sachdev HS, Gupta A. *J Appl Polym Sci*, **28**: 2125 (1983).
91. Mukherjee AK, Sachdev HS, Gupta A. *J Appl Polym Sci*, **28**: 2217 (1983).
92. Varma DS, Ravisankar S. *Angew Makromol Chem*, **28**: 191 (1973).
93. Varma DS, Ray ND. *Angew Makromol Chem*, **32**: 81 (1973).
94. Smirnova NV, Gabrielyan GA, Galbraikh LS. *J Appl Polym Sci*, **47**: 833 (1993).
95. Shalaby SE, El-Shahed MF, Hebeish A. *Acta Polym*, **35**: 321 (1984).
96. Shalaby SE, Al-Balakocy NG, Abo El-Ola SM. *J Appl Polym Sci*, **99**: 613 (2006).
97. Saihi D, El-Achari A, Vroman I, Caze C. *J Text Apparel Technol Management*, **4**: 1 (2004).
98. El-Salmawi K, El-Hosamy MB, El-Naggar AM, El-Gendy E. *Amer Dyestuff Repor*, **82**: 47 (1993).
99. Tobiesen FA, Michielsen S. *J Polym Sci Part A: Polym Chem Edn*, **40**: 719 (2002).
100. Michielsen S. INJ Fall, 41 (2003), http://www.inda.org/subscrip/inj03_3/p-41-44-michielsen.pdf. Accessed November 11, 2007.
101. Bhattacharya SD, Inamdar MS. *J Appl Polym Sci*, **103**: 1152 (2007).
102. Anjum N, Gupta B, Requet AM. *J Appl Polym Sci*, **101**: 772 (2006).
103. Sathe SN, Rao GSS, Devi S. *J Appl Polym Sci*, **53**: 239 (1994).
104. Mukherjee AK, Gupta BD. *J Appl Polym Sci*, **30**: 2643 (1985); *J Appl Polym Sci*, **30**: 2643 (1985), p. 2655.
105. Jain R, Gupta B, Anjum N, Revagade N, Singh H. *J Appl Polym Sci*, **93**: 1224 (2004).
106. Gawish SM, Gavet L, Chabert B. *American Dyestuff Reporter*, **78**: 54 (1989).
107. Shukla SR, Athalye AR. *J Appl Polym Sci*, **51**: 1567 (1994).
108. Mukherjee AK, Gupta BD. *J Appl Polym Sci*, **30**: 4455 (1985).
109. Mukherjee AK, Gupta BD. *J Soc Dyers Color*, **104**: 130 (1988).
110. Sunderdi F. *J Appl Polym Sci*, **22**: 3163 (1978).
111. Mukherjee AK, Gupta BD. *J Appl Polym Sci*, **30**: 3479 (1985).

112. Shukla SR, Athalye AR. *J Appl Polym Sci*, **49**: 2019 (1993).
113. Gupta B, Jain R, Anjum N, Singh H. *J Appl Polym Sci*, **94**: 2509 (2004).
114. Gupta B, Saxena S, Ray A. *J Appl Polym Sci*, **107**: 324 (2007).
115. Kubota H. *Eur Polym J*, **28**: 267 (1992).
116. Anjum N, Gulrez SKH, Singh H, Gupta B. *J Appl Polym Sci*, **101**: 3895 (2006).
117. Gupta B, Anjum N, Gulrez SKH, Singh H. *J Appl Polym Sci*, **103**: 3534 (2007).
118. Polowinski S. *J Appl Polym Sci*, **103**: 1700 (2007).
119. Halil UI, Can I, Oya SA. *Turk J Chem*, **27**: 403 (2003).
120. Öktem T, Ayhan H, Seventekin EP. *J Soc Dyers Color*, **115**: 274 (1999).
121. Jou Chi-H, Lee Jui-S, Chou Wen-L, Yu Da-G, Yang Ming-C. *Polym Adv Technol*, **16**: 821 (2005).
122. Jou Chi-H, Lin Shang-M, Yun L, Hwang Mou-C, Yu Da-G, Chou Wen-L, Lee Jui-S, Yang Ming-C. *Polym Adv Technol*, **18**: 235 (2007).
123. Huh MW, Kang Inn-K, Lee DH, Kim WS, Lee DH, Park LS, Min KE, Seo KH. *J Appl Polym Sci*, **81**: 2769 (2001).
124. Gupta B, Revagade N, Atthoff B, Hilborn J. *Polym Adv Technol*, **18**: 281 (2007).
125. Frushour BG, Knorr RS, in *Handbook of Fiber Science and Technology, Vol. IV. Fiber Chemistry*, M Lewin and EM Pearce, eds. (New York: Marcel Dekker Inc, 1985), p. 171.
126. Kisaichi A, Oguchi M, Takeda T, Aikawa A, Saito T. Method of Treating Synthetic Fibers or Synthetic Fiber Fabrics, U.S. Patent 4238193. Filed September 19, 1978.
127. Jia Z, Yang Y. *Polymer Bull*, **59**: 13 (2007).
128. Levine CA, Traylor TG. *J Appl Polym Sci*, **5**: 606 (1961).
129. Tsafack MJ, Levalois-Grutzmacher J. *Surf Coat Technol*, **200**: 3503 (2006).
130. Hafiz SAA, El-Rafie MH, Hassan SM, Hebeish A. *J Appl Polym Sci*, **55**: 997 (1995).
131. Ghosh P, Ganguly PK. *J Appl Polym Sci*, **52**: 77 (1994).
132. Haberme MB, Fena W, Abdellatif A, Bernard R, Brisson J. *Composite Interfaces*, **12**: 425 (2005).
133. Yang J, Haung P, Yang M, Lo SK. *J Biomed Mat Res*, **38**: 361 (1997).

8

IN AUTOMOBILES

JAMES W. RAWLINS AND JEREMY SWANSON

INTRODUCTION

Automobiles have undergone dramatic transformations in appearance, function, and comfort since their introduction. Globally, impetus for changes such as economic fluctuations, environmental concerns, and consumer demand are forcing automakers toward developing lighter, less expensive, and better-performing automobiles. In this quest, automobile producers are shifting toward higher performance with lighter and more cost-effective materials, particularly polymers. Many suggest that an entirely plastic automobile can now be realized and will not be too far off in the future. The demand for and use of polymers are growing rapidly throughout the automotive industry, often in the form of new composite developments. Moreover, plastics allow construction of complex integrated components as single-piece assemblies, thereby reducing manufacturing cost and weight from joining methods, which aids flexibility in design and enhanced functionality. Reduced material weight translates directly to fuel economy and emission improvements.

Automotive polymer applications have increased more than 600% by volume per vehicle since 1970, and current vehicles contain an astounding 50% by volume of polymers [1–4]. Polymer utilization was initiated in automotive interiors and grew outward, encompassing interior, exterior, and electrical components, primarily due to significant advantages in cost, process and production time, ease of installation, design freedom, and weight reduction. As materials' strength–to-weight ratios and service lives predictability continue

Polymer Grafting and Crosslinking, Edited by Amit Bhattacharya, James W. Rawlins and Paramita Ray

to improve, more metallic components will be replaced by efficient and customizable plastic options.

INTERIOR

Automobile interiors have benefited most from the incorporation of plastics. Heavy metal components originally used in vehicles required fasteners of various types that inevitably loosened with time, leading to vibrations and increased noise. The plastic systems that replaced these metal parts were not only lighter but also incorporated electrical, safety, or functional components into a single piece. Dramatic improvements in overall comfort have occurred due to reduction in vibration/noise and improvements in ergonomics, aesthetics, and durability. Among polymers, polyurethanes have had the greatest impact, in various roles such as seat cushioning, fabric and carpet backing, vibration and sound dampening, as well as energy-absorbing components throughout an automobile.

Auto Upholstery

Polyurethane foam is an important tool in the design and function of automobiles. Cushioning, sound dampening, and padding throughout the automobile are all served by the versatile nature of polyurethane foams. Polyurethanes comprise a variety of single-unit constructions within the automobile due to processing flexibility, adhesion characteristics, and overall performance. Controlled crosslinking is imperative for the ultimate polyurethane physical properties. Engineeering improvements in the understanding of blowing agents and, in combination with the microstructure, have produced urethane foams capable of addressing a multitude of automotive needs [5–8].

Polyurethane foams are typically synthesized via high temperature processes to ensure appropriate cell structure. Open cell polyurethane foams are the preferred material for automotive seat cushioning. Good seating is defined as giving support in all situations, reduced vibration, and maintenance of comfort in all temperature/humidity conditions [9]. Currently, polyurethane foams easily replace steel springs and early cellular foams due to their superior properties, all-in-one construction, and reduced cost. However, post-foam modifications require significant effort and often lead to performance reduction. Successful research and development to avoid post-foam modification has resulted in lower energy processes with enhanced foam performance.

Vegetable-oil-based foams are being investigated to enhance the use of natural and sustainable raw materials in polymers. Palm oil is a potential candidate for the replacement of petroleum-based polyurethane foam carpet backing [10]. Soybean oil is also attracting significant attention as a potential replacement for petroleum-based raw materials in foams. Incorporating these renewable resources into automotive applications will have a tremendous environmental impact as soon as the mechanical property challenges are overcome [10–12]. Unfortunately, some of the advantages of natural materials are offset by increased complexity: use of regional crops that might require increased pesticide use, reproducibility of raw materials, and property maintenance across the world.

Instrument Panels/Headliners

One of the most complex components of an automobile is the instrument panel. The all-metal technology used in earlier vehicles contributed significantly to the overall weight and required several-step fabrication processes and an enormous amount of post-fabrication work. Instrument panels have witnessed a significant amount of integration as the all-in-one approach facilitated by plastics rendered obsolete previous technologies that required the assembly of several individual components. Polyurethanes, acrylonitrile-butadiene-styrene (ABS), ABS/polycarbonate alloys, polycarbonates (PC), polypropylene (PP), modified polyphenylene ether (PPE), and SMA (styrene maleic anhydride) resins are commonly used for instrument panel applications. ABS and SMA graft copolymers are capable of addressing several structural and property needs due to the flexibility in raw material composition. These materials facilitate vibration reduction between plastic components in the automobile interior. All-in-one construction of polyurethane instrument panels enables the incorporation of several electrical components and computer chips without the use of multiple heavy metal fasteners. RIM and composite designs have resulted in improved integrated all-in-one headliner components with dramatically enhanced function while reducing weight and cost [13–15].

Air Ducts and Other Interior Components

Temperature control is an important part of all automobiles. ABS, SMA, and PP find applications in the control consoles as well as in ductwork. Plastics such as SMA and ABS graft copolymers are ideal for these applications as they easily achieve the complex shapes required for ducts along with lightweight durability [16, 17]. Ethylene-propylene-diene monomer (EPDM) rubbers find application in weather stripping and sealant applications due to their high thermal and chemical resistance, and good weathering characteristics.

Seat bases, flooring, and door trim applications, to name a few, utilize ABS, PP, PC, and polyurethanes to address the complex design needs while maintaining a low-cost approach. Environmental waste and sustainability concerns have prompted material development from recycled waste in interior automotive applications [17]. The emphasis on recycling has promoted the use of blended graft copolymers in place of crosslinked polymers.

AUTOMOTIVE ELECTRICAL SYSTEMS

Car electrical systems have also witnessed dramatic improvements via polymer utilization. Braking system regulation, fuel injection, oxygen sensors, navigation equipment, obstacle sensors, and high-performance audio systems are now made possible through a variety of polymers singularly and in combination.

Switches, Sockets, and Connectors

Electrical systems are required to perform consistently under extreme conditions and must survive lengthy exposure to high-temperature scenarios. Epoxy vinyl esters, unsaturated polyesters, vinyl acetates, irradiated polysulfones, and thermoplastic olefins (TPOs) have replaced their metal precursors in several applications. Performance improvements in properties such as heat resistance, flame retarding, and component lifetimes have dramatically improved the durability and reliability of electrical systems. Polyimides, for example, are used as interconnects, high-temperature adhesives, and fulfill damping requirements while maintaining excellent thermo-oxidative stability. Polyphenols are often used as insulators in electrical applications. These materials have facilitated innovations in sensors, all-in-one constructions, and dramatic reductions in production costs and times.

POWERTRAIN

Several plastics are now in routine use in engine systems. Although lower weight is the most obvious advantage of replacing conventional components, an important characteristic of polymeric parts is the ability for component integration while improving thermal properties, chemical resistance, and durability.

Engine Systems

Engine systems currently utilize a variety of high-temperature blends and graft copolymers to address the needs of rocker panels, intake manifolds, and heating/cooling hoses. Typical polymers for these applications include unsaturated polyesters, epoxy resins, irradiated polysulfones, and polyamides. The polymers are designed to maintain thermoplastic characteristics within the backbone of crosslinked networks. Properties such as toughness, ductility, heat resistance, moisture uptake, and strength have each been engineered through under-the-hood thermosetting applications. Additionally, these materials reduce weight and noise, resulting in a dramatic improvement in automobile acoustics and performance. EPDM rubbers are used for vibration reduction, seals, hoses, belts, and insulation. More innovative and aggressive applications employ crosslinked phenolics and nylons, completely replacing movable parts like pistons and resulting in dramatic weight savings with superior corrosion resistance.

Transmission

Phenolics, polyetherimides, nylons, and their composites find application in transmission components. Crosslinked phenolics, typically utilized as composites, are adequate replacements for metals in several stationary transmission parts that incorporate multiple components into a single unit. Phenolic composites are often used as bearings to adequately dissipate load energy while offering weight reduction over their metal counterparts.

EXTERIOR

The development of technologies that enhance safety has provided creative minds with near limitless possibilities for appearance that is both beautiful and functional. Replacement of metal components on the vehicle exterior provides dramatic weight and cost reduction while simultaneously reducing restrictions on vehicle design.

Bumpers and Fascia

As automobile safety evolved, so did the applications of polymers. In 1925, bumpers became standard components of all cars. Heavy metal beams have been replaced by complex plastic systems that provide significant protection against impact. The plastics used in these applications comprise a variety of TPOs, PC, polyesters, polyurethanes, and polyamides [18, 19]. Improvements in material stabilization via the use of hindered amine stabilizers have also increased their application potential dramatically [20].

Unsaturated polyesters and epoxy matrices in the form of composites now comprise a significant component of automotive fascia where weight reduction and performance are of more importance than cost [21–24]. These polymers now offer an additional advantage in the form of molded-in coloring [25, 26]. The ability to mold-in color provides a dramatic reduction in production and processing times and costs and also eliminates paint lines and the pollution associated with the painting process.

Car Lighting Systems

Car lighting systems are an area of constant innovation with plastics. Styling and functional improvements over traditional glass and metal systems have dramatically improved the overall function of vehicle lighting while simultaneously reducing cost and increasing design options. PC and acrylics such as polyacrylonitrile are the polymers of choice for these applications. Systems utilizing engineered hybrids of PC with TPOs and polyurethanes have resulted in improved safety and function. Major innovations in light-emitting diodes (LEDs) will further enhance the role of polymers in lighting systems while reducing cost and energy consumption.

Auto Trim

Exterior trim is a major component and styling feature of automobiles. Polymers such as EPDM rubber, nylons, polystyrene, PC, PVC, PP, polyesters, and polyurethanes are employed in mirror housings, side trim, wheel covers, and grilles. EPDM rubber is a major component of automotive weather stripping due to its versatile thermal characteristics and adequate weatherability. However, noise and vibration issues have created a desire to replace them in certain situations. ASA-AES (acrylonitrile-styrene-acrylate:acrylonitrile-ethylene-styrene) are weatherable graft copolymers that are used in auto trim applications because of their superior performance and mold-in color options. Moreover, the recycling potential of these materials surpasses those of most crosslinked systems.

As with the interior, polyurethanes find extensive applications outside the automobile as well. RIM polyurethane/metal hybrids have contributed to significant advances in exterior applications of plastics. These hybrid components are engineered to contain an aesthetically pleasing polyurethane shell with metal supports and polyurethane foam on the inside to provide energy, sound, vibration, and impact-absorbing benefits. Polyurethane and polyurea bedliner sprayable systems combine ease of application with great toughness, durability, and wear resistance soon after application. Exterior polyurethane applications are increasingly employing raw materials from renewable resources or recycled products [27, 28]. Further developments in recycling processes will enhance the use of these materials in automobiles and reduce their impact on the environment.

CHASSIS

Structural components offer significant potential for utilizing thermosetting polyesters. Both semistructural and fully structural applications are under investigation as heavy and costlier metal components continue to limit the efficiency of vehicles. However, current mass production processes for composite structural components are limited at best. Sheet, bulk, and injection molding provide some advances in medium volume applications, but the cost is still prohibitive for mass production of chassis-type structural materials.

CONCLUSION

Thermosetting and graft polymers have expanded their roles in automobiles by rapidly replacing costly fabricated metal components. The desire for environmentally responsible automobiles is driving the move toward the completely plastic car. Innovative approaches to the replacement of metal components coupled with the "green" revolution will further enhance the use of polymers. Composite materials provide ample opportunity for the construction of completely plastic automobiles in the near future. However, more advances are required in mass production processes for the development of a commercially viable, all-plastic automobile. The use of graft and crosslinked polymers will be instrumental in this pursuit as these chemistries facilitate the properties that will be tailored to multiple applications.

REFERENCES

1. American Chemistry Council, Plastics Division, Automotive, http://www.americanchemistry.com/s_plastics/sec_content.asp?CID=1080&DID=4234, accessed February 2008.
2. American Chemistry Council, Automotive Learning Center, "Automotive Markets: Vision and Technology Roadmap" www.plastics-car.org/s_plasticscar/index.asp, accessed February 2008.
3. Stewart R. *Plast Eng*, **63**: 9 (2007).
4. McCrum NG, Buckley CP, Bucknell CB. *Principles of Polymer Engineering, 2nd ed.* (New York: Oxford University Press, 1997), p. 7.
5. Kim SH, Lim H, Song JC, Kim BK. *J Macromol Sci, Part A*, **45(4)**: 323 (2008).
6. Sonnenschein M, Wendt BL, Schrock AK, Sonney JM, Ryan AJ. *Polymer*, **49(4)**: 934 (2008).
7. Sasaki T, Wada H, Horie A, Harada H, Murata Y, Takagi D. *Asahi Garasu Kenkyu Hokoku*, **53**: 67 (2003).
8. Park HH, Park JB. *J Cell Plast*, **39(4)**: 291 (2003).
9. Patten WN, Sha S, Mo C. *J Sound Vib*, **217(1)**: 145 (1998).
10. Shaari NZK, Lye OT, Ahmad S. *J Oil Palm Res*, **18**: 198 (2006).
11. Mielewski DF, Flanigan CM, Perry C, Zaluzec MJ, Killgoar PC. *Ind Biotech*, **1(1)**: 32 (2005).
12. Khalil H. *Can Chem News*, **58(9)**: 22 (2006).
13. Dolgopolsky I, Duley JA. *J Ind Text*, **30(1)**: 26 (2000).
14. Mkrtchyan L, Maier M. *Plast Rubber Compos*, **36(10)**: 445 (2007).
15. Lilley K, Mani A. *J Mater Eng Perform*, **7(4)**: 511 (1998).
16. Carter RO, McCallum JB. *Polym Degrad and Stab*, **45(1)**: 1 (1994).
17. Laverty JJ, Bullach RL, Ellis TS, McMinn TE. *Polym Recycl*, **2(3)**: 159 (1996).
18. Santoni A, Guo MC, Heuzey MC, Carreau PJ. *Int Polym Proc*, **22(2)**: 204 (2007).
19. Strebel JJ, Mirabella F, Blythe C, Pham T. *Polym Eng Sci*, **44(8)**: 1588 (2004).

20. Solera P, Capocci G. *Polym Polym Compos*, **7(8)**: 521 (1999).
21. Schubel P, Warrior N, Kendall K, Rudd C. *Composites Part A*, **37**: 1734 (2006).
22. Schubel P, Warrior N, Kendall K, Rudd C. *Composites Part A*, **37**: 1747 (2006).
23. Schubel P, Warrior N, Kendall K, Rudd C. *Composites Part A*, **37**: 1757 (2006).
24. Feraboli P, Masini A. *Composites Part B*, **35**: 323 (2004).
25. Modern Plastics, Feature Article, http://www.modplas.com/inc/mparticle.php?section=Feature%20Story&thefilename=feature01012008_01, accessed February 2008.
26. SABIC Innovative Plastics, www.geplastics.com, accessed Feb 2008.
27. Petrovic ZS, Guo A, Javni I, Zhang W, in *Natural Fibers Plastics and Composites—Recent Advances*, F Wallenberger and N Weston, eds. (Norwell, MA: Kluwer Academic Publishers, 2003), p. 167.
28. Zia KM, Bhatti HN, Ahmad BI. *React Funct Polym*, **67(8)**: 675 (2007).

9

IN CABLE TECHNOLOGY

ACHINTYA SEN

INTRODUCTION

Techniques to develop new classes of polymeric materials (insulating and sheathing) with improved specific properties through grafting and crosslinking of easily available commodity polymers like polyethylene (PE), ethylene propylene co- and ter-polymers and synthetic rubbers have gained worldwide acceptance and have opened new horizons in the field of material research. Judicious selection of these techniques can produce a tailor-made material possessing any attributed desirable properties by the simplest route and with a minimum involvement of manufacturing cost. The classical examples of the industrial applications of such techniques are power cables. Both grafting and crosslinking techniques are widely and successfully used in the manufacture of power cables.

A continuous supply of electricity is a prerequisite of today's life, and it can be interrupted by the failure of cables. Thus the demand for assured service performance is increasing. This has led to the use of the latest techniques of applying grafting and crosslinking for cables. It is essential to know and understand the different process variables, quality control, and characterization methods for useful applications of grafting and crosslinking in cables.

POWER CABLES

Power cables are the most essential practical media to convey electricity from generating stations to the final application/utilization/consumption points like

Polymer Grafting and Crosslinking, Edited by Amit Bhattacharya, James W. Rawlins and Paramita Ray

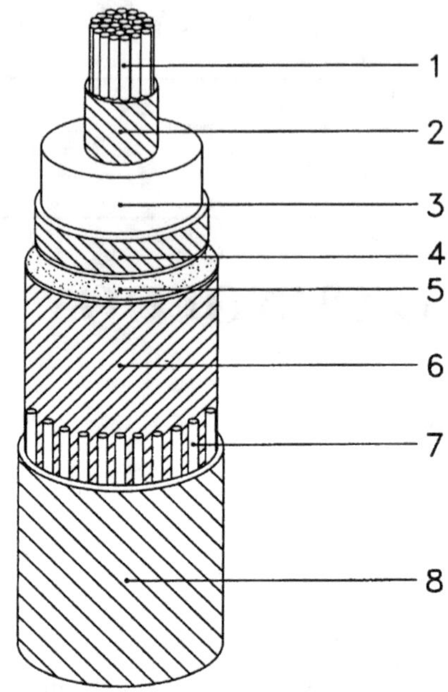

Figure 9.1. Typical construction of a peroxide crosslinked XLPE-insulated HV cable. (1) Stranded conductor. (2) Semiconducting conductor shield. (3) XLPE insulation. (4) Semiconducting insulation shield. (5) Shielding copper tape. (6) Inner sheath. (7) Wire armoring. (8) Outer sheath. Reproduced from *Rubber Products Manufacturing Technology*, AK Bhowmick, MM Hall, and HA Benarey, eds. (New York: Marcel Dekker Inc,) p. 682.

equipment, tools, and appliances of all types at factories, domestic premises, and other locations. The variety of uses and service requirements has resulted in a large number of designs, constructions, and above all the material systems, i.e., different types of cables. Typical construction of peroxide crosslinked polyethylene (XLPE) and silane XLPE insulated power cables are shown, respectively, in Figures 9.1 and 9.2.

TYPES OF CABLES

Generally, there are two types of cables: power cables and telecomunications cables. The use of grafting and crosslinking is limited to power cables. Power cables are normally classified on the basis of

- Voltage: low voltage (up to 3.3 kv), medium voltage (3.3–11 kv), high voltage (11–66 kv), and extra high voltage (above 66 kv) cables.
- Polymers: PVC (polyvinyl chloride) cable, XLPE (crosslinked polyethylene) cable, and elastomeric/rubber cable.

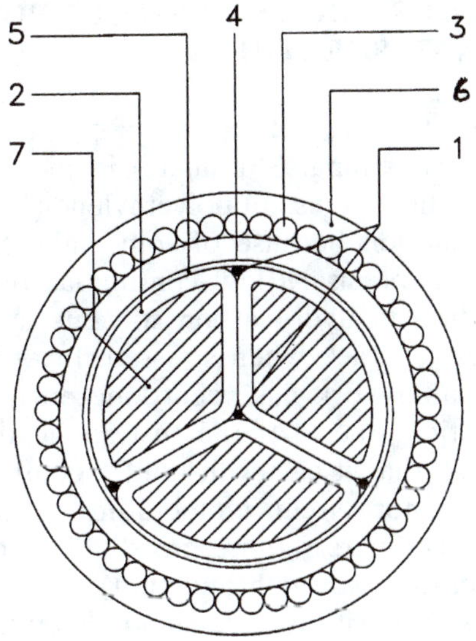

Figure 9.2. Typical construction of a silane crosslinked XLPE-insulated (three-core, stranded sector-shaped aluminum conductor) cable. (1) PVC filler (2) Silane cross-linked XLPE-insulation. (3) Galvanized round steel wire armoring. (4) Extruded PVC inner sheath. (5) Binder tape. (6) PVC outer sheath. (7) Stranded sector-shaped aluminum conductor. Reproduced from *Rubber Products Manufacturing Technology*, AK Bhowmick, MM Hall, and HA Benarey, eds. (New York: Marcel Dekker Inc,) p. 682.

- Applications and special properties: flame-retardant (FR) cable, fire-survival (FS) cable, radiation-resistant cable, submarine cable, etc.

POLYMERS USED IN CABLES INCLUDING GRAFTED POLYMERS

Mostly the polymers used in power cables are

- Crosslinkable polyethylene (XLPE), silane-grafted polyethylene (PE-g-Si), ethylene propylene rubber (EPR), and ethylene propylene diene (EPDM) rubber, silicone rubber. These are all used for insulation.
- Polyvinyl chloride (PVC), which is used for insulation as well as sheathing/jacketing.
- Ethylene copolymers such as ethylene vinyl acetate copolymer (EVA), ethylene ethyl acrylate (EEA), ethylene butyl acrylate (EBA), which are used for semiconducting layers (conductor and insulation screens/shields) and nonhalogenated flame-retardant sheaths.
- Elastomers like polychloroprene (PCP), chlorosulfonated polyethylene (CSM), acrylonitrile butadiene rubber (NBR), etc., which are used for sheathing or jacketing.

IMPORTANCE OF CROSSLINKING AND COMPOUND DESIGN FOR INSULATION AND SHEATHING

Polymers in their virgin state are not capable of meeting all the requirements of cable insulation and sheathing. Although it is one of the most important and versatile thermoplastics, the use of polyethylene (PE) has been restricted in its use as cable insulation because of some inherent defects such as a low-melting point [more precisely, low continuous operating temperature (70 °C)], rapid dissolution in hydrocarbon solvents at above-ambient temperature, and tendency to crack when stressed. In order to mitigate these disadvantages considerable work has been done on the crosslinking of PE [1]. Power cable insulation, one of the largest application areas of PE, also demands crosslinking so that it can withstand a continuous operating temperature of 90 °C and a short-circuit temperature of 250 °C. Similarly, EPR, EPDM, and silicone elastomers require crosslinking for their use as cable insulation. High-temperature strength and (hot) set properties of the cable insulation are not possible without crosslinking. In a similar way, special purpose elastomers such as CSM, NBR, or PCP require crosslinking or vulcanization to meet the properties of cable sheathing or jacketing. Crosslinking reactions actually bring about the formation of linkages between the long polymer chains, resulting in a network structure. The crosslinked polymers differ in many important aspects from that of uncrosslinked polymers. At elevated temperatures, crosslinked polymers generally behave like soft but elastic solids rather than viscous liquids [2]. Many authors refer to the improvement of stress cracking, solvent resistance, and weatherability [3], which are prerequisites for cable sheathings.

COMPOUNDING INGREDIENTS AND PROCESSES

In addition to the crosslinking agents, other ingredients such as antioxidants, and for sheathing, fillers, and process aids must be mixed to impart the desirable product properties after processing. Antioxidants are added to increase the service life by preventing polymer aging and degradation. Fillers are mainly used to improve the mechanical properties of the elastomers for sheathing and to improve the performance/cost ratio.

Elastomers compounding for both insulation and sheathing are done usually by internal mixers and two-roll mixing mills. Special care is taken for the mixing of crosslinking agents to avoid premature crosslinking during processing (compounding and extrusion). The temperature during mixing should not exceed the onset of decomposition temperature of the crosslinking agents. Mixing time and shear imparted during mixing should be optimum for achieving uniform mixing (distribution and dispersion) and avoidance of excessive heat generation.

CABLE CONSTRUCTION

Polymer-insulated power cable consists of low resistance metallic conductors (to carry current) covered with polymeric insulation to isolate the conductors from their surroundings and a polymeric jacket or sheath designed primarily as per the requirement of environmental conditions. Other major components include metal wire or belt armoring for mechanical protection, water blocking tape, and a metal sheath to prevent moisture ingress. In addition, high- and extra-high-voltage cables require a conductor and insulation shield to obtain a radial uniform field. The construction of a typical peroxide crosslinked XLPE-insulated HV cable is shown in Figure 9.1. Normally, the low- and medium-voltage cables (specifically XLPE insulated) are three cores with round or sector-shaped conductors, while the high- and extra-high-voltage cables are single core with a round-shaped conductor.

GRAFTING PROCESS TECHNOLOGY FOR CABLES

Unsaturated hydrolysable silanes with the general formula $R_nSiL_{(4n)}$, where R is a monovalent hydrocarbon radical containing olefinically unsaturated hydrocarbyl, alkoxyalkyl, acyloxy alkyl, and acyloxy alkyloxy alkyl, and each L is a hydrolysable organic radical, are grafted onto polyolefins for making it moisture crosslinkable [4]. Unsaturated organofunctional silanes are one of the most important classes of organofunctional monomers. These are grafted onto polyolefins to make them moisture crosslinkable. The technology of making a moisture curable polyolefin compound through silane grafting by bulk processing has been developed and used commercially since the early 1970s [5]. This method of making crosslinkable polyethylene through silane grafting has become increasingly important in recent years because of its various advantages such as easy processing, low cost and capital investment, and the favorable properties of the cables [6].

SILANE GRAFTING AND CROSSLINKING OF POLYETHYLENE

Chemistry

Vinyl trimethoxy silane (VTMS) is grafted onto polyethylene using dicumyl peroxide (DCP) as the initiator. The silane-grafted polyethylene in the presence of moisture and a Lewis acid catalyst undergoes hydrolysis, which converts the silane into silanol groups. These silanol groups then undergo condensation producing Si—O—Si linkages with the elimination of H_2O and produce crosslinked structure [7]. The mechanisms of silane grafting and crosslinking reaction through condensation reaction are shown in Schemes 9.1 and 9.2, respectively.

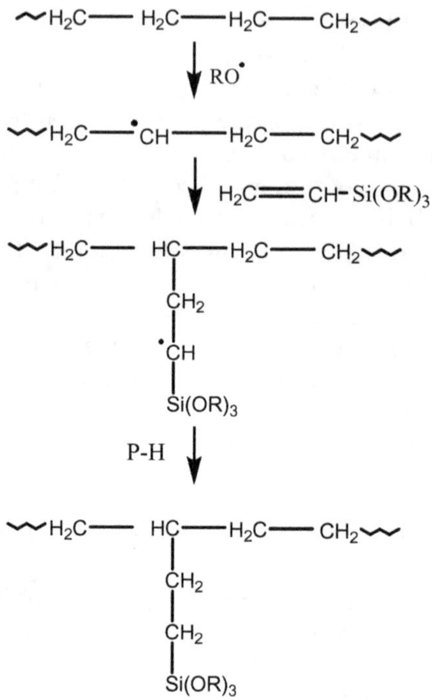

Scheme 9.1. Free-radical reactions producing silane-grafted PE. Reproduced from *J Appl Polym Sci*, **44**: 1153 (1992).

Technology

The VTMS-grafted polyethylene is produced usually by co-rotating twin screw extruder or co-kneader or a long continuous mixer, etc. A suitable grade of low-density polyethylene (preferably clean and dry) is fed and melted, and the mixture of DCP and VTMS is injected by a liquid-injection pump and processed at a temperature ranging from 160 °C to 190 °C. The processing conditions should be optimized as per the equipment used for grafting. Similarly, the concentration of DCP and VTMS should be optimized as per the requirements of the degree of grafting (which in turn will control the degree of crosslinking) and the grades of polyethylene. The liquid (solution of DCP in VTMS) injection rate is kept synchronized with polyethylene feed rate to maintain a uniform and consistent degree of grafting. The grafted polyethylene thus obtained should be dried and stored in sealed bags to avoid moisture ingress.

The catalyst masterbatch is produced and supplied separately. Tin compounds like dibutyl tin dilaurate is used as a catalyst. The catalyst type and concentration are again critical and should be selected judiciously to obtain the desired crosslink density with specific curing conditions but without risk

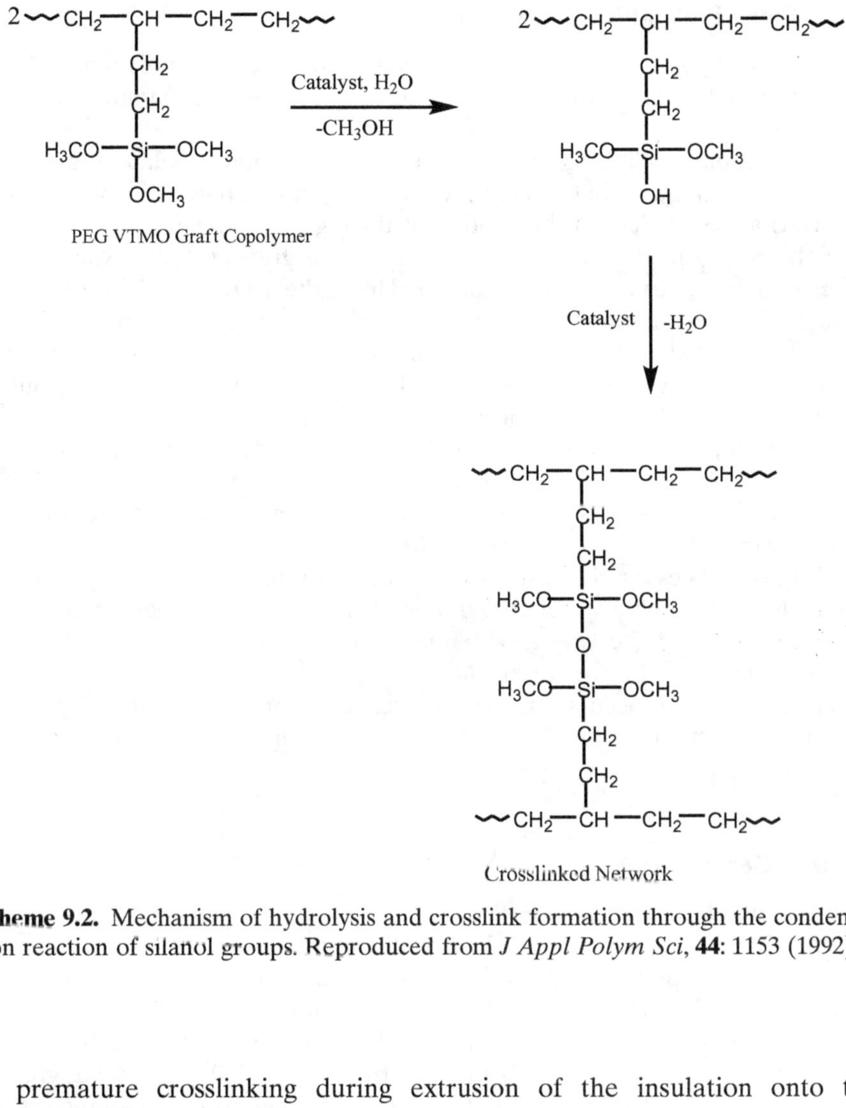

Scheme 9.2. Mechanism of hydrolysis and crosslink formation through the condensation reaction of silanol groups. Reproduced from *J Appl Polym Sci*, **44**: 1153 (1992).

of premature crosslinking during extrusion of the insulation onto the conductor [7].

CROSSLINKING TECHNOLOGY

Peroxide Cure

Peroxide mixed with respective polymers decomposes when exposed to high temperature (heat) and pressure, generating radicals that abstract the hydrogen radicals (H*) from the polymer chain, producing polymer macroradicals. The polymer macroradicals combine to form crosslinks [8].

Silane-Extrusion and Moisture-Cure Process

For silane crosslinking applications in cables, two methods are followed and are popularly known as the Sioplas technology process and Monosil technology process.

In the Sioplas technology process, the core compounds (silane-grafted PE plus a catalyst masterbatch along with color masterbatch if coloration is required) are extruded on the conductor through cross-head dye.

In the Monosil technology process, the silane grafting takes place during the cross-head extrusion on conductor. Here, the LDPE or LLDPE is fed through the main hopper, and a mixture of silane, peroxide, and catalyst is injected in the extruder. The temperature and residence time are maintained in such a way that grafting is completed before the insulation comes out of the extruder. This can be well accomplished by using a single screw extruder with a length:diameter (L:D) ratio of 25:1. A higher degree of grafting can be achieved and scorching can be avoided if the polyethylene (LDPE or LLDPE) is dried at the hopper by dehumidified dry air to a moisture content of less than 50 ppm and then fed to the extruder.

For both processes, the insulated conductor is immediately water cooled by passing it through a water trough and then wound on a reel. Crosslinking is done by placing the reel at a temperature of 60 °C–70 °C in a steam or high-humidity chamber or water bath.

As crosslinking occurs only in presence of moisture and a catalyst, this curing system is largely immune to thermally activated premature crosslinking.

Electron-Beam Curing

High-energy electron-beam (EB) curing is used to crosslink insulation based on polyethylene for wire and cable as well as for heat-shrinkable tubing used for cable joints. In radiation crosslinking process, heavy radiation (β rays) creates radicals by the abstraction of hydrogen, and carbon-carbon bonds form in a similar manner to the chemical crosslinking process [9, 10]. Because of very short wave lengths ($10^{-7\text{ to }-3}$) and high frequencies (10^{18-21}), EB has greater penetrating power than other radiation processes like infrared and microwave and can be used for crosslinking relatively thick insulation. Unlike microwave and ultrasonic heating, the EB process needs no curing agent like sulfur or peroxides for vulcanization/crosslinking. Although peroxide is not required for the EB crosslinking, the presence of this accelerator enhances the crosslinking process. Similarly, addition of a di- or trifunctional monomer to the curing system enhances gel formation because the monomer end-capped macroradicals have greater reactivity [11]. The depth of penetration possible with EB is also very important for cable insulation crosslinking. In general, the depth of penetration is directly proportional to the accelerator voltage. However, as the voltage increases, the potential for polymer degradation also increases. Hence, the possibility of using very high voltage accelerators to cure thick insulation (i.e., high voltage) is limited [11].

It has been reported that high-energy radiation produces crosslinks preferentially in the amorphous phase and in the fold surfaces [12]. Moreover, irradiation crosslinking produces an ~20% chain scission whereas chemical crosslinking with DCP produces a 5% chain scission [13].

Ultrasonic Curing

Another addition in the crosslinking process is through ultrasonic energy. Mori and Isshika used ultrasound in addition to conventional heat for polyethylene crosslinking [14]. Unlike microwaves, ultrasonic waves are not electromagnetic radiation. They are essentially elastic waves created by the cyclic vibratory movement of an ultrasonic transmitter. Two types of transmitters are frequently used: piezoelectric crystals and magnetostrictive devices [15].

CROSSLINKING PROCESSES USED IN INDUSTRY

The technology of crosslinking has been significantly improved in past two decades. The technology has passed through several stages during this period. A number of procedures and process variants have been tried to learn the limits and efficiency of the individual processes, enabling a comprehensive evaluation to select the most technically and commercially suitable one [16]. Any assessment of the different crosslinking processes must take into account the size and voltage spectrum of the cables to be manufactured, the design and construction of the cables, the technical requirements, and the overall economy of the plants and processes.

Batch Curing

Batch curing is normally done for elastomeric cables. The cable or core in its uncured state is spirally overlapped with rubberized cotton tape or polyester film for better consolidation, then coiled in trays or reeled on drums. Where maximum consolidation and improved surface finish are required, a metal sheath is applied. In a batch or discontinuous method, these drums or creels are placed in autoclaves in steam at a pressure corresponding to temperature of 130 °C–170 °C; curing cycles vary between 15 and 90 min.

Continuous Curing—HCV, CCV

A continuous vulcanization (CV) plant consists of a curing tube attached to the die face of the extruder head. There are three types of continuous vulcanization plants: horizontal continuous vulcanization (HCV), catenary continuous vulcanization (CCV), and vertical continuous vulcanization (VCV). The

HCV line is not advantageous and for core diameters above 15 mm; CCV and VCV are more appropriate.

In the peroxide-cure process, heat and pressure can be applied by circulating high-temperature, high-pressure, and generally, dry nitrogen through the steel curing tube. The nitrogen gas temperature is of the order of 300 °C–450 °C at a pressure of 10 kg/cm^2. The heat required to raise the temperature in the curing zone can be obtained from high-temperature nitrogen, or radiant heaters fixed on outside the tube. The elevated temperature causes the peroxide to react and form the crosslinked structure. The heat provided to the insulation causes the peroxide to decompose, generating volatiles such as acetophenone, methane, water vapor, and alpha-methyl styrene. The high pressure causes the gasses released in the crosslinking process to remain in the molten polymer so they do not form voids that might lead to partial discharge and deterioration of the cable insulation. This pressure needs to be maintained until the entire insulation (crosslinked) gets solidified sufficiently to exit the CV tube [17].

To determine temperature profile and line speed, the following factors are to be considered:

- The heating and cooling are accomplished by heat transfer through insulation.
- The line speed affects the temperature profile.
- The time/temperature profile of the insulation should be sufficient to crosslink the insulation.
- Curing is not always completed in the heating zone. Part of the curing occurs in the cooling zone.
- The time in the cooling zone should be sufficient to reduce both the surface and the conductor temperature. Generally, when cable exits the CV tube, the conductor remains hotter than the surface and insulation. Care should be taken to maintain the quality.

As noted previously, heat transfer occurs through the outer surface; therefore, to achieve the optimum throughput, the surface achieves very high temperature. An important consideration is that this temperature should not be so high as to cause problems for the performance of the cable. Thus, it is universal practice to set a maximum permitted surface temperature. This becomes more critical due to semiconducting layer. Well-designed materials are able to withstand temperatures as high as 275 °C. But the normal range is from 250 °C to 275 °C.

A typical CCV line is shown schematically in Figure 9.3.

The vulcanizing tube of a CCV line is generally set at an angle between 12 ° and 25 °. The conductor is fed from a payoff through an accumulator. This allows a new reel of conductor to be loaded onto the payoff. Before it runs out, it is joined with the old one by welding, thus maintaining the continuity of the process. The accumulator allows time for the two conductors to be

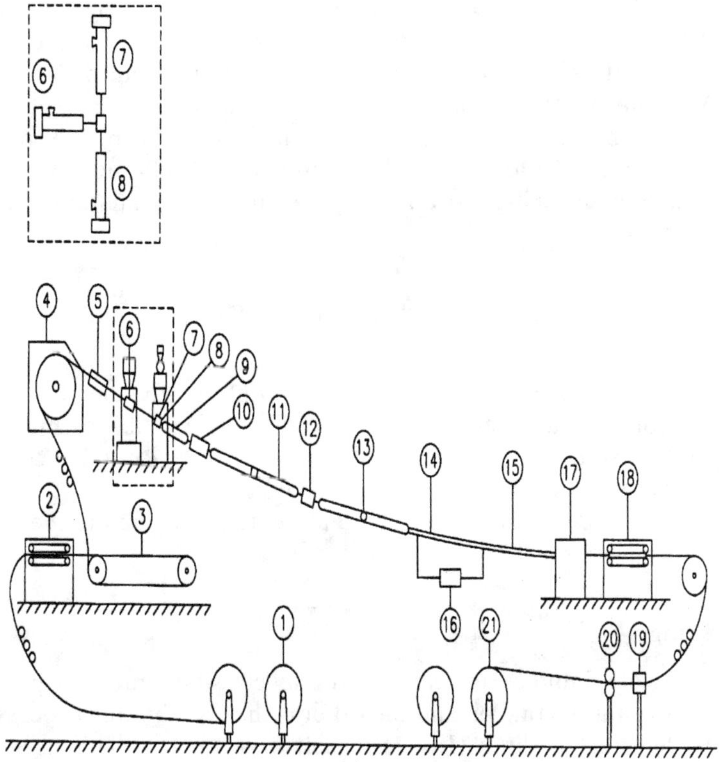

Figure 9.3. Layout of a CCV plant. (1) Payoff stand. (2) Helper capstan. (3) Payoff accumulator. (4) Payoff capstan. (5) Heater. (6) No. 1 extruder. (7) No. 2 extruder. (8) No. 3 extruder. (9) Splice box. (10) X-ray monitor. (11) Heating tube. (12) Catenary sensor. (13) Curing tube. (14) Precooling tube. (15) Cooling tube. (16) Water level controller. (17) Exit seal. (18) Take-up capstan. (19) Diameter gauge. (20) Length counter. (21) Take-up stand. Reproduced from Sen AK, in *Rubber Products Manufacturing Technology*, AK Bhowmick, MM Hall, and HA Benarey, eds. (New York: Marcel Dekker Inc,) p. 682.

welded. The tube consists of a heating zone and cooling zone. The cable is made to conform to the longitudinal axis (center) of the tube by maintaining a tension appropriate to its weight per unit length. This has been possible with sophisticated automatic control systems. The catenary sensor is set at a point before which the crosslinking of the surface is completed, and thus any possibility of any surface scratches or damage is eliminated. After passing through the heating zone, and upon completion of the crosslinking reactions, the cable enters the cooling tube, and after cooling, it is wound up. The total length of the tube, including the catenary part, the straight part, and the cooling part, is about 150–160 m. However, when the insulation wall thickness becomes 20 mm or the outer diameter reaches 80 mm, tension control in the catenary becomes critical, and cable extruded with insulation gets deformed by gravity before the crosslinking reactions are completed. Therefore, manufacture of XLPE cable with diameters exceeding 80 mm is usually carried out by VCV.

Since VCV has a vertical tube, the problem of deformation by gravity does not arise, even in the case of high-insulation thickness cables. Processing is thus easier than CCV. One of the disadvantages of VCV, however, is that a special building with a a high tower is required for the installation of the vertical equipment.

Mitsubishi Dainichi Continuous Vulcanization (MDCV) Long Land Die Process

In the MDCV process, the equipment is horizontal, and the die of the extruder is extended for several meters, forming a vulcanizing tube to maintain the pressure and prevent void formation. The uncrosslinked extruded core exiting the die enters the tube, and its surface is coated with a lubricant. The core is then slide and moved through the tube and heated to complete the crosslinking reactions.

Liquid Curing

Commercial-scale liquid curing can be broadly classified into two groups. One is the process using silicone oil and the other is the crosslinking process known as the liquid-curing medium (LCM) or salt bath process. This uses sodium, potassium, and a molten eutectic mixture as the heating medium.

In the Fuzikura continuous vulcanization (FZCV) process, the entire system remains filled with silicone oil when in operation; the oil is used for curing and cooling the cable core under pressure. Two circulatory circuits are used, the first via a heat exchanger to give the necessary heat to enable the crosslinking followed by a second circuit, which includes a chiller to cool the cable before it leaves the line.

Infrared Curing

Another important crosslinking method used for the cable manufacture is infrared curing. With this technique, the elastomers are continuously crosslinked by the infrared energy source after extrusion at ambient pressure and at relatively low temperature. Silicone rubber cables, especially, are made by applying this technique.

Electron-Beam Curing

Crosslinking of polyethylene for a variety of wire and cable insulations is one of the largest applications of electron-beam processing. After extruding the polyethylene insulation over the conductor through the cross-head of the extruder, it passes under the electron scanner at a speed of 150–9000 m/min with 10–20 mA beam currents or dose values ranging from 1 to 10 Mrad. The EB process has a few advantages like compact configuration, high speed, and noncritical extrusion conditions. However, electron-beam curing can be used and suitable for limited thicknesses only because high-energy electrons cannot penetrate beyond a certain thickness. EB curing has been successfully used to cure polyethylene cable up to 33 kV, ethylene propylene diene terpolymer for use in control cables, and to cure thermocouple wires.

Ultrasonic Crosslinking

In the early 1970s, Fujikura Electric developed an ultrasonic process to continuously crosslink polyethylene-insulated power cables [18]. They used a 400 kHz, 310 W barium titanate cylindrical ultrasonic transducer. They coupled the ultrasonic transducer to a conventional continuous vulcanization (CV) tube and used pressurized water to couple the ultrasonic power to polyethylene. This process not only provided uniform curing throughout the cable but also increased the manufacturing speed.

CONDUCTOR AND INSULATION SCREENS/SHIELDS

To maintain a uniformly divergent electric field and to smooth out the surface of irregularities of the conductor strands, a layer of semiconductive material is extruded over the conductor. Conductor screens are usually used on cables rated above 2 kV. The interface between the screens and the insulation must be smooth and free from voids to minimize water and electrical tree initiation and partial discharge. The polymers, semiconducting carbon black, and crosslinking characteristics are of paramount importance to meet the requirements of the conductor screen. Generally, EBA and EEA polymers, clean furnace black, antioxidants, and organic peroxides are used for making the conductor screen compounds. The outer layer of the cable core is the insulation screen, which is a layer of crosslinked semiconducting material that is extruded in tandem with the conductor screen and insulation. For medium- and high-voltage cables, the compound used for bonded insulation screens is often the same compound used for the conductor screen. When a strippable screen is required, the specially engineered compound should be used.

RECENT DEVELOPMENTS IN XLPE CABLES

Tree Retardant Crosslinked Polyethylene

Water Trees Dendritic treelike structures that grow within the XLPE insulation from the voids and/or defects in presence of both an electric field and moisture are called water trees. Vented water trees generally grow from either the conductor or insulation screen into the insulation. Water trees are micro-cavities that contain moisture within them. They tend to grow at localized field enhancement points such as voids or contaminants within the insulation material. Water trees reduce the insulation dielectric strength in the immediate area of the tree.

Electrical Trees Treelike growth that produces partial discharges within the insulation is called electrical trees. Partial discharges occur in voids and may eventually grow to a treelike permanent and carbon-coated channels causing the breakdown of the cables. Electrical trees initiate and propagate under dry conditions at much higher electrical stresses than those required for water trees [19].

Water Tree Retardant Crosslinkable Polyethylene (WTR-XLPE)

Water trees grow relatively slowly in XLPE cables over a period of months or years of service. As they grow, the electrical stress can increase to the point that an electrical tree is generated at the tip of water tree [20]. Once initiated, electrical trees grow rapidly until the insulation is weakened to the point that it can no longer withstand the applied voltage, and an electrical fault occurs at the water/electrical tree location leading to the failure of the cable. Many actions have been tried to reduce water tree growth. The following two approaches to insulation technology have been widely adopted to limit water tree growth. Both of these are modification of classic XLPE materials:

- Modification of the polymer structure, called "polymer" WTR-XLPE (sometimes called copolymer modified XLPE)
- Modification of the additive package, "additive" WTR-XLPE (sometimes called TR-XLPE)

These insulation compounds, combined with the use of super clean semiconducting screens (free from any type of ionic contaminants) and well-designed manufacturing processes can produce cables with a long assured service life. In both instances, the compounds maintain the excellent electrical properties of standard XLPE (high dielectric strength and very low dielectric losses). WTR-XLPE insulated cables were commercialized in the early 1980s and have now been performing reliably in service for over 20 years [21, 22].

CROSSLINKING OF ELASTOMERIC CABLES

Silicone Cables

Silicone is normally crosslinked by peroxide through free-radical generation and forms carbon-carbon linkage similar to the way PE is crosslinked. Silicone is used as insulation for special purpose cables such as radiation-resistant cables, fire survival cable, etc. Normally, saturated silicone elastomers and vinyl modified silicone elastomers are used. Both types of silicone elastomer compounds are heat-cured by one of the organic peroxides:

- Bis (2,4-dichlorobenzoyl) peroxide
- Dibenzoyl peroxide
- Dicumyl peroxide
- 2,5-dimethyl 2,5-bis (t-butyl peroxy hexane)
- Ditertiary butyl peroxide

The last three are usually considered to be vinyl specific. It indicates that these peroxides will result in a good cure only with vinyl that contains silicone elastomers. The first two will cure both vinyl and non vinyl containing silicone elastomers. Bis (2,4-dichlorobenzoyl) peroxide is normally used industrially and requires a curing temperature range of between 110 °C and 130 °C. Compounds containing this peroxide may be cured without external pressure.

As mentioned, the best way of curing silicone rubber cable is through the infrared-curing technique. Immediately after extrusion through the cross-head, the insulated core enters to the infrared chamber, and the silicone insulation gets crosslinked by the infrared energy source. In this technique, high pressure is not required, and crosslinking takes place at ambient pressure.

In addition, through peroxide, the silicone insulation can be crosslinked by hot air or the nitrogen vulcanization technique at a high-temperature range 315 °C–350 °C in a few seconds.

The chemical structure of silicone rubber imparts tolerance to radiation and also offers several advantages: long service life, improved circuit integrity when cables are exposed to fire, and no evolution of potentially corrosive halogen or sulfur containing gases as by-products of combustion. The basic requirements of cables for nuclear applications are given in the Standard IEEE 323—1974.

EPR/EPDM Cables: Peroxide Curing and Sulfur/Accelarator Curing

Normally EPR (saturated backbone) and EPDM (diene-modified) elastomers for high-voltage insulation are widely cured by using peroxides, mostly dicumyl peroxide and 2,5-dimethyl 2,5-bis (t-butyl peroxy hexane) under a nitrogen atmosphere in a CCV or VCV line similarly to XLPE cables. The crosslinking of EPR follows the same mechanism as followed by PE, but because of presence of a tertiary carbon atom (of propylene unit), chain scission may occur in the case of EPR, resulting less curing efficiency. The presence of unsaturation (diene) in EPDM increases the cure efficiency, and chain scission becomes almost negligible.

The special purpose elastomeric cables where EPDM is used as insulation is cured by sulfur accelerator systems. The most suitable ones are the combination of thiazole and sulfenamide type accelerators, and sometimes peroxide depending on the requirements with respect to insulation thickness, extrusion and curing temperature, process safety, curing rate, curing technique (media), and cure state to be achieved. The insulation cores are cured by the batch-curing process in an autoclave as mentioned previously. It is also possible to cure it in CV line under nitrogen pressure.

Chlorosulfonated Polyethlene (CSM) Sheath

A combination of sulfur and accelerator is a common and versatile method for the curing of a CSM sheath of the cables. The common system is sulfur:1 part plus tetramethyl thiuram disulfide (TMTD):2 parts. The high state of cure is achieved economically even in highly compounded CSM sheaths. Nickel dibutyl dithiocarbamate (NBC) is used to improve the heat resistance of vulcanizates in the presence of mixed metal oxides, but it reduces the scorch safety. A CSM can also be cured with peroxide, but it is necessary to add an acid acceptor. A coagent or cure promoter such as triallyl cyanurate is added to improve the cure efficiency.

N,N′-m-phenylene dimaleimide is a primary curing agent for CSM. Maleimide cures also require the presence of an acid acceptor and a butyraldehyde-amine condensation product as accelerator (HVA-2) [23]. In addition to acting as an effective secondary cure accelerator for CSM, which improves the state of cure, HVA-2 improves the electrical properties of the vulcanizates.

Nitrile Butadiene Rubber (NBR) Sheath

NBR (nitrile butadiene rubber) as a cable sheath is normally cured by sulfur accelerator system. Depending on the requirements of state and rate of cure, aging resistance, etc., a combination of accelerators such as thiuram disulfides, thiazole or sulfonamide, and sometimes a sulfur donor like DTDM (dithiodimorpholine) is used. The NBR sheath is normally cured by a batch-curing technique in an autoclave at a temperature ranging from 130 °C to 160 °C. NBR-sheathed cables are used where oil resistance is required and are widely used in oil refineries.

Polychloroprene (PCP) Rubber Sheath

For effectively regulating the scorch time, cure rate, cure state, and vulcanizate properties of PCP cable sheaths, a combination of sulfur, accelerator, and metal oxides is used. Sulfenamides, thiazoles, and thioureas are common accelerators. Thiourea accelerators [in particular, ethylene thiourea (ETU)] have been used for years. Handling precautions are required for using ETU. The combination of MgO and ZnO are used effectively as metal oxides. In addition to regulating the cure rate and state, these metal oxides serve as acceptors of acid (HCl), which is released in trace amounts during processing, curing, and service aging. PCP rubber sheath is also cured normally in an autoclave (batch curing) similar to NBR and CSM sheaths at a temperature ranging from 130 °C–150 °C.

Ethylene Vinyl Acetate (EVA) Cables

EVA (with high VA content) is especially used for nonhalogenated flame-retardant (FR) cables because of its capacity to hold high amounts aluminium trihydrate (ATH,$Al_2O_3{\cdot}3H_2O$) and magnesium hydroxide $Mg(OH)_2$ in the compound. EVA is cured by peroxides such as dicumyl peroxide and 2,5-dimethyl 2,5-bis (t-butyl peroxy hexane), and sometimes a coagent like triallyl cyanurate is added to achieve higher state of cure. The curing temperature is kept within a range of 160 °C–180 °C in batch curing. Above that temperature, H_2O may be eliminated from ATH. $Mg(OH)_2$ has better stability than ATH. The peroxide cure mechanism follows the same route as PE.

CHARACTERIZATION AND QUALITY CONTROL OF GRAFTED AND CROSSLINKABLE CABLE COMPOUNDS

Rheometry is the best way of checking compounds to see whether they will achieve the desired and optimum curing state with the optimum rate of cross-linking. The Monsanto rheometer is the most suitable one as it provides a continuous measurement of dynamic shear modulus where the compounds are cured, i.e., transformed from a thermoplastic state to a thermoset (crosslinked) state, in the rheometer, under heat and pressure. The study is carried out by measuring the torque on a conical disc rotor embedded in a matrix as it sinusoidally oscillates through a small arc (+3 °). The torque is a linear function of crosslink density determined by swelling measurements through the proportional constant, which varies with the compounds. Moreover, from the rate of rise of torque, the rate of crosslinking, and from the onset of rise in torque (i.e, onset of crosslinking) scorch safety, i.e., the process window before premature crosslinking, can be predicted. It is a widely accepted tool for determining crosslinking characteristics and regular quality checking of the compounds. However, this method is not suitable for ascertaining the crosslinking characteristics of silane-grafted compounds.

Another rapid but precise method, particularly for peroxide crosslinkable XLPE, is to measure (i) the crosslinking exotherm and (ii) the melting point reduction by a differential scanning calorimeter [8, 24]. The curing exotherm is directly related to the amount of peroxide present and degree of crosslinking achieved. The melting point of PE is reduced upon crosslinking, and the extent of reduction bears a linear relationship with the degree of crosslinking within the range practiced in the industry. From the measurement of crosslinking exotherm and melting point reduction, the uniformity of peroxide distribution and dispersion and its crosslinking efficiency can be precisely determined by this technique.

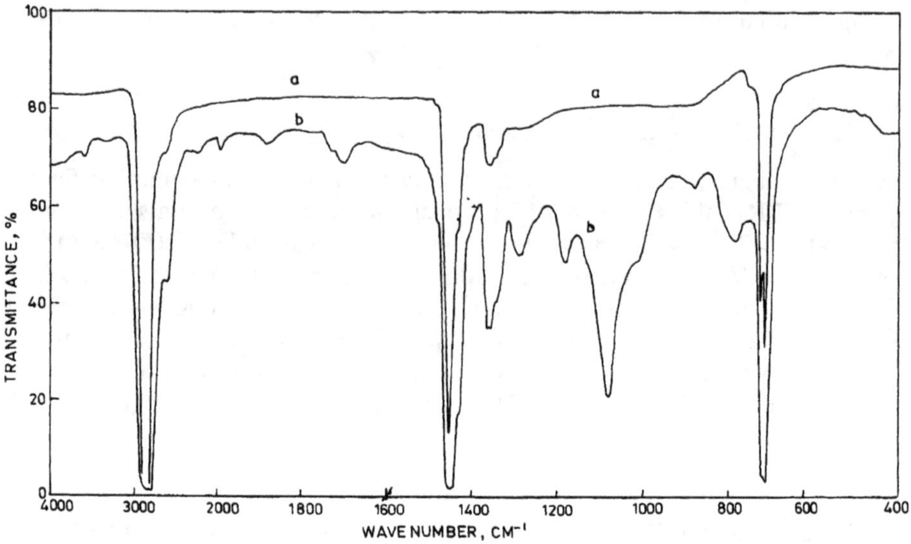

Figure 9.4. Infrared spectra of (a) PE and (b) silane-grafted PE. Reproduced from *J Appl Polym Sci*, **44**: 1153 (1992).

The degree of crosslinking of silane-grafted polyethylene is mostly dependent on the degree of grafting (i.e., the number of moles silane grafted on polyethylene) and the rate of crosslinking, including scorch safety, is dependent on the concentration of catalyst and moisture in the system.

The degree of silane grafting (the presence of the silane group) in polyethylene can be determined by infrared spectroscopy. The absorbance of Si–O–C-group stretching of the grafted silane at at 1090 cm^{-1} is used for characterization [7]. The infrared spectra of polyethylene and silane-grafted polyethylene are shown in Figure 9.4.

The degree of grafting can be determined from the intensity/depth of absorption of the Si–O–C-group stretching. In practice, thin films of PE and silane-grafted PE are made by molding in a hydraulic press at a temperature of 175 °C; free silane is removed from the films by acetone extraction, dried in vacuum, and then spectroscopy is performed.

Moisture (H_2O) content is determined by a Karl Fisher type of titrator. About 1.5–2.0 g samples are heated to 180 °C in a sealed glass oven at 180 °C; the liberated moisture is carried out by dry nitrogen gas and automatically titrated by the Karl Fisher reagent.

SWELLING RATIO AND GEL FRACTION

To measure the degree of crosslinking in industry, normally gel fraction or swelling ratio is measured, in particular for XLPE cable insulations.

Swelling Ratio

Previously weighed samples are allowed to swell in xylene at 120 °C for 24 h to reach swelling equilibrium, after which the test pieces are dried on the surface and rapidly weighed into stoppered bottles. The samples are then dried to constant weight in a vacuum oven at 60 °C. The weight difference between the swollen and dry test pieces gives the weight of the solvent, the value of which divided by the density of the solvent to give v_1. In the same manner, v_2 is obtained from the weight and density of the dry test piece. The degree of swelling or swelling ratio q is calculated as $q = v_2/(v_1 + v_2)$. An average of three measurements is taken.

Gel Fraction

Previously weighed samples (cut into a number of smaller pieces) are allowed to swell, and the soluble portions are extracted in xylene at 120 °C for 24 h. The samples are then dried to constant weight in a vacuum oven at 60 °C. The percentage of the weight remaining with respect to the initial weight gives the gel fraction. An average of three measurements is taken.

Gel fraction is easy to carry out and regularly practiced in industry to ascertain the curing state for XLPE compounds and cables. As per the industry norms, the gel fraction for peroxide XLPE should be 80±2% to attain the required continuous operating temperature, short circuit temperature, hot elongation, and set properties and thermal aging properties. For silane-crosslinked XLPE, a gel fraction of 65%–70% can meet the above requirements. This difference is due to the differences in the crosslinked network structure that results from peroxide and silane crosslinking. Peroxide crosslinking of PE forms planar C–C crosslinks whereas silane crosslinking forms three-dimensional –Si–O–Si– linkages through trialkoxy functionality [25].

TROUBLESHOOTING IN CABLE MANUFACTURING (WITH SPECIAL REFERENCE TO GRAFTING AND CROSSLINKING)

- Low degree of grafting (resulting lower degree of crosslinking): to increase the concentration of silane with optimum concentration of peroxide, to optimize the temperature of grafting and residence time in the extruder after silane injection.
- Scorching during extrusion on conductor: to reduce the moisture content of the silane-grafted PE, to optimize the catalyst concentration. Color masterbatch (if used) should have no or minimum catalytic effect.
- Reduction of MFI (melt-flow index) on grafting: to optimize the peroxide concentration and silane concentration. Excessive concentration may result in partial crosslinking in addition to grafting, resulting in a reduction in MFI.

- Deviation from concentricity in CCV line: to adjust the temperature profile and rotation of the conductor and insulated core during extrusion.
- Failure in electrical breakdown (high voltage) test: to check the insulation thickness, contamination in insulation, and protrusion. Special care should be taken during the production of compounds and cables. Both insulation and semiconducting shielding compounds should be free from any type of contamination to enable long life of XLPE-insulated cables [26, 27].

SCOPE OF FUTURE STUDIES AND DEVELOPMENTS

The importance of power generation, distribution, and conveyance of the electricity to the application points will continue in the future. The success of delivery largely depends on the performance of power cables. Performance comes from quality, and for cables it is based largely on the quality of compounds and the manufacturing process, as XLPE (both peroxide and silane crosslinked) cables are most important for conveying the electricity. Though grafting and crosslinking technology has progressed considerably, there is still room for further study and developments as requirements are being more and more stringent. For EHV cables and their performance and long-term reliability, it is of paramount importance to make defect-free insulation. Hence, studies are required to develop crosslinking systems, technology, and base polyethylene, so that there will be no voids, defects, and even irregularities that will result from crosslinking. There is particular interest in developing efficient catalytic systems so that the scorch problem can be eliminated without sacrificing a reasonable rate of cure and degree of crosslinking, and they can be made industrially usable. The relationship of crosslinked structures (morphology) with long-term aging and service life of the cables must be established; accordingly, standards need to be prepared and set in place.

In the case of silane crosslinking, it is necessary to develop more efficient catalyst systems to avoid scorching. To date, the major constraint for making EPR silane crosslinkable is its poor scorch-resistance properties. It is of particular interest to develop efficient catalytic systems so that the scorch problem can be eliminated without sacrificing reasonable rate of cure and degree of crosslinking and to make it industrially usable.

REFERENCES

1. Manley TR, Qayyum MM. *Polymer*, **12**: 176 (1971).
2. Labana SS, in *Encylopedia of Polymer Science and Engineering, Vol. 4*, HF Mark, NM Bikales, CG Overberger, G Manges, and JI Kroschwitz, eds. (New York: John Wiley & Sons, 1986), p. 350.
3. Andreopoulos AG, Kampouris EM. *J Appl Polym Sci*, **31**: 1061 (1986).

4. British Patent 1,542,543 (Dec. 31, 1979).
5. Scott HG. U.S. Patent 3,646,155 (Feb. 29, 1972).
6. Sen AK. "Grafting and Crosslinking of Polyethylene, Ethylene Propylene Rubbers and Their Blends," Ph.D. dissertation, 1991, p. 63.
7. Sen AK, Mukherjee B, Bhattacharyya AS, De PP, Bhowmick AK. *J Appl Polym Sci*, **44**: 1153 (1992).
8. Sen AK, Bhattacharyya AS, De PP, Bhowmick AK. *J Thermal Anal*, **37**: 19 (1991).
9. Sorio A, Gale GM, in *Proceedings International Conference, Polymer Extrusion III* (London: Plastics and Rubber Institute, 1985), p. 27/1.
10. Sano K, Ishitani H. *Rad Phys Chem*, **25**: 849 (1985).
11. Bhowmick AK, Mangaraj D, in *Rubber Products Manufacturing Technology*, AK Bhowmick, MM Hall, and HA Benarey, eds. (New York: Marcel Dekker Inc., 1994), p. 386.
12. Ungar GJ. *J Matr Sci*, **16**: 2635 (1981).
13. Kao YH, Phillips PJ. *Polymer*, **27**: 1669 (1986).
14. Mori E, Isshiki S. *Fujikura Technol Rev* (1979) 41.
15. Bhowmick AK, Mangaraj D, in *Rubber Products Manufacturing Technology*, AK Bhowmick, MM Hall, and HA Benarey, eds. (New York: Marcel Dekker Inc., 1994), p. 376.
16. Sen AK, in *Rubber Products Manufacturing Technology*, AK Bhowmick, MM Hall, and HA Benarey, eds. (New York: Marcel Dekker Inc., 1994), p. 682.
17. Orton H, Hartlein R. *Long-Life XLPE-Insulated Power Cables* (Orton Consulting Engineers International, Vancouver, Canada, and National Electric Energy Testing Research and Application Center, Atlanta, Georgia, USA, 2006), p. 88.
18. Bhowmick AK, Mangaraj D, in *Rubber Products Manufacturing Technology*, AK Bhowmick, MM Hall, and HA Benarey, eds. (New York: Marcel Dekker Inc., 1994), p. 383.
19. Orton H, Hartlein R. *Long-Life XLPE-Insulated Power Cables* (Orton Consulting Engineers International, Vancouver, Canada, and National Electric Energy Testing Research and Application Center, Atlanta, Georgia, USA, 2006), p. 16.
20. Orton H, Hartlein R. *Long-Life XLPE-Insulated Power Cables* (Orton Consulting Engineers International, Vancouver, Canada, and National Electric Energy Testing Research and Application Center, Atlanta, Georgia, USA, 2006), p. 41.
21. Campus A, Ulrich M. *Jicable '03*, 2003, Paper B9.5, p. 350.
22. Mendelshon A, Person T, Shattuck G, Hartlein R. *Jicable '03*, 2003, Paper B9.5, p. 556.
23. Graff RS, in *Rubber Technology*, M Morton, ed. (London: Chapman & Hall, 1995), p. 339.
24. Sen AK, Mukherjee B, Bhattacharyya AS, De PP, Bhowmick AK. *Thermochimica Acta*, **157**: 45 (1990).
25. Sen AK, Mukherjee B, Bhattacharyya AS, De PP, Bhowmick AK. *Polym Degrad Stab*, **36**: 281 (1992).
26. Bostrom J, Marsden E, Hammpton R, Nilson U, Lennartsson H. *International Symposium Electrical Institute* (2002), p. 91.
27. Sarma H, Cometa E, Densely J, *IEEE*, **18**: 15 (2002).

10

IN SEPARATION AND PURIFICATION

Mohamed Nasef

INTRODUCTION

Grafted and/or crosslinked polymers obtained by the modification of polymers are attractive materials because of their new or enhanced properties compared to corresponding conventional polymers. These new or enhanced properties arise principally from the introduction of new functionalities and/or interlinkages to the original polymers structures by grafting and/or crosslinking reactions. Because their chemical and physical properties may be tailored over a wide range of characteristics, the use of grafted and/or crosslinked polymers has found a permanent place in various fields. This includes separation processes [1, 2], energy conversion [3–6], solid state [3], biomedical [7], biological [8, 9], and environmental applications [10, 11]. Among all polymers, the contribution of grafted and/or crosslinked polymers to the field of separation and purification has been tremendous due to their versatile involvement in a wide spectrum of commercial applications today; many more are being developed for the future in the chemical, environmental, pharmaceutical, and biotechnological industries [12].

The primary objective of this chapter is to review the status of the applications of grafted and/or crosslinked polymers in the field of separation and purification. It starts with a brief review of basic principles covering classifications and descriptions of grafted and/or crosslinked polymers on the basis of the separation/purification functions together with their engineering configurations and operating systems and followed by intensive discussion on a wide range of applications. The format of the discussion follows the diversity of

Polymer Grafting and Crosslinking, Edited by Amit Bhattacharya, James W. Rawlins and Paramita Ray

applications in a single separation field not only in the variation in the forms and the types of grafted and/or crosslinked polymers but also as to their configurations and operating systems.

CLASSIFICATION OF GRAFTED AND/OR CROSSLINKED POLYMERS

Grafted and/or crosslinked polymers are available in different forms that vary in their classifications as shown in Figure 10.1. These polymers can be found in various physical forms including gels, resins, fibers, films, and fabrics that have widely differing chemical and physical properties and can be chemically active. The majority of these forms are of synthetic origin; i.e., they are made of modified synthetic polymers such as polyethylene, polystyrene, and polyvinyl fluoride, while some of them are obtained from modified natural polymer sources including chitosan, starch, and cellulose. Considering the separation function, grafted and/or crosslinked polymers can be classified into various categories including ion exchangers, chelating adsorbents, hydrogels, and affinity polymers, as shown in Figure 10.2.

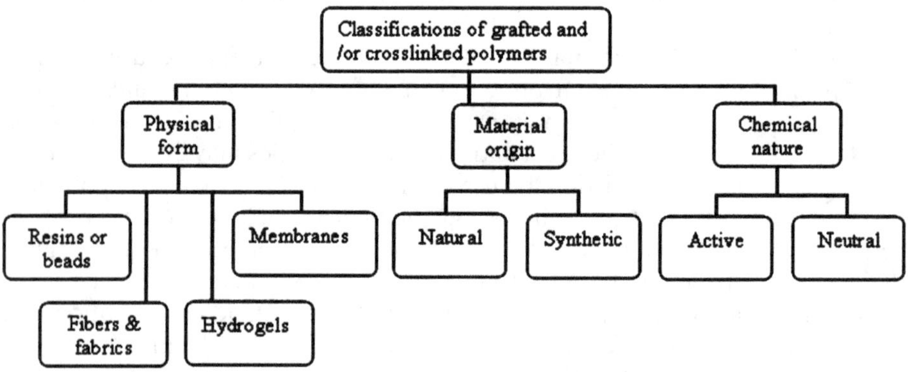

Figure 10.1. A schematic representation for various classifications of grafted and/or crosslinked polymers.

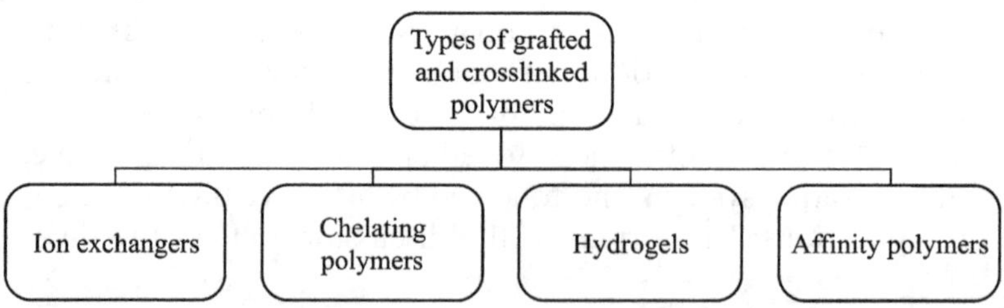

Figure 10.2. A schematic representation for various types of grafted and/or crosslinked polymers.

Ion Exchangers

Ion Exchange Membranes In principle, an ion exchange membrane (IEM) is a highly swollen polymer film carrying fixed charged chemical groups covalently bonded to the polymer backbone that can exchange their mobile ions for ions of similar charges from the surrounding medium. There are three types of IEMs based on the type of the fixed charged groups: cation, anion, and combined cation/anion exchange (bipolar) membranes. The separation behavior of each type of membrane is determined by the chemical nature of attached charged group. Cation exchange membranes contain negatively charged groups, such as $-SO_3^-$, $-COO^-$, $-PO_3^{2-}$, $-PO_3H^-$ and, $-C_6H_4O^-$, that are fixed to the membrane backbone and allow the passage of cations but reject anions. Anion exchange membranes contain positively charged groups, such as $-NH_3^+$, $-NRH_2^+$, $-NR_2H^+$, $-NR_3^+$, $-PR_3^+$ and, $-SR_2^+$, that are fixed to the membrane backbone and allow the passage of anions but reject cations. Since IEMs separate cations from anions and vice versa from aqueous solution of similarly charged ions, they should have a high permselectivity, i.e., a high transport number for counterions.

IEMs are commonly prepared using various routes based on hydrocarbon substrates such as poly(styrene-co-divinylbenzene) and polysulfone (PS), or fluorocarbon ones such as poly(tetrafluoroethylene) (PTFE) and poly(ethylene-*co*-tetrafluoroethylene) (ETFE), as well as other conjugated polymers that undergo chemical treatment to introduce functional groups [1–3, 13]. Figure 10.3 shows the chemical structures of poly(styrene-*co*-divinylbenzene) cation and anion exchange membranes. Grafted and/or crosslinked IEMs reportedly prepared using various modifications that can be conducted by chemical, photochemical, radiochemical, or plasma modification methods [3]. Membrane preparation starts by the grafting of active acrylic monomers, e.g., acrylic acid (AAc) or an inactive vinyl monomer such as styrene (Sty) followed by the desired chemical treatment: sulfonation to confer cation

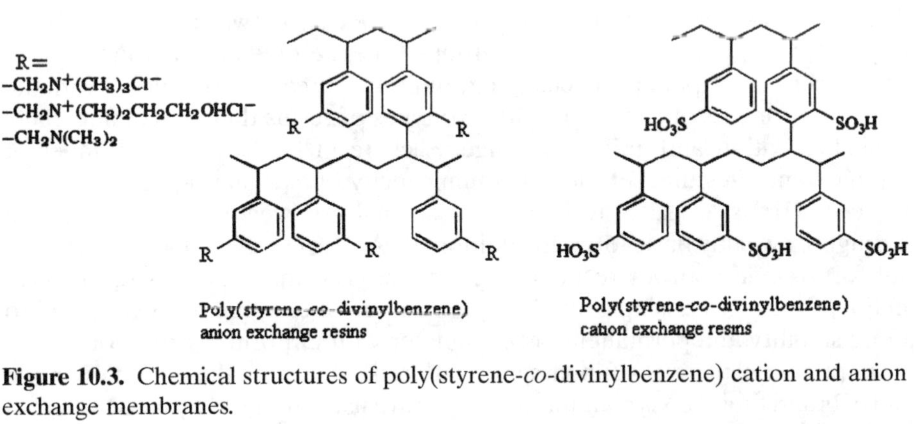

Figure 10.3. Chemical structures of poly(styrene-*co*-divinylbenzene) cation and anion exchange membranes.

exchange functionality or chloromethylation and amination to confer anion exchange characters. Crosslinking of the membranes is commonly performed by adding a small amount (up to 10%) of a crosslinking agent such as divinylbenzene (DVB) during the grafting reaction. Various aspects related to preparation and applications of radiation-grafted IEMs have been recently reviewed by Nasef and Hegazy [3].

IEMs are broadly classified into strong or weak acid cation exchange membranes and strong or weak base anion exchange membranes depending on the strength of the functional groups. Strongly acidic cation exchange membranes contain $-SO_3H$ groups, whereas $-COOH$ groups are present in weakly acidic cation exchange membranes. Similarly, strongly basic anion exchange membranes contain quaternary ammonium exchange groups, and weakly basic anion exchange membranes have groups derived from tertiary (=N–), secondary (=NH), or primary ($-NH_2$) amines. In the bipolar form of IEMs, one side contains a weakly or strongly anion exchange layer and the other side possesses a weakly or strongly cation exchange layer.

IEMs can be also classified depending on their structure and preparation procedure, into two major categories: (a) homogeneous and (b) heterogeneous. They can be further divided depending on the degree of heterogeneity into: (i) homogeneous IEMs, (ii) interpolymer membranes, (iii) microheterogeneous graft- and block-copolymer membranes, (iv) snake-in-the-cage IEMs, and (v) heterogeneous IEMs [1].

IEMs have been widely used in separations in various industries based on systems that apply various driving forces (concentration, potential, or pressure gradient). These systems include diffusion dialysis (DD), electrodialysis (ED), pervaporation (PV), and reverse osmosis (RO). To meet the requirements for such applications, IEMs have to essentially possess low electrical resistance combined with high ionic permselectivity, moderate degree of swelling, chemical stability, and high mechanical strength. The diverse characteristics of ion exchange membranes make them suitable for different separation applications [1–3, 13–16].

Ion Exchange Resins Ion exchange resin (IER) is a network polymer of high molecular weight having fixed ionic groups. Like their membrane counterparts, these resins are capable of exchanging counterions (cations or anions) with the ionic components of a solution with two types of resins that are commercially available: cation and anion exchange. Each resin has a distinct number of mobile ionic sites that set the maximum capcity of exchanges per unit weight of resins. IERs are also available in a strongly or weakly acid form and a strongly or weakly basic form. In addition, IERs are also available in two physical forms: macroporous resins and gel resins. The macroporous resins have major advantages arising from their high degree of crosslinking, which lead to more stability and permanent pore diameter with uniform distribution.

In principle, there is a fundamental difference between the operation mode of IERs and their IEMs counterparts. IERs are used in a flow-by mode whereas IEMs are used in a flow-through mode. Therefore, kinetics of the ion exchange for IEMs is highly favorable relative to IERs, which heavily rely on the diffusion to achieve ion capture in their pores.

Ion exchange resins have found widespread applications in various fields including water and waster water treatment; oil and gas, electronics, electroplating, and tanning industries; sugar decolorization; starch hydrolysated deionization; acid catalysis of sucrose; steam flooding; and the improvement of soil water retention [17, 18].

Chelating Polymers

Chelating or ligand exchange polymers refer to polymers that bind metal ions by coordinating interaction and sometimes by ionic interactions. The chelating polymers are insoluble in water (they have a crosslinked matrix) and are characterized by two components: the polymer backbone, which provides the stability, and the functional groups, which represent the ligands required for metal complexation. The most common coordinating atoms present in the main or side chain are N, O, P, and S. It is possible to make chelating polymers that have a selective absorption capacity for specific metal ions by fixing the desired ligand groups on the polymer backbone. For example, polymers containing amidoxime groups have high affinity for uranyl ions (UO_2^{2+}) and their use has been attempted for the recovery of uranium from seawater [19].

Upon use, the selectivity of chelating resins relies on the stability of metal complexes at the appropriate pH values. Practically, chelating polymers are very useful for selective adsorption of certain metal ions from their mixtures and recovery of soluble metal ions from solutions [20]. Adsorption of toxic metal ions using chelating polymers has the advantage of high efficiency, easy handling, availability of different adsorbents, and cost effectiveness. Therefore, it is of great importance, especially in environmental applications [21, 22]. Chelating polymers can be divided into two classes depending on the origin of their backbone: graft adsorbents and graft biosorbants.

Graft Adsorbents Graft adsorbents are a class of chelating polymers that have a modified backbone of synthetic polymers containing immobilized functional groups. These adsorbents have the flexibility of being prepared with various combinations of functional-group-containing grafts (e.g., polyamines, polyacrylonitrile, polyacrylamides, and polyacrylic acid) and polymer substrates such as polyethylene (PE), polypropylene (PP), and PTFE, thus opening a broad spectrum of prospects for tailor-made applications such as the removal of toxic heavy metals and recovery of precious metals. The adsorption mechanisms in these polymers have generally been attributed to the formation of complexes between the functional groups on the adsorbents and the metal ions to be removed [23].

Grafted and Crosslinked Biosorbants Grafted and crosslinked biosorbants are a class of chelating polymers that have a modified backbone of biopolymers (polysaccharides) such as chitosan, alignate, starch, and cellulose. These biosorbents are much more selective yet cheaper than traditional ion exchange resins, although their native form shows inferior biosorbance performances caused the instability of physical and chemical properties [24]. Therefore, modification of these polysaccharides has been explored as a way of combining their best attributes with those of synthetic graft polymers [25]. Several functional groups such as amino, phosphate, amido, sulfonate and carboxyl, and hydroxyl groups can be grafted onto the polysaccharides to give them additional ionic characters with their stability further enhanced by crosslinking [26].

The biosorption mechanism in these grafted and crosslinked materials is a complex process involving (i) ion exchange, (ii) chelation, (iii) complexation, and (iv) adsorption by physical forces such as H-bonding, entrapment in inter- and intrafibrilliar capillaries, and spaces of the structural polysaccharides network as a result of the concentration gradient and diffusion through the sorbent [27–28].

Particularly, chitosan-based biosorbents are efficient selective materials that have been frequently used in various separation applications including removal of heavy metals and dyes. For example, beads obtained by the casting of crosslinked and grafted chitosan have high porosity and large surface areas, high stability in acidic media, and are most suitable for industrial applications [29–30]. Other biopolymers such as starch [31–37] and cyclodextrins [38–42] have been also proposed as low-cost biosorbents. The recent developments in the synthesis of biosorbents obtained from modified polysaccharides (starch, cyclodextrin, chitin, and chitosan) were reviewed by Cirni [43].

Hydrogels

Hydrogels are highly swollen, hydrophilic three-dimensional network structures of polymers capable of absorbing significant amount of water (few tens to thousands percent of their dry weight) (also discussed in Chapter 6) and therefore can have the free exchange of water within the external contaminated water preparations. Depending on the nature of their network, hydrogels have two main classes: (i) permanent hydrogels having entangled covalently crosslinked networks formed by UV or ionizing radiation [44–47], and (ii) reversible hydrogels with networks formed by secondary interactions, i.e., molecular entanglements and/or ionic and H-bonding, or hydrophobic forces, making them lack homogeneity, and having free chain ends and loops that causes transit network defects [48].

Hydrogels can have various crosslinked or entangled network macromolecular structures including linear homopolymers, linear copolymers, block copolymers, graft copolymers, polyions-multivalent ions, polyion-polyion, or bonded complexes. The physicochemical properties of the hydrogels depend not only on the molecular gel structure, and the degree of crosslinking but also on the content and the state of water in the hydrogel. A number of interesting grafted hydrogel materials have been reported in recent years particularly for water treatment [49], heavy-metal removal and extraction of uranium [50–56], and dyes [57]. More details on the synthesis of hydrogels, and their and applications can be found elsewhere [58].

Affinity Polymers

Affinity polymers consist fundamentally of porous structures having selective affinity ligands (immobilized biological molecules such as antigens, and antibody or synthetic molecules such as dye, ion exchange ligands, and amino acids) attached to their pores. The affinity polymers can be found in membranes and bead (resin) forms. Preparation of affinity polymers usually invovles preparation and activation of the basic polymer, and coupling of affinity ligands to the activated polymers [59]. The chemical structures of the matrices and how various immobilized ligands can be utilized for specific classes of separation was reviewed by Charcosset [60].

Affinity Membranes Affinity membranes are the most commonly used form of affinity polymers. These membranes were basically introduced to overcome the shortcomings (e.g., fouling) associated with microfiltration and ultrafiltration membranes. They allow high flow rates to be achieved at low pressures and enable the processing of large volumes within short processing times compared to affinity beads [61]. Affinity membranes are made of highly hydrophobic porous polymers, mostly in hollow-fiber form, such as PE, PP, and poly(vinylidene fluoride) (PVDF), and PS having entrapped reactive groups (–OH, $-NH_2$, –SH, or –COOH) for further coupling of spacer arms or ligands. These membranes can be obtained by chemical activation, coating, or grafting in addition to the copolymerization of two functional monomers [62]. The first commercial affinity PS-based hollow-fiber membranes were produced by Monsanto in 1986. Later, different amine groups modified-microporous nylon-6 hollow-fiber membranes were reported by Klein and coworkers [63]. Saito and coworkers have introduced various functional groups including iminodiacetate, amidooxime, phosphoric acid, alchohol hydroxyl, phenylalanine, and chlorovinylpyridine to polyethylene (PE) hollow fibers by the radiation grafting of monomers such as glycidylmethacrylate (GMA) followed by specific chemical treatments [64–69]. Affinity membranes were also prepared by photochemical grafting [70, 71].

Affinity membranes have been used for heavy-metal removal [72, 73] and proposed as alternatives to stationary beads for affinity membrane chromatography mainly for the separation of biomolecules [59]. Nonetheless, adoption of this technology has been slow because membrane chromatography has been limited by lower binding capacity than that of conventional columns, even though the high flux advantages provided by membrane adsorbers would lead to higher productivity [74]. The latest advances in preparation of affinity membranes and their applications were reviewed by Klein [61].

Affinity Beads Like their corresponding membranes, affinity beads are porous particles having selective affinity ligands, including immobilized biological or synthetic molecules attached to their pores. These beads have been found widespread application in the separation and purification of biological macromolecules in both analytical and preparative scales in a variety of bioprocesses [75]. Affinity beads are used as separation media in adsorption chromatography, an alternative efficient separation technique to conventional chromatography; adsorption chromatography is often used for the isolation and purification of fermentation products [76, 77]. Column operations are commonly adopted for such applications. The selectivity of beads is largely a function of the three-dimensional geometry of the pores of the beads and the occluded functional/ligand groups; therefore, both of them dictate the separation and adsorption capacity of the desired compounds [78].

SEPARATION AND PURIFICATION SYSTEMS

The feasibility and success of separation and purification processes depend on the nature of the separating polymer and the operating conditions. In other words, the morphology of the polymer dictates the selection of the separation process and its operating mode. As mentioned earlier, the basic morphologies of grafted and/or crosslinked polymers include membranes, resins or beads, and hydrogels. To bring such polymer forms into applications, they have to be hosted in a proper engineering system that provides a set of adjustable operating parameters to control their performance. Accordingly, the separation systems take various configurations depending on the physical form of separating polymers and the volume of the solution to be treated. When polymers are in a resins form, the system takes a configuration best represented by ion exchange systems. Figure 10.4 shows a schematic representation of various

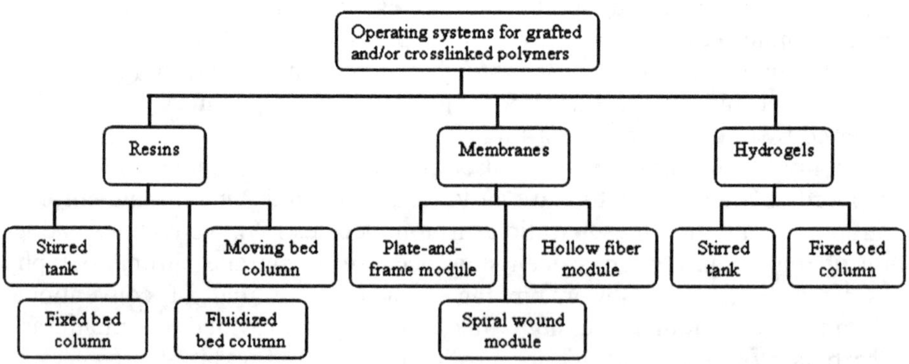

Figure 10.4. A schematic representation of various morphologies of grafted and/or crosslinked polymers and their respective operating systems.

morphologies of grafted and/or crosslinked polymers and their respective operating systems.

There are four basic operating configurations for resins forms: (i) continuous-stirred tank, (ii) fixed-bed column, (iii) fluidized-bed column, and (iv) a moving-bed (closed-loop) system [79].

Similarly, putting grafted and/or crosslinked polymers in a film or a membrane form into any practical application requires a housing vessel called a "module." A membrane module provides support and protection against operating pressures and the daily wear and tear of the production environment and allows the application of a control strategy over the system. More details on the requirements for efficient module designs to house various membrane types can be found elsewhere [80].

The most common configurations of membranes modules are (i) plate-and-frame, (ii) spiral wound, (iii) tubular, and (iv) hollow fiber. Membrane modules can be combined to form cascade systems of different configurations (parallel, series, and tapered). Selection of the membrane module for particular membrane separation must balance a number of parameters, including the cost of the module, volume to be treated, concentration polarization control, and resistance to fouling [81].

Various processes utilizing grafted and/or crosslinked polymers in the field of separation and purification such as ED, DD, RO, PV, ultrafiltration (UF), adsorption chromatography, and affinity membrane chromatography are available [81, 82].

APPLICATIONS IN SEPARATION AND PURIFICATION

Grafted and/crosslinked membranes have been found to meet the requirements for many separation processes in a large number of industrial applications. The purpose of separation dictates the selection of the type of the grafted/crosslinked polymers form and the system configuration for a particular application. The most conventional media treated with functional grafted/crosslinked polymers are aqueous solutions. Therefore, the separation with these polymers can be classified depending on the purpose, into three main categories: (i) purification of a solvent, (ii) purification of a solution, and (iii) extraction of ions. More details on these categories can be found in Zagorondni [79]. The next subsection covers the various applications of grafted and/or crosslinked polymers under four main categories: water production, chemical industry, environmental, and biotechnological and biomedical applications.

Water Production Applications

The interest in water production has always been high and recently renewed due to the shortage of water in many remote areas and diminishing natural fresh water resources. Grafted and/or crosslinked polymers, mainly in form of neutral membranes, IEMs, and their corresponding resins, accomplish a great deal in water purification systems dealing with ion, particulate, organic compounds, and microrganism removal. Membranes vary dramatically in pore size, molecular weight cutoff, and ion rejection. Various membrane separation processes including ED, ion exchange, RO, and UF have been developed for water production applications [81, 82].

Desalination of Brackish Water Production of water by desalination methods continues to receive attention despite four decades of commercialization. The essential requirement that determines the viability of any scheme to provide fresh water suitable for human use is the cost. Desalination of brackish water by electrodialysis (ED) is considered to be the most important large-scale application in terms of installation number (>2000 plants worldwide with a total production capacity of >1,000,000 m^3 of water per day are installed with a >1.5 million m^2 membrane area), which is competing with RO and fractional distillation [1, 2, 83–86]. This process is also the most economical for water with a low salt concentration (<5000 ppm). ED is also used for the production of potable water [87].

In principle, ED is a significant membrane separation process where selective separation of certain ions from water takes place by migration across IEMs under the influence of an electric field applied on two electrodes bordering a number of cationic and anionic membranes placed in an alternating pattern. The degree of separation is mainly determined by the properties of IEMs used in the process, whereas the economics of the process is determined by the operating costs that involve the energy consumption. The various process design parameters, e.g., flow turbulence, limiting current density, effective cell pair area, and cell dimensions also have significant effect on the energy consumption and thus on the total operating cost. Recent developments in IEM properties have led to tremendous improvement in the efficiency of ED [1, 2].

Various grafted and/or crosslinked IEMs prepared by chemical grafting and crosslinking are commercially available. Sty-DVB-based membranes (shown in Figure 10.3) are best-known in the literature for the application in ED.

The use of grafted and crosslinked anionic and cationic membranes for sea water desalination by ED has not been genuinely addressed. Only few articles are reported in the literature. Aminabhavi et al. have developed homogeneous anion exchange membranes for the conversion of sea and brackish water into potable water by simultaneous polymerization of 4-vinylpyridine (4-VP) and crosslinking with epichlorohydrin (EPI) and aniline on nonwoven support cloth at different ratios [89]. The resulting membranes were quarternized with methyl iodide using hexane as a solvent.

Choi et al. studied the desalination of water by ED with two types of IEMs prepared by the radiation grafting of GMA onto PE and PP nonwoven fabrics followed by treatment of the obtained grafts with triethylamine (TEA) and phosphoric acid [90]. Currently, research is focusing on addressing the remaining unsolved issues such as membranes organic fouling, selective permeability for specific ions, and instability of quaternary ammonium exchange groups in alkaline solution in anion exchange membranes [84, 91, 92].

Desalination of Sea Water Water desalination by RO is another field where membranes have served since their early days [88, 93]. In this application, membrane material plays a significant role in affecting the system performance [94, 95]. Neutral membranes and IEMs have been utilized for such applications. However, the industrial application of IEMs is far smaller compared to neutral membranes. A limited number of studies reported the use of radiation-grafted IEMs for RO application [3].

Radiation-grafted cation exchange membranes prepared by the grafting of acrylic monomers, i.e, acrylic acid (AAc), methacrylic acid (MAc) and acrylamide (AAm) and vinyl monomers (vinyl acetate, VAc) onto PE and PTFE and FEP (copolymers of PTFE with hexafluoropropylene) films were tested for water desalination with a salt rejection >90% [96–99]. The performance toward salt rejection was further improved by using membranes obtained by the grafting of combined (binary) monomer systems of Sty/AAm onto various polymer films [100].

Anionic exchange membranes prepared by grafting and subsequent quaternization (treatment with methyl bromide) of 2-VPn and 4-VPn onto poly(vinyl chloride) (PVC) and poly[3,3-bis(chloromethyl)oxetane], (penton) films demonstrated very good RO properties including sufficient salt rejection [101, 102]. Similar quaternized poly(4-VPn) grafted-PE and -PTFE membranes also demonstrated a combined good flux and salt rejection properties [97, 103–106].

Cation exchange membranes having phosphoric acid groups obtained by radiation-induced grafting of GMA onto PE hollow fibers followed by phosphorylation of the epoxy rings were also reported to be capable of maintaining the salt splitting capacity of 0.68 mol/kg in their Na form [64].

Commercial microfiltration PE hollow-fiber membranes modified by radiation-induced grafting of diethylaminoethyl methacrylate (DEAEMA) followed by amination with DEA recorded a high pure water flux of $1.0\,m\cdot h^{-1}$ at a filtration pressure of 0.1 MPa and an ion exchange capacity of $1\,mol\cdot kg^{-1}$ [68].

Softening of Hard Water Water softening is one of the most classical applications of ion exchangers. This treatment involves the removal of hardness caused by the presence of Ca^{2+} and Mg^{2+} ions from water solution by exchanging them with the Na^{+} ion available in ion exchangers. Conventional softening technologies are replaced by nanofiltration where the membrane operation does not produce by-products to the finished water and therefore eliminates the by-product disposal cost.

IEMs prepared by photografting of porous UF membranes have been tested for water softening. Bequet et al. [106] reported a composite nanofiltration membrane for water softening using a route consisting of photografting AAc onto the surface of UF-porous PS membranes with UV. The surface modified membrane was stabilized by crosslinking it with *N,N'*-methylene-bis-acrylamide. The obtained membranes showed softening properties of 95% recovery when measured in dead-end mode.

Similar membranes prepared by the photografting of AAc and sodium allyl sulfonate (SAS) onto polyetherketone cardo (PEK-C) with UV without a photoinitiator showed unexpectedly large permeation flux without loss of salt retention [107].

The treatment of natural nonpotable and irrigation-unsuitable water by novel cellulose-based hydrogels synthesized from reactive networks and graft copolymers of either AAm or AAc and a mixture of them in the presence of benzoyl peroxide as an initiator and glutaraldehyde (GL) as a crosslinker was also recently reported [49]. The water treated with these hydrogels became potable and could be used for agricultural purposes.

Chemical Industry Applications

Production of Caustic Soda Production of caustic soda and chlorine is a large IEM-dependent industry that produces about 48 million tons of NaOH and about 42 million tons of Cl_2 starting from NaCl and H_2O [1]. In the chloroalkali ED cell, the anode and cathode are separated by means of a cation-exchange membrane [108]. A saturated NaCl is introduced at the anode compartment while the water is fed at the cathode. ED of both feeds results in the evolution of Cl_2 and H_2 at the anode and the cathode, respectively. At the same time, the Na^+ ions migrate toward the cathode, forming NaOH by combining with the OH^- that results from H_2O hydrolysis at the cathode while Cl_2 gas is produced at the anode. The use of IEMs has enabled the technology to overcome the two major challenges of the conventional process, i.e., the explosive reaction of Cl_2 with H_2 and the dissolving of Cl_2 in NaOH to form a hypochlorite solution. The common cation exchange membranes applied in this process are perfluorinated polymers containing sulfonic acid fixed groups, carboxylic acid groups or mixture of the two groups (e.g., Nafion membranes from DuPont, USA, and Aciplex from Ashahi Chemical Industry, Co., Japan). Commercially grafted and crosslinked membranes such as Flemion (Asahi Glass Co. Ltd., Japan) and Morgane CDS (Solvay S.A., Belgium) are also good substitutes for Nafion membranes in this application.

Alternative radiation-grafted strongly acidic cation exchange membranes for ED have been reported in the literature [109–115]. These membranes are commonly prepared by radiation-induced grafting of Sty or Sty/DVB onto films such as PTFE and PE films followed by sulfonation reaction.

Another version of weakly acidic radiation-grafted membranes having –COOH were tested for the production of NaOH by ED of NaCl solution. These membranes maintained a current efficiency in the ED cells in a level of

96–97% at a constant cell voltage of 3.7V [116]. More membranes of this type were developed by Hayashi and Kotani [117] and McRae [118].

Momose et al. developed highly stable cation exchange membranes for the chloro-alkali industry by radiation-induced grafting of α,β,β-trifluorostyrene onto ETFE films under controlled reaction conditions and subsequent sulfonation [119, 120]. The membranes obtained were found to have a combination of physicochemical properties that meet the requirements of caustic soda production [121]. This is believed to have led to commercial products such as Flemion membranes from Ashai Glass Co. Ltd. (Japan). Similar versions of these membranes are produced by RAI Research Inc., USA, and are also believed to have been developed based on a patent by D'Agostino et al. [122]. Another version of these membranes that contains weakly acid characters and is prepared by the grafting of fluorinated monomers such as methyl trifluoroproproenoate onto PTFE films was also found to be promising for ED processes in general and the chloro-alkali industry in particular [123].

Recently, the research on radiation-grafted perfluorinated sulfonic acid membranes has been exploited to great extent after these membranes were found to be potential alternative solid polymer electrolytes for fuel cell applications [124, 125]. Reviews of a large number of articles on the preparation, characterization, and testing of polystyrene sulfonic acid based on fluorinated polymers and their derivatives have been published [3, 4, 6].

Production of Acids The electrochemical splitting of salts by ED to their original constituents is highly desirable so as to minimize chemical consumption and effluent treatment costs as well as to enable the reuse of acids and bases. The ED of sodium sulfate (Na_2SO_4) has long been known, and in recent years it has earned increasing interest because of the mounting demand for H_2SO_4 and NaOH production.

Tzanetakis et al. reported the ED of Na_2SO_4 using anion exchange membranes prepared by the radiation grafting of vinylbenzyl chloride (VBC) onto PVDF films followed by amination using trimethylamine (TMA) [126]. The membranes showed acceptable selectivity and low electrical resistance together with high chemical, thermal, and mechanical stability. However, their performance in terms of current efficiency and transport properties was inferior to Nafion® 117 and sulfonated-poly- (Sty)-grafted PVDF membranes.

Cation exchange membranes obtained by radiation-induced grafting of styrene onto PVDF films followed by a sulfonation reaction were also tested for the ED of Na_2SO_4. These membranes outperform Nafion® 117 membranes in terms of current efficiency, achievable acid and base concentrations, and cell voltages, and offer an economical alternative process [127].

Similar commercial radiation-grafted membranes (Pall R1010 and Pall R1030) were also tested for ED of Na_2SO_4 in comparison with Nafion® 117 membrane [128]. The Pall R1010 gave higher current efficiency and NaOH generation than Nafion® 117.

Water Electrolysis The splitting of water into hydrogen and oxygen by electrolysis is a process that is conventionally conducted by applying high voltage in a cell containing an anode and cathode and an alkaline water solution. Crosslinked sulfonic acid membranes have been used to replace the liquid electrolyte. The use of these membranes in the electrolyzer has improved the degree of separation and increased the energy efficiency compared to the classical alkaline process with a simplification of cell design and a low temperature production capacity. Furthermore, the electrolyte membrane electrolyzer can work at higher pressures, thus avoiding the compression of the H_2 required for its transport [129–131]. The electrolyte membrane plays a vital role in this cell as it acts as a selective separator; i.e., it transfers H^+ formed at the anode to the cathode side and prevents the mixing of H_2 and O_2 ions again [132]. Nafion membranes (DuPont) are the most common materials used for this application. The first membrane electrolyzers were developed by General Electric Co. in 1966 for space applications [133]. Extensive research and development on electrolyte membrane electrolyzers has been also performed within the Japanese WE-NET program [134].

Commercial radiation-grafted sulfonic acid membranes were tested for electrolysis of water and found to be a suitable cost-effective alternative for Nafion membranes [135]. These membranes were developed by Chlorine Engineers Co. through the radiation grafting of α,β,β-trifluorostyrene (TFS) onto ETFE films; the subsequent sulfonation displayed a performance stability up to 10,000 h operation.

Dehydration of Solvents Grafted/crosslinked polymers in the form of membranes can be effectively used to separate liquid mixtures in competition with conventional chemical processes such as distillation, absorption, liquid extraction, and fractional crystallization through a process called pervaporation (PV). This process has the advantage of creating cost savings and making environmentally friendly technology; therefore, it has attracted attention from the chemical, biochemical and petrochemical industries [136].

In the PV process, a liquid mixture is brought into contact with the membrane, which preferentially permeates one of the liquid components (more volatile) as a vapor. The vapor of the more permeable component is cooled and condensated under the influence of the vapor pressure difference created by evacuation of the permeate side. The performance characteristics, i.e., permeability and the selectivity of the membrane for various components in organic-organic mixtures are functions of the solubility and the diffusivity of each component. Removal of water from solvents and separation of the organic solvents mixture are the most common applications of pervaporations [136]. Particularly, the removal of water from a concentrated alcohol solution is one of the most important industrial PV applications.

Membranes of both synthetic polymers and derivatives of natural polymers have been tested for separation of liquid mixtures. Particularly, extensive efforts have been made to develop grafted or crosslinked PV membranes with good performance and stability using various preparation methods, including radiation-induced grafting [117, 136–139].

Ping et al. have reported a high PV performance for crosslinked PVA-poly(AAc) blend membranes for separation of alcohol/water mixtures [140]. An increase in the permeability was observed with the crosslinked membranes at low poly(AAc) contents. Other blend membranes of grafted poly(AN) membranes were also tested for separation of water/ethanol mixtures by PV [141]. These membranes were developed by blending poly(AAc) and poly(AAc)-g-poly(ethylene oxide) (PEO).

Wang et al. (1998) developed 2,3-(epoxy propyl)-methacrylate (EPMA) chemically grafted polyurethane (PU) membranes for separation of an alcohol/water mixture by PV separation [125]. The grafted membranes showed a degree of grafting-dependent water permselectivity with the performance of grafted membranes higher than that of the ungrafted PU membrane for all water/alcohol mixtures. A grafted membrane with an 83.0% degree of grafting recorded a maximum performance in terms of the separation factor and permeation rate with values of 209 and 203 $g{\cdot}m^{-2}h^{-1}$, respectively.

Poly(acrylate) membranes have been shown to have high flux and selectivity toward water in alcohol/water mixtures [142]. For example, radiation-grafted membranes prepared by grafting AAc onto various films such as PE, FEP, and ETFE exerted superior PV properties toward ethanol/water mixtures [143, 144]. AAc-grafted membranes also showed superior permeability toward methanol compared to MAc-grafted counterparts. The permeability was found to be a function of the degree of grafting in either membrane [145].

PVA membranes modified by radiation grafting of AAc and MAc were also tested for pervaporation of methanol/water mixtures [146]. Similar membranes based on crosslinked PVA modified with monochloroacetic acid were tested for PV separation of ethanol/water mixtures [147]. Modification of the surface of the crosslinked membranes enhanced the permselectivity by almost twofold when compared to that for the only crosslinked PVA membrane, while the total flux remained unchanged. In another study, chemically crosslinked PVA membranes crosslinked with amic acid followed by imidization at 150 °C showed an optimum PV results at a 12 wt. % crosslinking agent in the membranes, giving separation factors of 70–380 and permeation rates in range of 30 to 1600 $g(m^2{\cdot}h)^{-1}$ depending on the operating temperature and feed mixture composition [148].

The permeation of water and acetone was studied using a porous hollow-fiber membrane grafted with GMA treated by DEA [149]. The pure water flux curve exhibited a sharp decrease at a conversion ratio of the epoxide to the DEA group between 40 and 60%. The acetone flux curve had a maximum value near the 20% conversion point. The DEA-group-containing membrane, with a conversion below 20%, showed a higher swelling in acetone than in water, suggesting that these membranes are acetone selective.

It has long been recognized that hydrophilic natural polymers such as sodium alginate (Na–Alg) and chitosan are good materials for PV separation

separation, namely, dehydration of solvents due to their strong affinity to water molecules [136]. Huang et al. tested ionically crosslinked Na-Alg membranes for dehydration of ethanol/water and isopropanol/water mixtures by pervaporation. The membranes crosslinked with Ca^{2+} ions showed the highest pervaporation performance in terms of flux and separation factors [150]. Similar Na–Alg membranes crosslinked with glutaraldehyde (GA) showed an improved performance toward the separation of ethanol/water by pervaporation [151]. The performance of GA crosslinked Na–Alg membrane was enhanced by blending it with different hydrophilic polymers like PVA [152, 153] and cellulose [154].

In another study, GA crosslinked membranes of high and low viscosity grades Na–Alg were used in pervaporation of acetic acid/water and isopropanol/water mixtures at 30 °C for feed mixtures containing 10–50 wt. % of water. Membranes prepared from low-viscosity-grade Na–Alg showed the highest separation selectivity of 15.7 for 10 mass % of water in the feed mixture, whereas membranes prepared with high-viscosity-grade Na–Alg exhibited a selectivity of 14.4 with a slightly higher flux than that observed for the low-viscosity-grade Na–Alg membranes [155].

Toti and coworkers prepared blend membranes of Na–Alg with polyacrylamide-*grafted*-guar gum (PAAm-*g*-GG) and studied their pervaporation performance for acetic acid/water and isopropanol/water mixtures [156, 157]. The membranes obtained from neat Na–Alg and a 1:1 blend of Na–Alg and PAAm-*g*-GG showed the highest selectivity for 10 wt. % water in the feed mixture, whereas membranes prepared with the 3:1 blend ratio of Na–Alg to PAAm-*g*-GG showed the highest separation selectivity of 20 wt. % water in the feed.

Alg composite membranes crosslinked with 1,6-hexanediamine (HDM) or PVA were prepared by casting an aqueous solution of Alg and HDM or PVA on a hydrolyzed microporous PAN membrane and tested for pervaporation separation of acetic acid/water mixtures [158]. The flux was higher for the K^+ membrane than for the Na^+ membrane while the separation factor showed an opposite tendency. This HDM-crosslinked Na–Alg composite membrane showed a high separation factor of 161 and a good permeation rate of $262\,g{\cdot}m^{-2}h^{-1}$ for pervaporation of 85 wt. % acetic acid aqueous solutions at 70 °C. The crosslinked Alg composite membrane with PVA has much lower separation properties (particularly lower selectivity for the acetic acid/water mixture) as compared to the alcohol/water system.

The PV performance of modified chitosan membranes obtained by surface carboxylation of GA crosslinked chitosan membranes by maleic anhydride (MAn) was tested for separation of ethanol/water mixtures [159]. The modified membranes were found to have a remarkably improved separation factor and flux. In separation of 90 wt. % ethanol/water mixture at 60 °C, the separation factor and the flux of the MAn-treated chitosan-GA membrane were, respectively, found equal to 634 and $300\,g{\cdot}m^{-2}h^{-1}$, compared to 105 and $250\,g{\cdot}m^{-2}h^{-1}$ in the untreated crosslinked membrane.

Chen et al. reported crosslinked chitosan–silk-fibroin complex membranes and their applications in the pervaporation of an isopropanol/water mixture. The flux was significantly improved while high selectivity was maintained when suitable amounts of $AlCl_3$ were added to the feed mixture [160].

Water-soluble blocked diisocyanate (DIC)-incorporated chitosan membranes prepared using a solution technique were tested for separation of isopropanol/water mixtures at various temperatures (30 °C–60 °C) [161]. The membranes with higher amounts of DIC showed the best separation ability. A membrane containing 40 wt. % of crosslinker showed the highest separation selectivity of 5918 with a flux of 2.20×10^{-2} kg g·m^{-2} h^{-1} at 30 °C for 5 wt. % of water in the feed.

Radiation-grafted poly (AAc)/nylon6 membranes were tested for pervaporation of acetic acid/water mixtures [162]. These membranes gave a high separation factor of 300 for acetic acid/water with flux rates in the range of 40 to 60 g·m^{-2} h^{-1}.

Separation of Solvent Mixtures PV separation using grafted and/or crosslinked membranes offer an alternative energy saving process to separate azeotrope mixtures, such as benzene and cyclohexane mixtures, posing a challenge to the chemical industry. Particularly, composite membranes combining a benzene-soluble polymer (having an affinity for benzene) with a benzene-nonsoluble one (forming the membrane backbone) became popular for such applications. The selectivity of these membranes was enhanced by crosslinking, graft polymerization, and/or microphase separation [163].

Uragami and coworkers tested the PV performance of poly(alkylmethacrylate) (PAMA) crosslinked with ethylene glycol dimethacrylate (EGDMA) for separation of benzene/cyclohexane mixtures [164]. The crosslinked PAMA membranes exhibited a benzene permselectivity, and the permeation rate increased with increasing benzene in the feed solution.

2-hydroxyethyl methacrylate (HEMA) (branch)–methyl acrylate (backbone) comb-type graft copolymer membranes having microphase-separated structures were investigated for pervaporation of benzene/cyclohexane mixtures [165]. Benzene was found to preferentially permeate through the membranes with its permeation rate increased with increasing MA mole fraction in the graft copolymers. It was found that the continuous phases of poly(MA) domains functioned as permeating pathways for benzene, and that the selectivity of the graft polymeric membranes occurred by dissolution of the solvent into the membranes.

Yanagishita et al. prepared crosslinked radiation-grafted membranes based on asymmetric polyimide (PIM) films obtained on nonwoven fabric by the phase inversion process fabric followed by electron-beam (EB) irradiation and subsequent grafting with 4-VPn. The membranes showed high benzene selectivity when used for separation of benzene/cyclohexane mixtures [166].

Pervaporation of methanol/toluene mixtures were investigated using ungrafted poly(ethylene-terephthalate) (PET) and radiation-grafted poly(ethylene-terephthalate)-*graft*-polystyrene (PET-*g*-PSty) membranes. It was found that PET-*g*-PSty membranes exhibited better toluene selectivity than the ungrafted PET membrane, while the permeation fluxes of the grafted membranes were lower. Grafted PET-*g*-PSty membranes with degrees of grafting up to 35% were found to be better than the ungrafted PET for PV of toluene/methanol systems [167, 168].

Pore-filled membranes composed of porous substrate filled with polyelectrolyte polymers have been proposed for organic-liquid pervaporation separation [169]. Grafted membranes prepared by photoinduced graft-filling polymerization of porous films of poly(AN)and PIM grafted with methyl methacrylate (MMA) from the vapor phase showed benzene permselectivity for a benzene/cyclohexane mixture (50/50, v/v) by PV [170, 171]. The composite poly(AN)-based membrane showed a total flux of 0.12 $kg\,g{\cdot}m^{-2}\,h^{-1}$ with a separation factor of 9.1 at 30 °C compared to 0.71 $kg\,g{\cdot}m^{-2}\,h^{-1}$ and 7.4 for composite PIM-based membranes at the same temperature.

Pore-filling membranes composed of porous PE or PP filled with grafted poly(MMA) prepared by plasma copolymerization have found their application in pervaporation separation of the benzene/cyclohexane mixture [169, 172]. The membranes showed high permselectivity for benzene.

Similar pore-filling membranes composed of poly(AN) membranes combined with photografted poly(EGMA) were tested for the pervaporation separation of organic mixtures such as methanol/cyclohexane. High methanol selectivity (2000) and extraordinarily high permeate fluxes ($\leq 8\,kg\,g{\cdot}m^{-2}\,h^{-1}$) were achieved with membranes having a grafted selective layer of <1 μm thick [173].

The use of pore-filling-type membranes for separation of chloroform/*n*-hexane mixtures by reverse osmosis (RO) was also reported. The membranes obtained by grafting of MMA into porous PE and crosslinked with vinyl acrylate (VA) and *N,N′*-methylene- bis-(acrylamide) (MBAAm) [174]. The HDPE-*g*-MMA/VA membranes showed RO separation capability, but they deteriorated after the RO experiments were repeated several times. The HDPE-*g*-MMA/MBAAm membranes exhibited an improved stability and a reasonable separation performance at pressures as high as 12 MPa.

Earlier, Chapiro et al. developed radiation-grafted membranes FEP-*g*-poly(4-VPn) capable of separating ethanol from a model mixture of ethanol/diproperyl ether bearing a composition of 20:80 (v/v) by pervaporation [175].

Recovery of Precious Metals Due to increasing demand, precious metals, including silver (Ag), gold (Au), palladium (Pd), and platinum (Pt), are recovered from a wide variety of mining and industrial waste streams [176]. Biosorption with natural polymers such as modified chitosan offers promising alternative technologies to conventional ones (solvent extraction and chemical precipitation) for the recovery of precious metals [177, 178].

Chitosan modified by GA crosslinking is an efficient sorbent for Pt recovery from dilute acidic solutions. The sorption capacity reaches levels as high as

300 mg Pt g^{-1} (~1.5 mmol Pt g^{-1}) at pH range of 2.0–2.5. Uptake values of 100 mg Pt g^{-1} and higher were obtained at concentrations as low as 0.2 mg Pt L^{-1} [179].

Guibal et al. improved the sorption capacity of chitosan-based sorbents by the grafting of thiourea and rubeanic acid (dithiooxamide) for the extraction of Pd from dilute solutions [180]. The added chelating sulfur functions were found to give the sorbent a dual anion exchange/chelating structure, enhancing the metal sorption capacity, which makes these chitosan derivatives highly efficient in the extraction of Pd from dilute acidic solutions.

Crosslinked chitosan derivatives prepared by crosslinking with GA and hexamethylene diisocyanate (HMDIC) followed by grafting of sulfur compounds showed high efficiency for removal of Au from dilute acidic solutions: maximum uptake capacity reaches 600 mg Au g^{-1} (~3 mmol g^{-1}). The grafting of sulfur compounds and HMDIC crosslinking was found to enhance the sorption kinetics of Au [181].

Environmental Applications

Removal of Heavy Metals from Industrial Waste Water Removal of toxic heavy metals such as Cu^{2+}, Pb^{2+}, Ni^{2+}, Cd^{2+}, Zn^{2+}, Mn^{2+}, Co^{2+}, Hg^{2+}, and Fe^{3+} from industrial waste water is essential from the standpoint of environmental pollution control. This is to eliminate the direct toxic effect of such heavy metals on human and animal health. Various methods for the removal of heavy metals such as ion exchange, chelation, reverse osmosis, coagulation precipitation, electrochemical operation, filtration, and adsorption were adopted [182]. Among many chelating agents, various forms of grafted and crosslinked synthetic or natural polymers containing functional groups (ion exchange resins and membranes, affinity adsorbents, and hydrogels) have been found to be attractive materials for these applications [183, 184]. This was enhanced by the availability of a variety of cheap host polymers and a large number of grafting monomers.

The grafting of hydrophilic monomers onto a variety of polymer films offers various types of membranes that have ion-binding capacity [185–188]. To further enhance the metal adsorption capacity of the membranes, the grafting of binary monomer mixtures was introduced. Cation/anionic membranes prepared by radiation grafting of AAc and 4-VPn onto low-density PE films showed a good potential for separation of Fe^{3+} and other heavy metals such as Pb^{2+}, Cd^{2+}, and Cu^{2+} from contaminated water resources [189]. The stability of similar membranes was improved by grafting of AAc and N-VPn onto PTFE films. These membranes were found to have potential for separation of Cu^{2+}, Ni^{2+}, Pb^{2+}, Cd^{2+}, Zn^{2+}, Mn^{2+}, Co^{2+}, Cr^{2+}, and Fe^{3+} ions [186, 190].

Double (strongly and weakly) acidic membranes prepared by the radiation grafting of Sty/AAc onto low-density PE films followed by sulfonation and alkaline treatment have shown good affinity for adsorption of Pb^{2+} and Fe^{3+} ions [10].

The binary grafting mixture of Sty/AAc was also used to modify PVA films, converting them to hydrogel membranes that showed high selectivity toward Ni^{2+} and Co^{2+} as reported by Abd El-Rehim et al. [191]. They also investigated another membrane by grafting Sty and MAn monomers onto PE films, followed by treatment that includes conversion of the membranes to carboxylic acid and amino groups by alkaline hydrolysis and treatment with different alkylamines [192]. It was found that the grafted films treated with thiosemicarbazide prefer Cu^{2+}, the hydroxylamine hydrochloride treatment prefers Cr^{3+}, and the NaOH treatment prefers Fe^{3+} in a mixture of Cu^{2+}, Cr^{3+}, and Fe^{3+} at 70 °C.

Chelating membranes prepared by radiation grafting of epoxy-group-containing monomers such as GMA onto porous PE films followed by conversion of epoxy groups to iminodiacetate groups have shown great potential for removal of Cu^{2+} from $CuCl_2$ solution [193].

Hollow-fiber membranes prepared by radiation grafting of GMA onto PE hollow fibers followed by amination reaction have shown great potential for the removal of Pb^{2+} and Pd^{2+} ions [194, 195]. Similar PE-*g*-poly(GMA) hollow-fiber membranes activated by iminodiacetation and sulfonation were proposed for the removal of Co^{2+} and Cs^{+} [11, 194], and Pd^{2+} ions [196] and germanium oxide [197] from effluents produced by processing spent catalysts and the production of PET, respectively.

Strongly acidic cationic membranes prepared by radiation grafting of a Sty/DVB mixture onto FEP films followed by sulfonation were originally proposed for acid recovery (chromic acid) from the electroplating industry by the separation of Cu^{2+}, Ni^{2+}, and Fe^{3+} ions from the chromium plating solution [195]. Weakly acidic cation exchange membranes prepared by grafting AAc onto PE, PTFE, FEP, and PFA films have shown a remarkable capability to separate Cu^{2+}, Ni^{2+}, Mo^{2+}, Zn^{2+}, Mn^{2+}, Co^{2+},Cr^{2+}, and Fe^{3+} ions from low and intermediate level waste water [10].

Removal of Hg^{2+} from solutions by radiation-grafted PE-*g*-poly(AAm) membranes was reported by Gupta et al. [188]. These membranes have achieved 99% separation when used to treat a solution containing 200 ppm of Hg^{2+} ions.

Sugi et al. prepared resins to remove heavy metals by the immobilization of phenylalanine (Phe) spherical macroreticular Sty/DVB beads. The obtained resins were found to be capable of separating Cu^{2+} ions from its mixture with Co^{2+} and Ni^{2+} ions. These resins were also capable of removing Cu^{2+} ions from seawater [198].

Photografted membranes composed of poly(2-hydroxyethylmeth- acrylate-co-methacrylamindophenylalanine) were tested for Cu^{2+} adsorption [199, 200]. The membranes showed metal affinity in the order of $Hg^{2+} > Ni^{2+} > Cu^{2+}$, with metal adsorption increasing with increased pH, levelling off at a pH of 5. Grafted Phe/PE hollow-fiber membranes prepared by radiation-induced grafting were also investigated for removal of Cu^{2+} ions [201].

Kumar et al. studied the removal of Co^{2+}, Ni^{2+}, Mn^{2+}, and Cd^{2+} ions from aqueous solution using a grafted adsorbent obtained by radiation grafting AN onto a nonwoven thermally bonded PP followed by the amidoximation of nitrile groups to a chelating amidoxime (AMO) group. The adsorption capacities of the adsorbent for the metal ions were found to follow the order $Cd^{2+} > Co^{2+} > Ni^{2+} > Mn^{2+}$. The kinetics of adsorption of these ions indicated that the rate of adsorption of Cd^{2+} ions was faster than that of other ions studied [202].

Trochimczuk and Streat prepared effective chelating resins composed of a copolymer of AN/EA/DVB for removal of Cd^{2+} ions from a solution in the presence of Ni^{2+} ions. The initial polymer is modified by means of 7.5% w/w ethylenediamine thiophosphonate. The chelating resins showed high selectivity for Cd^{2+} ions in the presence of Ni^{2+} ions in aqueous solutions, independent of the pH which was set in the range of 3.5 to 5.6 [203].

Grafted and/or crosslinked natural polymers such as chitosan and starch offer very promising materials for removal of toxic metal ions from waste water. Particularly, chitosan has been subjected to a tremendous number of studies because of its excellent adsorption capacity for heavy-metal removal. For example, the saturation adsorption capacity of Cd^{2+} ions on chitosan crosslinked with GA decreased exponentially from 250 to $100\,mg \cdot g^{-1}$ as the extent of crosslinking increased from 0 to $1.3\,mol\,GA \cdot mol^{-1}$ of amine [204]. A similar study was conducted by Rorrer et al. [205] with a saturation adsorption capacity of $518\,mg \cdot g^{-1}$ for Cd^{2+} ions on GA-chitosan 1 mm diameter beads.

Beside GA, other crosslinking agents including EPI [206], ethylene glycol diglycidyl ether (EGDGE), iminodiacetic acid [207], nitriloacetic acid, and organic diisocyanates [208] have been used to crosslink chitosan adsorbents. For example, chitosan crosslinked with EGDGE, DEGDGE ether, and PEGDGE showed an interesting selectivity for Cu^{2+} over Ni^{2+} and Co^{2+} ions [209–211].

Polyaminated highly porous resins have been prepared from porous beads of chitosan by crosslinking them with EGDGE and then reacting the crosslinked beads with EPI and polyethyleneimine (PEI), respectively. The order of selectivity of adsorption of metal ions onto the resins at pH 7 was $Hg^{2+} > UO^{2+} > Cd^{2+} > Zn^{2+} > Cu^{2+} > Ni^{2+}$ ions. Interestingly, Ca^{2+} Ga^{3+} As^{3+}, Sr^{2+} ions were not adsorbed on the resins at all [212, 213].

GA- or EPI-crosslinked chitosan membranes were tested for the removal of Hg^{2+} ions from aqueous solutions in a batch reaction under various conditions. The maximum adsorbed amounts, at pH 6.0, were 30.3 and $75.5\,mg \cdot g^{-1}$, respectively, to EPI-crosslinked and GA-crosslinked chitosan membranes. The GA-crosslinked chitosan membranes showed a lower percentage of recovery, but they were chemically stable at lower values of pH, allowing their use in acidic solutions and continuous stages of adsorption and desorption [214]. Similar crosslinked chitosan beads were used for the removal of Fe^{2+} and Fe^{3+} ion from aqueous solutions. The capacity of Fe^{3+} ion adsorption using chitosan and crosslinked chitosan beads was greater than Fe^{2+} ion adsorption [215].

Chitosan grafted with poly(AN) has been further modified by the reaction of the cyano group of chitosan bead-*g*-poly(AN) copolymer with HA, after crosslinking the chitosan with GA and AN grafting to yield amidoximated crosslinked chitosan [216], a derivative that has higher adsorption for Cu^{2+}, Mn^{2+}, and Pb^{2+} ions as compared to crosslinked chitosan. The adsorption capacity had a linear dependence on pH in the cases of Cu^{2+} and Pb^{2+} ions. However, a slight decrease in the adsorption capacity was observed in the case of Zn^{2+} and Cd^{2+} ions [217].

Chitosan-*g*-PVPn resins synthesized by homogenous grafting of 4-VPn onto chitosan were effectively used to bind to Cu^{2+} and then became soluble in dilute HCl [218]. Amphoteric flocculent, which can function as a polycationic flocculent in acidic media and as a polyanionic flocculent in basic media, having much higher adsorption properties, was also synthesized by grafting alkanedioic, alkenedioic, and alkenoic acids [219].

Good adsorbents for Hg^{2+} and Cd^{2+} ions were also obtained by crosslinking chitosan with diisocyanates such as nitriloacetic acid. More details on the prospective of crosslinked chitosan adsorbents can be found in the review by Verma et al. [30].

Recently, other polysaccharide derivatives such as starch, cyclodextrins (CD), cellulose, and biomass have been under consideration as materials to use for developing biosorbents for heavy-metal removal. For example, crosslinked and carboxymethylated starches (CMSs) have been proposed as sorbents for the removal of divalent heavy-metal ions (Cu^{2+}, Pb^{2+}, Cd^{2+}, and Hg^{2+} ions). The removal efficiency of these metal ions depends on the carboxymethyl group and the starch content in the sorbent.

Chan studied the adsorption dynamics of Cu^{2+} ions onto insoluble amphoteric starch that contained quaternary ammonium and phosphate groups [220]. The Cu^{2+} ion sorption process was found to occur in two stages: external mass transport occurs in the early stage and intraparticle diffusion occurs in the long-term stage [221–223].

Zhang and Chen investigated crosslinked poly(dimethylamino- ethyl methacrylate) starch grafted copolymers containing tertiary amine groups as adsorbents for Pb^{2+} and Cu^{2+} ions [224]. The crosslinked starch graft copolymer with a 60% degree of grafting achieved an equilibrium adsorption capacity of 2.09 and 2.12 mmol g^{-1} (dry weight) for Pb^{2+} and Cu^{2+} ions, respectively.

Guo et al. investigated the adsorption of Cu^{2+} ions on crosslinked starch phosphate carbamates adsorbent prepared from an aqueous solution [225]. The adsorbent achieved a maximum adsorption capacity of 1.60 m mol·g^{-1} for Cu^{2+} ions with a regeneration capacity of above 96% when treated with 1 N HCl solution for 1 h.

The use of biomass has been also reported for the development of biosorbent chelating materials for heavy-metal removal. The biomass of *Penicillium chrysogenum* was modified by the grafting of AAc or polyethylenimine (PEI) on the surface of ozone pretreated biomass [226, 227]. However, the preparation process was very complex and not cost effective.

Yu et al. reported a simple process to prepare a biosorbent with high sorption capacity by grafting of poly(amic acid), which was obtained through the reaction of pyromellitic dianhydride (PMDA) and thiourea, onto the biomass (baker's yeast) surface at 50 °C for 4 h. The adsorption capacity of the modified biomass increased 15- and 11-fold for Cd^{2+} and Pb^{2+} ions, respectively, compared with the pristine biomass. The regenerated biomass can be used for at least four times with little loss of uptake capacity [228]. The same authors extended their work by modifying the surface of the dried biomass of baker's yeast by crosslinking the cystine with GA. The adsorption capacity of the modified biomass for Cd^{2+} and Pb^{2+} ions showed an increasing trend compared with the pristine biomass due to the presence of cystine on the biomass surface. The adsorption capacities for Cd^{2+} and Pb^{2+} ions were 11.63 and 45.87 mg g^{-1}, respectively [229].

The use of metal-binding proteins (amino acids) such as glycine (Gly) immobilized onto a solid support (crosslinked resins) as a metal chelating agent has been investigated. The metal-binding capabilities of poly(AAm) crosslinked with *N,N*-methylene-bis-acrylamide (NNMBA) immobilized with Gly was reported for heavy metals: Co^{2+}, Cu^{2+}, Fe^{2+}, Fe^{2+}, Ni^{2+}, and Zn^{2+} ions. The metal-binding capacity increased with the increase in crosslinking up to 8% beyond which it declined [230, 231].

In another study, George and Mathew studied the metal-binding ability of DVB-crosslinked poly(AAm) supported Gly toward Co^{2+}, Cu^{2+}, Ni^{2+}, and Zn^{2+} ions [232]. Interestingly, as the degree of crosslinking increases from 2 to 20%, the metal complexation decreases due to a decrease in the available carboxylate ligands for metal binding with an increase in DVB content. The comparison between the metal-ion complexation characteristics between Gly functionalities supported on DVB-crosslinked poly(AAm) and NNMBA-crosslinked poly(AAm) toward Co^{2+}, Cu^{2+}, Ni^{2+}, and Zn^{2+} ions showed the latter to be more effective at metal complexation than the former, while inclusion of DVB showed an increased metal selectivity over NNMBA [233].

Hydrogels of synthetic or natural polymer origins have been also used for heavy-metal removal. In particular, chelating radiation-grafted poly(VPn/AAc) copolymer hydrogel was investigated for the removal of Fe^{3+}, Cu^{2+}, and Mn^{2+} ions from aqueous solution in a batch equilibration technique under various conditions. The removal of the metal ions followed the order of $Fe^{3+} > Cu^{2+} > Mn^{2+}$ ions [55].

Merrifield et al. investigated the uptake of Hg^{2+} ions by thiol-grafted chitosan gel beads [234]. The gel beads were prepared by dissolving chitosan and converting it into spherical beads using a phase-inversion technique followed by crosslinking to improve the beads' porosity and chemical stability. Cystine was grafted onto the beads in order to improve the adsorption affinity of Hg^{2+} ions to the beads. The adsorption capacity was approximately 8.0 mmol-Hg^{2+} g^{-1} dry beads at pH 7, and decreased with decreasing pH.

Radiation-grafted acid hydrogels composed of PVA/citric acid (CA)/water and PVA/succinic acid (SA)/water were investigated for heavy-metal removal. The incorporation of SA and CA groups onto PVA networks remarkably improved the affinity of these hydrogels for Co^{2+} and Ni^{2+} ion uptake [50].

Radiation-grafted PVA/ethylenediaminetetraacetic acid (EDTA) hydrogels were tested for removal of Pb^{2+}, Cd^{2+}, and Hg^{2+} ions. The incorporation of EDTA moieties onto PVA networks resulted in a hydrogel, which has a strong absorption affinity for Pb^{2+} and Cd^{2+} ions. The maximum chelating capacities of the hydrogel were found to be 8.5 mg/g for Pb^{2+} ions and 4.2 mg/g for Cd^{2+} ions. No affinity for Hg^{2+} ions by the hydrogel was observed [235].

Decontamination of synthetic Pb^{2+}, Zn^{2+}, Cd^{2+}, and Ni^{2+} ion solutions was investigated using crosslinked silica gels chemically modified with poly(ethyleneimine) (PEI) [236].

Hydrogels from cellulose water-soluble derivatives such as chloromethylated (CM) cellulose and from crosslink CM-chitin and CM-chitosan, crosslinked by high-energy radiation at high concentrated pastelike conditions, have shown improved metal adsorption capacity and can be suitable for heavy-metal removal [46, 237, 238].

Removal of Dyes from Industrial Waste Water Removal of dyes generated from industries such as dye manufacturing, textile, pulp and paper production, tanneries, chemical production, and paints and varnishes poses a real challenge for environmental waste control. This is primarily due to dyes' diverse and continuously changing character, complex chemical nature, persistent color, inhibitory and nonbiodegradable nature, and toxicity [239]. At present, various methods are available for color removal, including coagulation flocculation with agent adsorption over activated carbon, charcoal or silica and membrane-filtration processes, oxidation via ozonation or hydrogen peroxide, and treatment with fungi or bacteria. Each one of these methods has advantages and drawbacks when chosen for application. Due to cost consideration, there is renewed research interest in the production of alternative sorbents to replace the costly dye-removing materials and related processes. Attention has been focused on various natural solid supports, which are able to remove pollutants from contaminated water at low cost. Natural polymers offer potentially economical alternative sorbents for dye removal. These natural polymers are abundant, renewable, biodegradable resources and can be easily modified. For example, chitosan molecules have a high content of amine and hydroxyl functional groups, which gives them an extremely high affinity toward many classes of dyes including disperse, direct, reactive, anionic, sulfur and naphthol [240]. Thus, chitosan- and chitin-modified biosorbents have been the subject of a large number of studies, recently reviewed by Crini [26].

Grafting and crosslinking of chitosan have also been made to improve its removal performance and selectivity for dyes, to control its diffusion properties, and to decrease the sensitivity of sorption to environmental conditions. Both batch contacting and column processes are available for chitosan materials with solution containing dyestuffs [241]. The use of crosslinked chitosan for the removal of dyes has been intensively investigated [242–244].

Several crosslinked chitosan sorbents were prepared according to a procedure in which chitosan beads were crosslinked with GA, EPI, or EGDE [62].

The obtained crosslinked chitosan-EPI beads showed higher adsorption capacity than chitosan crosslinked with GA or EGDE, respectively. It was found that 1 g chitosan adsorbed 2498 mg of reactive blue 2. Chitosan-based biosorbents have also demonstrated outstanding removal capabilities for direct dyes. In comparison with commercial activated carbon, the chitosan-EPI beads exhibited excellent performance for adsorption of anionic dyes: the adsorption values were 3–15 times higher at the same pH [244]. The adsorption capacities for some crosslinked chitosan and grafted chitosan biosorbents are summarized in Table 10.1.

Chao et al. have used four kinds of phenol derivatives, namely 4-hydroxybenzoic acid (BA), 3,4-dihydroxybenzoic acid (DBA), 3,4-dihydroxyphenylactic acid (PA), and hydrocaffeic acid (CA) individually as substrates for tyrosinase to be grafted onto chitosan [245]. The modified chitosan gel beads were found to be effective in adsorption of cationic dyes such as crystal violet (CV) and Bismarck brown Y (BB). The adsorption of CV and BB by chitosan modified by carboxyl groups showed a chemical, equilibrated, and saturatable mechanism.

Crosslinked starch containing amine groups were proposed for dye removal by Delval et al. [34–36]. The sorption results showed that the adsorption rate was high and the sorption capacities were significant. Several hundreds ppm of dyes could be removed from water effectively in a few minutes using column experiments.

Radiation-grafted adsorbent starch beads prepared by radiation grafting AN onto maize starch were tested for the removal of basic, acid, and direct dyes from aqueous solution [246]. Similar radiation-grafted beads were evaluated for the removal of basic violet 7, basic blue 3, direct yellow 50, and acid red 37 from aqueous solutions after the nitrile group was converted into amidooxime by reaction with hydroxylamine [247].

Crosslinked cyclodextrin (CD) polymers also have interesting diffusion properties and possess an amphiphilic character. This biosorbent demonstrated an efficient extraction of dyes [39, 40, 248].

TABLE 10.1. The adsorption capacities for some crosslinked chitosan and grafted chitosan biosorbents.

Biosorbent	Dye type	Adsorption capacity ($mg \cdot g^{-1}$)
Crosslinked chitosan beads	Anionic/reactive blue 2	2498
Crosslinked chitosan beads	Anionic/reactive red 2	2422
Crosslinked chitosan beads	Anionic/direct red 81	2383
Crosslinked chitosan beads	Anionic/reactive red 189	1936
Crosslinked chitosan beads	Anionic/reactive yellow 86	1911
–COOH grafted chitosan beads	Cationic/crystal violet	N/A
–COOH grafted chitosan beads	Cationic/Bismarck brown	N/A

[Reproduced from *Dyes Pigments*, **60**: 69(2004)].

Dye biosorbents coupling chitosan with CD such as (CD-*graft*-chitosan showed excellent sorption properties toward different classes of dyes present in textile waste water with a superior rate of sorption and high efficiency as compared to those of parent chitosan and of the well-known CD-epichlorohydrin gels [240]. A maximum adsorption capacity of 77.4 mg·g^{-1} was recorded for acid blue 25.

Treatment of Radioactive Waste Water Ion exchange is an effective treatment method for liquid radioactive waste in nearly all phases of the nuclear fuel cycle, including the early stage of uranium ore treatments, the chemical control of primary circuit coolants during nuclear power plant operations, and polishing water effluents at spent fuel reprocessing plants [249]. Organic and synthetic ion exchangers in membranes and resins forms have found their specific fields of application in different purification and liquid radioactive waste treatment processes.

PE-*g*-poly(AAc) membranes prepared by the radiation grafting of AAc on PE were tested for the recovery of uranium from simulated waste solution. The membranes were found to be highly selective toward Zr, which is the most important contaminating element in uranium recovery [250].

Sulfonated hollow-fiber membranes prepared by radiation grafting of Sty onto hollow PE fibers followed by chemical treatment with chlorosulfonic acid have shown good separation properties for Co^{2+} ions [251]. Yamagishi et al. earlier removed Co^{2+} ions from their solution using similar chelating radiation-grafted porous hollow-fiber membranes [252].

Grafted poly(AAm) membranes prepared by radiation grafting of AAm onto polyvinylchloride (PVC) showed a great tendency to separate Co-60 from liquid radioactive waste containing a mixture of Co-60 and Cs-137 [253].

Amidoxime containing fibrous hydrogels prepared by radiation-induced grafting of AN onto PP fibers and nonwoven fabrics followed by subsequent amidoximation by HA have been found to be very promising for the recovery of uranium and vanadium from sea water [254]. The adsorption for uranium and vanadium was found to be, respectively, 576 and 1800 μg·g^{-1} fiber, from sea water having concentrations of 3.3 and 1.5 μg·g^{-1} sea water, respectively. Similar potential adsorbent materials were prepared by graft copolymerization of an AN/N-VPn mixture followed by amidoximation reaction with HA. An improved adsorption capacity of 0.75 g $UO_2^{2+}g^{-1}$ dry amidoximated adsorbent from its aqueous solution (e.g., 1400 ppm) was obtained [255].

Amidoximated AN/*N*-vinylimidazole (VIm) copolymer hydrogels prepared by γ-radiation followed by amidoximation of the AN moieties of copolymers with HA hydrochloride in a basic medium were used to separate UO_2^{2+} from aqueous solutions by a complexation process. The adsorption capacity was found as high as 0.64 g $UO_2^{2+}g^{-1}$ dry amidoximated copolymer, indicating that that amidoximated poly(AN/VIm) hydrogels are very suitable chelating agents for the adsorption of UO_2^{2+} ions from aqueous systems [53].

Separation of UO_2^{2+} was also attempted by Choi and Nho [256] using chelating membranes prepared by grafting AN, AAc, or AN/AAc mixture. The AN unit of the grafted film was transformed into amidoxime groups. This membrane was found to be good material for UO_2^{2+} separation. In another study, GMA and AN were grafted onto PE and subsequently transformed into IDA and oximes, respectively [194]. These membranes were effective for complexation with UO_2^{2+} and Cu^{2+} ions. The separation of UO_2^{2+} ions by oximes has also been demonstrated by such membranes [257].

Earlier, chitosan derivatives with ketoacids were found to be effective in removal of Co-60 from nuclear plant waste water and for separation of uranium from dilute solutions as well as from saline water [258].

Recovery of Waste Acids Recovery of spent acids from waste acids produced by several industries such as pickling and surface treatment, pigments, and hydrometallurgy has become of major environmental concern. Conventional technology for spent acid recovery includes either neutralization, which leads to large amounts of sludge, or thermal techniques such as distillation and evaporation or pyrolysis, which are economically viable only for large volumes of effluents. Among the possible alternatives, ED with IEMs is an attractive technique for treating waste acids [259].

The recovery of sulfuric acid from the zinc hydrometallurgy industry has proved to be viable in laboratory and pilot scale ED cells when tested with three crosslinked commercial IEMs: Neosepta CMS (Tokuyama Soda Co.), Morgane CRA (Solvay), Selemion CHV (Asahi Glass Co.) in comparison with Nafion membranes [260].

Recovery of sulfuric acid produced in the diamond manufacturing process by the diffusion dialysis (DD) method using a commercial crosslinked anion exchange membrane was conducted by Jeong et al. [261]. About 80% of the acid could be recovered from waste sulfuric acid, which contained 4.5 M free-acid at a flow rate of $0.26 \times 10^{-3}\,m^3 \cdot h^{-1} m^{-3}$. The concentration of recovered sulfuric acid was 4.3 M, with impurities as low as 2000 ppm. This study has revealed that DD has high economic prospects for such an application.

Sulfonated and crosslinked poly(Sty)-*grafted* ETFE membranes prepared by radiation grafting showed an excellent performance in acid recovery [262]. Poly(4-VPn)-*grafted* ETFE membranes also showed a comparable performance to commercial IEMs (De Morgan) in acid recovery by dialysis [263].

Also, membranes obtained by radiation grafting of vinyl monomers such as AAc, MAc, AAm, and N-VPn, 2-HEMMA onto various films (PE, PP, and ETFE) were found to be promising for acid recovery by dialysis [117, 264, 265].

Biotechnology and Biomedical Applications

Purification of Protein Rapid development in biotechnology requires reliable and efficient methods to purify commercial-scale quantities of proteins for use in the pharmaceutical and food industries [266]. Selective grafted resins and hydrogels are widely used in separation and purification of proteins based on (i) ion exchange, (ii) affinity, and (iii) metal complexation that forms various chromatographic modes and techniques developed with packed-bed columns. The major limitations associated with conventional packed-bed chromatography for protein separation and purification are mainly high pressure drop, diffusion limitation, and long processing time. Such drawbacks can be overcome by using adsorptive microporous membranes as chromatographic media. These membranes offer higher throughput and shorter processing times, with an easy scaling up when compared with packed bed systems [267, 268]. Membrane adsorbers combine the selectivity of chromatographic resins with the high productivity of filtration membranes [269].

Membrane adsorbers can exist in a variety of configurations [270, 271] with a variety of adsorptive mechanisms (e.g., ion exchange, hydrophobic, reversed-phase, or affinity interactions). The principle of selective or specific collection of ions and proteins during permeation of the liquid across a microporous membrane has been verified using various systems [73, 268, 272].

Grafting has been used to increase the binding capacity of rigid polymeric membranes through reduced steric hindrance associated with the binding of large molecules on rigid surfaces, as reported by Koguma et al. [273].

The use of grafted porous affinity and chelating membranes for the collection of proteins has been reported in the literature. For example, lysozyme adsorption was examined during the permeation of the lysozyme solution (pH 6) through a microporous cation-exchange hollow fiber containing sulfopropyl (SP) groups prepared by radiation-induced grafting of GMA followed by hydrolysis of the resulting epoxide group into a diol, and then converting the diol into the SP groups. The SP group density of the resulting hollow fiber ranged from 0.21 to 0.84 $mol \cdot kg^{-1}$ of dry fiber with a pure water flux of 2.7 $m \cdot h^{-1}$ at a filtration pressure of 0.1 MPa. These membranes showed a negligible diffusional resistance of lysozyme to the SP group. The binding capacity of lysozyme to the fiber was constant in this range of SP group density [274]. Adsorption of lysozyme was also investigated with similar membranes containing sulfonic acid groups (SO_3H) anchored onto grafted microporous PE hollow- fiber membranes [275].

A hydrophobic amino acid having a pseudoaffinity ligand, such as tryptophan or -phenylalanine, immobilized into a microfiltration PE hollow-fiber membrane was used to remove disease-causing proteins from blood. These affinity membranes were prepared by radiation-induced grafting of GMA onto porous PE hollow fibers and subsequent coupling of the polymer brush grafted epoxide group with tryptophan or -phenylalanine. The remaining epoxide groups were hydrolyzed into a diol group with sulfuric acid [65, 66]. Bovine γ-globulin as a model protein was adsorbed onto the pseudoaffinity microporous PE hollow fiber membranes. The equilibrium binding capacity was found to be between two extremes of oriented bindings: end-on and side-in bindings onto the interface between the polymer brush and the liquid.

Hydrogel membranes prepared by grafting monomers such as *N*-vinyl-2-pyrrolidone onto PE, PP, PTFE, and FEP were found to have potential for biomedical applications [276–279]. Membranes prepared by grafting ethylene glycol methacrylate onto PE films were also found to have surfaces that show low levels of adsorption of protein and adhesion of platelets [280].

Chromatographic supports for the use in protein purification using affinity chromatography were also prepared by grafting polar and ionic monomers onto porous polymer substrates [268, 271, 281]. For example, immobilized metal ion (Cu^{+2}) affinity hollow-fiber membranes prepared by radiation-induced grafting of GMA onto hydrophilized PE membranes followed by conversion of the epoxy group by iminodiacetic acid treatment have shown potential for protein adsorption [282, 283].

High-capacity membrane adsorbents composed of commercial cellulose hydrate grafted with selective ligands were used as a stationary phase for chromatographic purification of human serum albumin (HSA) from desalted human plasma. The membranes showed high productivity not only for albumin, but also for products of higher molecular values such as human immunoglobulin G (HIgG) and set to be promising for commercial-scale separation and purification [284].

Separation of HSA, the most abundant protein present in human plasma, from HIgG, the second-most abundant plasma protein, was also investigated by membrane chromatography using a novel macroporous gel-filled membrane under various operating conditions. The membranes were prepared by anchoring a quaternary ammonium salt gel within the pores of a nonwoven fabric of PP. The membranes displayed an HSA-binding capacity range as high as 290–300 $mg \cdot ml^{-1}$ (under saturating conditions), which was not only significantly higher than the binding capacities reported for other chromatographic membranes, but also higher than the binding capacities of conventional gel-based chromatographic media. The suitability of using this membrane for plasma fractionation was demonstrated by the separation of a simulated feed solution consisting of HSA and HIgG [285].

Blood Purification (Haemodialysis) Blood purification using the dialysis process (haemodialysis) is one of the most important and oldest membrane biomedical applications where membranes are used as an artificial kidney for patients suffering from renal failure. The membrane replaces the kidney in removing toxic low-molecular components such as urea, creatinine, phosphate, and uric acid.

Membranes prepared by radiation-induced grafting have been for years attractive alternative biocompatible materials for haemodialysis application [3]. Haemodialysis membranes prepared by radiation-induced graft copolymerization of vinyl alcohol and *N*-VPn onto PE films were early reported in the literature [286]. These membranes showed a tenfold increase in the solute permeability compared to commercial dialysis membranes.

TABLE 10.2. List of various applications of grafted and/or crosslinked polymers.

No.	Categories	Areas of applications	References
1	Water production applications	—Desalination of brackish water	1, 2, 81–92
		—Desalination of seawater	64, 68, 88, 93–106
		—Softening of hard water	7, 49, 106
2	Chemical industry applications	—Production of caustic soda	3, 4, 6, 108–125
		—Production of mineral acids	126–128
		—Water electrolysis	129–135
		—Dehydration of solvents	>117, 125, 136–162
		—Separation of solvent mixtures	163–175
		—Recovery of waste acids	176–181
3	Environmental applications	—Removal of heavy metals from industrial waste water	10, 11, 30, 46, 50, 55, 182–238
		—Removal of dyes from industrial waste water	26, 34, 36, 39, 40, 62, 239–248
		—Treatment of radioactive waste water	53, 194, 249–258
		—Recovery of waste acids	117, 259–265
4	Biotechnological and biomedical applications	—Purification of proteins	65, 66, 73, 266–285
		—Blood purification	3, 146, 185, 264, 286
5	Other applications	—Gas separation	287, 288
		—Chiral separation of enantiomers	289

Other membranes of PVA modified by the radiation grafting of AAc and MAc monomers were tested for dialysis [146]. The permeability of these membranes toward NaCl, urea, creatinine, and uric acid was higher than that of commercial PVA and cellephane membranes.

Membranes made of poly(2-HEMMA) grafted chloroprene rubber and γ-methyl-L-glutamate obtained by grafting 2-HEMMA onto chloroprene rubber and γ-methyl-L-glutamate showed good dialytic capabilities when they were tested for the permeability of urea, creatinine, and creatine in a dialysis cell [185, 265].

Similar membranes obtained by radiation grafting 2-HEMMA onto PP films showed permeation rates of 4.9 and 19 times higher than the original PP film for urea and uric acid, respectively. These values are close to the commercial cuprophane membranes [264]. The various applications of grafted and/or crosslinked polymers covered in this chapter are summarized in Table 10.2.

Other Applications

Asymmetric membranes prepared by radiation-induced grafting a from vapor phase of several AAc monomers onto poly(vinyltrimethylsaline) films were investigated and showed improved gas-separation properties [287]. Radiation crosslinking of these membranes in the presence of EGDMA or radiation-induced grafting with allyl methacrylate from a vapor phase was found to improve the resistance of the membrane toward hydrocarbon solvents beside enhancing the gas separation properties for O_2 and N_2 [287].

Complex membranes of polyl[(1-trimethylsilyl)-1-propyne]-*graft*-poly(AAc)-Ag^+ prepared by the radiation-induced grafting of AAc onto polyl(1-trimethylsilyl)-1-propyne followed by treatment with $AgNO_3$/glycerol, was found to have great potential for the separation of C_4 olefin/paraffin gas mixtures (isobutene/isobutane) and other linear gas mixtures having close boiling temperature (butene/butane). A complex membrane having a 121.5% degree of grafting showed a typical selectivity of 7.7 and 7.3 for isobutene/isobutane and trans-2-butane/n-butane gas mixtures, respectively [288].

Radiation grafting was also used to prepare enantioselective membranes for chiral separation of racemic mixtures of two enantiomers, a process commonly used in the pharmaceutical industry to separate effective therapeutic agents [289]. Nakamura et al. prepared enantioselective membranes by grafting GMA onto irradiated PE porous hollow fibers followed by amination. Then, the membrane was immobilized with bovine serum albumin. The membrane was crosslinked with GA to suppress protein leaching. Evaluation of this membrane for tryptophane under chromatographic conditions produced an excellent enantioseparation factor of 12.

CONCLUDING REMARKS AND FUTURE DIRECTIONS

Grafted and/or crosslinked polymers continue to play very significant roles in the fields of separation and purification. A large number of chemically, radiochemically, photochemically or plasma-modified polymers have been developed in various forms and functionalities for specific separation applications, and many more are under development. Special attention is given to natural polymer derivatives, as they are economically attractive materials. Nevertheless, greater effort is needed to bring many more of grafted/crosslinked polymers from the experimental level to the pilot scale production and extend their practical applications. This can be achieved by addressing several aspects, which include boosting the selectivity without compromising flux and stability, optimizing polymer modification techniques, understanding the mechanism of transport, establishing structure-property relationships for polymer forms, using the proper engineering configurations, testing separation with actual samples under various conditions, and using cost-effective materials.

REFERENCES

1. Nagarale RK, Gohil GS, Shahi VK. *Advan Colloid Inter Sci* **119**: 97 (2006).
2. Xu T. *J Membr Sci*, **263**: 1 (2005).

3. Nasef MM, Hegazy E-SA. *Prog Polym Sci*, **29**: 499 (2004).
4. Renaud S, Bruno A. *Prog Polym Sci*, **30**: 644 (2005).
5. Smitha B, Sridhar S, Khan AA. *J Membr Sci*, **259**: 10 (2005).
6. Gubler L, Gürsel SA, Scherer GG. *Fuel Cells*, **5**: 317 (2005).
7. Jagur-Grodzinski J. *React Funct Polym*, **39**: 99 (1999).
8. Hoffman A. *Radiat Phys Chem*, **9**: 207 (1977).
9. Clough RL. *Nucl Inst Meth Phys Res B*, **185**: 8 (2001).
10. Hegazy E-S, Kamal H, Maziad N, Dessouki A. *Nucl Instr Meth Phys Res B*, **151**: 386 (1999).
11. Choi S-H, Nho YC. *J Appl Polym Sci*, **71**: 999 (1999).
12. Abrams IM, Millar JR. *React Funct Polym*, **35**: 7 (1997).
13. Kariduraganavar MY, Nagarale RK, Kittur AA, Kulkarni SS. *Desalination*, **197**: 225 (2006).
14. Sata T. *Pure Appl Chem*, **58**: 1613 (1986).
15. Strathmann H, in *Membrane Separation Technology—Principles and Applications*, RD Nob and SA Stern, eds. (Amsterdam: Elesevier Science BV, 1995), p. 214.
16. Strathmann H, *Ion Exchange Membrane Separation Processes* (Amsterdam: Elsevier, 2004), p. 114.
17. Korkisch J. *Handbook of Ion Exchange Resins: Their Application To Inorganic Analytical Chemistry*, Vol. **1** (Boca Raton, FL: CRC Press, 1989), p. 36.
18. Dardel F, Arden TV, in *Ullmann's Encyclopedia of Industrial Chemistry*, F Ullmann, W Gerhartz, YS Yamamoto, FT Campbell, R Pfefferkorn, and JF Rounsaville, eds. (Weinheim: Wiley-VCH Verlag GmbH & Co., 2005), p. 1.
19. Nonaka T, in *The Polymeric Materials Encyclopedia*, JC Salamone, eds. (Boca Raton, FL: CRC Press, Inc., 1996), p. 130.
20. Geckeler KE, in *The Polymeric Materials Encyclopedia*, JC Salamone, eds. (Boca Raton, FL: CRC Press, Inc., 1996), p. 4101.
21. Chanda M, Rempel GL. *React Funct Polym*, **35**: 197 (1997).
22. Saglam A, Bektas S, Patir S, Genc O, Denizli A. *React Funct Polym*, **47**: 185 (2001).
23. Yantasee W, Lin Y, Fryxell GE, Alford KL, Busche BJ, Johnson CD. *Ind Eng Chem Res*, **43**: 2759 (2004).
24. Chen JP, Yang L. *Ind Eng Chem Res*, **44**: 9931 (2005).
25. Pal S, Mal D, Singh RP. *Colloids Surf A: Phys Eng Aspects*, **289**: 193 (2006).
26. Crini G. *Bioresour Technol*, **97**: 1061 (2006).
27. Ravi Kumar MNV. *React Funct Polym*, **46**: 1 (2000).
28. Lee ST, Mi FL, Shen YJ, Shyu SS. *Polymer*, **42**: 1879 (2001).
29. Rodrigues CA, Laranjeira MC, De F, Valfredo T, Sadler E. *Polymer* **39**: 5121 (1998).
30. Verma AJ, Deshpande SV, Kennedy JF. *Carbohydr Polym*, **55**: 77 (2004).
31. Khalil M, Abdel-Halim M. *Starch/Stärke*, **53**: 35 (2001).
32. Lutfor MR, Sidik S, Wan Yunus WMZ, Rahman MZA, Mansour A, Haron MJ. *J Appl Polym Sci*, **79**: 1256 (2001).

33. Delval F, Crini G, Janus L, Vebrel J, Morcellet M. *Macromol Symp*, **166**: 103 (2001).
34. Delval F, Crini G, Morin N, Vebrel J, Bertini S, Torri G. *Dyes Pigments*, **53**: 79 (2002).
35. Delval F, Crini G, Vebrel J, Knorr M, Sauvin G, Conte E. *Macromol Symp*, **203**: 165 (2003).
36. Delval F, Crini G, Bertini S, Morin-Crini N, Badot PM, Vebrel J, Torri G. *J. Appl Polym Sci*, **93**: 2650 (2004).
37. Delval F, Crini G, Bertini S, Filiatre C, Torri G. *Carbohydr Polym*, **60**: 67 (2005).
38. Crini G, Cosentino C, Bertini S, Naggi A, Torri G, Vecchi C, Janus L, Morcellet M. *Carbohydr Res*, **308**: 37 (1998).
39. Crini G, Morcellet M. *J Sep Sci*, **25**: 1 (2002).
40. Crini G. *Bioresour Technol*, **90**: 193 (2003).
41. Kitaoka M, Hayashi K. *J Incl Phenom Macrocyclic Chem*, **44**: 429 (2002).
42. Yu JC, Jiang ZT, Liu HY, Yu L, Zhang L. *Anal Chim Acta*, **477**: 93 (2003).
43. Crini G. *Prog Polym Sci*, **30**: 38 (2005).
44. Kume T, Nagasawa N, Yoshii N. *Rad Phys Chem*, **63**: 625 (2002).
45. Yoshii F, Zhao L, Wach RA, Nagasawa N, Mitomo H, Kume T. *Nuclear Inst Methods Phys Res B*, **208**: 320 (2003).
46. Zhao L, Mitomo H, Nagasawa N, Yoshii F Kume T. *Carbohydr Polym*, **51**: 169 (2003).
47. Wach RA, Mitomo H, Nagasawa N, Yoshii F. *Rad Phys Chem*, **68**: 771 (2003).
48. Hoffman AS. *Advan Drug Deliv Rev*, **43**: 3 (2002).
49. Chauhan GS, Lal H. *Desalination*, **159**: 131 (2003).
50. Bodugoz H, Pekel N, Guven O. *Rad Phys Chem*, **55**: 667 (1999).
51. Pekel N, Sahiner N, Guven O. *Rad Phys Chem*, **59**: 485 (2000).
52. Pekel N, Sahiner N, Güven O. *J Appl Polym Sci*, **81**: 2324 (2001).
53. Pekel N, Güven O. *Colloids Surf A: Phys Eng Aspects*, **212**: 155 (2003).
54. Kanto lu M en, Güven O. *Nucl Instr Meth Phys Rev B*, **151**: 218 (1999).
55. El-Hag AA, Shawky HA, Abd El-Rehim HA, Hegazy E-SA. *Euro Polym J*, **39**: 2337 (2003).
56. Francis S, Varshney L. *Rad Phys Chem*, **74**: 310 (2005).
57. Duran D , Güven O. *Nucl Instr Meth Phys Rev B*, **151**: 196 (1999).
58. Coviello T, Matricardi P, Marianecci C, Alhaique F. *J Control Release*, **119**: 5 (2007).
59. Zou H, Luo Q, Zhou D. *J Biochem Biophys Methods*, **49**: 199 (2001).
60. Charcosset C. *J Chem Tech Biotech*, **71**: 95 (1998).
61. Klein E. *J Membr Sci*, **179**: **1** (2000).
62. Zeng XF, Ruckenstein E. *Ind Eng Chem Res*, **35**: 4169 (1996).
63. Kugel K, Moseley A, Harding G, Klein E. *J. Membr Sci*, **74**: 115 (1992).
64. Saito K, Kaga T, Yamagishi H, Furusaki S, Sugo T, Okamoto J. *J Membr Sci*, **43**: 131 (1989).
65. Kim M, Saito K, Furusaki S, Sugo T, Ishigaki I. *J Chromatogr*, **585**: 45 (1991).

66. Kim M, Saito K, Furusaki S, Sugo T, Ishigaki I. *J Chromatogr*, **586**: 27 (1991).
67. Kawai T, Sugita K, Saito K, Sugo T. *Macromolecules*, **33**: 1306 (2000).
68. Kobayashi K, Tsuneda S, Saito K, Yamagishi K, Furusaki S, Sugo T. *J Membr Sci*, **76**: 209 (1993).
69. Tsuneda S, Kagawa H, Saito K, Sugo T. *J Colloid Inter Sci*, **176**: 95 (1995).
70. Thom V, Jankova K, Ulbricht M, Kops J, Jonsson G. *Macromol Chem Phys*, **199**: 2723 (1998).
71. Ulbricht M, Riedel M. *Biomater*, **19**: 1229 (1998).
72. Tsuneda S, Saito K, Furusaki S, Sugo T, Okamoto J. *J Membr Sci*, **58**: 221 (1991).
73. Iwata H, Saito K, Furusaki S, Sugo T, Okamoto J. *Biotechnol Prog*, **7**: 412 (1991).
74. Boi C. *J Chromatogr B*, **848**: 19 (2007).
75. Casey JT, O'Cleirigh C, Walsh PK, O'Shea DG. *J Microbiol Meth*, **58**: 327 (2004).
76. Guo W, Ruckenstein E. *J Chromatogr B* **795**: 61 (2003).
77. Gordon NF, Moore CMV, Cooney CL. *Biotechnol Adv*, **8**: 471 (1990).
78. Adachi T, Isobe E. *J Chromatogr*, **1063**: 33 (2004).
79. Zagorondni AA. *Ion Exchange Materials: Properties and Application* (Amsterdam: Elsevier, 2007), p. 263.
80. Baker RW. *Membrane Technology and Applications*, (Chichester: John Wiley and Sons Inc., 2004), p. 139.
81. Baker RW. *Membrane Technology and Applications*, (Chichester: John Wiley and Sons Inc., 2004), p. 355.
82. Mulder M. *Basic Principles of Membrane Technology*, (Dordrecht: Kluwer Academic Publishers, 1996), p. 231.
83. Veza JM. *Desalination*, **133**: 259 (2001).
84. Lee H-J, Moon S-H, Tsai SP. *Sep Purif Technol*, **27**: 89 (2002).
85. Demircioglu M, Kabay N, Kurucaovali I, Ersoz E. *Desalination*, **153**: 329 (2002).
86. Tanaka Y. *J Membr Sci*, **215**: 265 (2003).
87. Haung TC, Wang JK. *Eng Ind Chem Res*, **32**: 133 (1993).
88. Sata T. *J Membr Sci*, **93**: 117 (1994).
89. Aminabhavi TM, Vijaya Kumar, Naidu B, Sridhar S. *J Appl Polym Sci*, **94**: 1827 (2004).
90. Choi S-H, Han JY, Jeong RJ, Lee K-P. *Rad Phys Chem*, **60**: 503 (2001).
91. Lindstrand V, Sundström G, Jönsson A. *Desalination* **128**: 91 (2000).
92. Lindstrand V, Jönsson A-S, Sundström G. *Desalination*, **130**: 73 (2000).
93. Bohdziewicz J, Bodzek M, Wasik E. *Desalination*, **121**: 139 (1999).
94. Tanaka Y. *J Membr Sci*, **163**: 277 (1999).
95. Matsuura T. *Desalination*, **134**: 47 (2001).
96. Hegazy E-S, Ishigaki I, Okamoto J. *J Appl Polym Sci*, **26**: 3117 (1981).
97. Hegazy E-S, El-Dessouky M, El-Sharabasy S. *Rad Phys Chem*, **27**: 323 (1986).
98. Hegazy E-S, El-Assy N, Dessouki A, Shaker M. *Rad Phys Chem*, **33**: 13 (1989).
99. Higuchi A, Iijima T. *Polymer*, **26**: 1207 (1985).

100. Vigo F, Capannelli G, Munari S. *Desalination*, **37**: 313 (1981).
101. Bittencourt E, Stannett V, Williams JL, Hopfenberg HB. *J Appl Polym Sci*, **26**: 879 (1981).
102. Bittencourt E, Stannett V, Hopfenberg HB, Williams JL. *J Appl Polym Sci*, **26**: 2897 (1981).
103. Canepa G, Munari S, Rossi C, Vigo F. *Desalination*, **13**: 159 (1973).
104. Kaur I, Gupta A, Misra B, Chauhan G. *Desalination*, **110**: 115 (1997).
105. Kaur I, Misra B, Kohli A. *Desalination*, **139**: 357 (2001).
106. Bequet S, Abenoza T, Aptel P, Espenan J-M, Remigy J-C, Ricard A. *Desalination*, **131**: 299 (2000).
107. Qiu C, Nguyen QT, Ping Z. *J Membr Sci*, **295**: 88 (2007).
108. Balster J, Stamatialis DF, Wessling M. *Chem Eng Process*, **43**: 1115 (2004).
109. Chen WKW, Mesobian RB, Ballantian DS, Metz DJ, Glines, A. *J Polym Sci*, **23**: 903 (1957).
110. Chen W-W. *Method of Forming Graft Copolymer Ion Exchange Membranes*, U.S. Patent 3,257,334, June 21, 1966.
111. Lee L, Liu K-J. *Preparation of High Performance Polyelectrolyte Membrane*, U.S.Patent 4,083,768, April 11, 1978.
112. Dobo J, Somogyi A, Czvikovszky T. *J Polym Sci, Part C*, **4**: 1173 (1964).
113. Munari S, Vigo F, Tealdo G, Rossi C. *J Appl Polym Sci*, **11**: 1563 (1967).
117. Machi S, Kamel I, Silverman J. *J Polym Sci: Polym Chem Ed*, **8**: 3329 (1970).
115. Xu Z, Gen-hu W, Han-ing W, Gyn C, Min-hua N. *Rad Phys Chem*, **22**: 939 (1983).
116. Suhara M, Arai K. *Electrolysis of Aqueous Alkali Metal Chloride Solution*. German Patent De, 2,916,111 (1980).
117. Hayashi K, Kotani N. *J Polym Sci: Polym Chem Ed*, **18**: 191 (1980).
118. McRae W, in *Effective Membrane Processes: New Perspectives*, R Paterson, ed. (London: Mechanical Engineering Publications Ltd., 1993), p. 149.
119. Momose T, Yoshioka H, Ishigaki I, Okamoto J. *J Appl Polym Sci*, **37**: 2165 (1989).
120. Momose T, Yoshioka H, Ishigaki I, Okamoto J. *J Appl Polym Sci*, **38**: 2091 (1989).
121. Momose T, Tomie K, Harada H, Miyachi H, Kato H. *Method for Preparation of Graft Polymeric Membranes*, U.S. Patent 4,605,685, August 12, 1986.
122. D'Agostino V, Lee J, EH CJ. *Trifluorostyrene Sulfonic Acid Membranes*, U.S. Patent 4,012,303, March 15, 1977.
123. Kostov G, Matsuda O, Machi S, Tabata Y. *J Membr Sci*, **68**: 133 (1992).
124. Guzman-Garcia AG, Pintauro PN, Verbrugge MW, Schneider EW, *J Appl Electrochem*, **22**: 204 (1992).
125. Wang H, Capuano GA. *J Electrochem Soc*, **145**: 780 (1998).
126. Tzanetakis N, Varcoe J, Slade RS, Scott K. *Electrochem Comm*, **5**: 115 (2003).
127. Tzanetakis N, Taama WM, Scott K, Varcoe J, Slade RS. *Desalination*, **151**: 275 (2002).

128. Tzanetakis N, Taama WM, Scott K. *Filtr Separ*, **40**: 33 (2003).
129. Millet P, Pineri M, Durand R. *J Appl Electrochem*, **19**: 162 (1989).
130. Rasten E, Hagen G, Tunold R. *Electrochim. Acta* **48**: 3945 (2003).
131. Slavcheva E, Radev I, Bliznakov S, Topalov G, Andreev P, Budevski E. *Electrochimica Acta*, **52**: 3889 (2007).
132. Stucki S, Scherer GG, Schlagowski S, Fischer E. *J Al Electrochem*, **28**: 1041 (1998).
133. Kreuter W, Hofmann H. *Hydrogen Ener Progr XI*, **1**: 537 (1996).
134. Mitsugi C, Harumi A, Kenzo F. *Int J Hydrogen Energy*, **23**: 159 (1998).
135. Scherer GG, Momose T, Tomiie K. *J Electrochem Soc*, **135**: 3071 (1988).
136. Shantora V, Huang R, Shao P, Huang RYM. *J Membr Sci*, **287**: 162 (2007).
137. El-Nesr E, Dessouki A, Abdel-Bary E. *Polym Int*, **46**: 150 (1998).
138. Matsuda O, Kostov G, Tabata Y, Machi S. *J Appl Polym Sci*, **24**: 1053 (1979).
139. Gupta B, Mukherjee AK. *Polym Plast Eng Technol*, **29**: 299 (1990).
140. Ping Z, Nguyen Q-T, Essamri A, Neel J. *Polym Adv Technol*, **5**: 320 (1994).
141. Garcia FE, Nobrege R, Oliveira CMF. *Polym Bull*, **37**: 519 (1996).
142. Semenova SI, Ohya H, Soontarapa K. *Desalination*, **110**: 251 (1997).
143. Zhigong R, Guixiang L, Sugo T, Okamoto J. *Int. J Radiat Appl Instr Part C: Rad Phys Chem*, **39**: 421 (1992).
144. Ping Z, Nguyen Q, Chen S, Ding Y. *J Membr Sci*, **195**: 23 (2002).
145. Niemoeller A, Scholz H, Goetz B, Ellinghorst G. *J Membr Sci*, **36**: 385 (1988).
146. Shantora V, Huang R. *J Appl Polym Sci*, **26**: 3223 (1981).
147. Yong SK, Sang WL, Un YK, Jyong SS. *J Membr Sci*, **51**: 215 (1990).
148. Huang RYM, Yeom CK. *J Membr Sci*, **51**: 273 (1990).
149. Tsuneda S, Saito K, Furusaki S, Sugo T, Ishigaki I. *J Membr Sci*, **71**: 1 (1992).
150. Huang RYM, Pal R, Moon GY. *J Membr Sci* **160**: 101 (1999).
151. Yeom CK, Lee K-H. *J Membr Sci*, **135**: 225 (1997).
152. Yeom CK, Lee K-H. *J Appl Polym Sci*, **67**: 949 (1998).
153. Kurkuri MD, Toti US, Aminabhavi TM. *J Appl Polym Sci*, **86**: 3642 (2002).
154. Yang G, Zhang L, Peng T, Zhong W. *J Membr Sci*, **175**: 53 (2000).
155. Toti US, Tejraj MA. *J Membr Sci*, **228**: 199 (2004).
156. Toti US, Aminabhavi TM. *J Appl Polym Sci*, **85**: 2014 (2002).
157. Toti US, Kariduraganavar MY, Soppimath KS, Aminabhavi TM. *J Appl Polym Sci*, **83**: 259 (2002).
158. Wang XN. *J Membr Sci*, **170**: 71 (2000).
159. Zhang W, Li G, Fang Y, Wang X. *J Membr Sci*, **295**: 130 (2007).
160. Chen W, Li W, Zhong W, Yu T. *J. Macromol Sci: Pure Appl Chem*, A**34**: 2451 (1997).
161. Choudhari SK, Kittur AA, Kulkarni SS, Kariduraganavar MY. *J Membr Sci*, **302**: 197 (2007).
162. Huang R, Xu Y. *J Membr Sci*, **43**: 143 (1989).
163. Garcia-Villaluenga JP, Tabe-Mohammadi A. *J Membr Sci* **169**: 159 (2000).
164. Kuniaki I, Okumura H, Miyata T, Uragami T. *J Membr Sci*, **132**: 193 (1997).

165. Terada J, Hohjoh T, Yoshimasu S, Ikemi M, Shinohara I. *Polym J*, **145**: 347 (1982).

166. Yanagishita H, Arai J, Sandoh T, Negishi H, Kitamoto D, Ikegami T, Haraya K, Idemoto Y, Koura N. *J Membr Sci*, **232**: 93 (2004).

167. Khayet M, Nasef MM, Mengual JI. *J Membr Sci*, **263**: 77 (2005).

168. Khayet M, Nasef MM, Mengual JI. *Desalination*, **193**: 109 (2006).

169. Yamaguchi T, Nakao S, Kimura S. *Macromol*, **24**: 5522 (1991).

170. Wenzel A, Yanagishita H, Kitamoto D, Endo A, Haraya K, Nakane T, Hanai N, Matsuda H, Koura N, Kamusewitz H, Paul D. *J Membr Sci*, **179**: 69 (2000).

171. Yanagishita H, Kitamoto D, Endo A, Haraya K, Nakane T, Hanai N, Arai J, Matsuda H, Idemoto Y, Koura N. *J Membr Sci*, **203**: 191 (2002).

172. Yamaguchi T, Nakao S, Kimura S. *Ind Eng Chem Res*, **32**: 848 (1993).

173. Ulbricht M, Schwarz H-H. *J Membr Sci*, **136**: 25 (1997).

174. Kai T, Goto H, Shimizu Y, Yamaguchi T, Nakao S-i, Kimura S. *J Membr Sci*, **265**: 101 (2005).

175. Chapiro A, Schmidt N, Randriamahefa S, Misra B, in *ACS 16th Biennial Symposium on Advances in Polymerization and Polymeric Materials* (New Jersey: ACS, 1992).

176. Brooks CS. *Metal Recovery from Industrial Wastes* (Boca Raton, FL: CRC Press, 1991), p. 267.

177. Byerley JJ, Schareri JM, Rioux S, in *Biohydrometallurgy*, J Salley, R McCready, and PL Wichlacz, eds. (Montreal: CANMET, 1989), p. 301.

178. Göksungur Y, Üren S, Güvenç U. *Bioresour Technol*, **96**: 103 (2005).

179. Guibal E, Larkin A, Vincent T, Tobin JM. *Ind Eng Chem Res*, **38**: 4011 (1999).

180. Guibal E, Sweeney, NVO, Vincent T, Tobin JM. *React Funct Polym*, **50**: 149 (2002).

181. Arrascue ML, Garcia HM, Horna O, Guibal E. *Hydrometallurgy*, **71**: 191 (2003).

182. Dabrowski A, Hubicki Z, Podkoscielny P, Robens E. *Chemosphere*, **56**: 91 (2004).

183. Lee W, Saito K, Furusaki S, Sugo T, Makuuchi K. *J Membr Sci*, **81**: 295 (1993).

184. Kawai T, Saito K, Sugita K, Katakai A, Seko N, Sugo T, *Rad Phys Chem*, **59**: 405 (2000).

185. Fang Y-E, Lu X-B, Cheng Q. *J Appl Polym Sci*, **68**: 83 (1998).

186. Hegazy E-S, El-Rehim H, Shawky H. *Rad Phys Chem*, **57**: 85 (2000).

187. Gupta B, Anjum N. *J Appl Polym Sci*, **82**: 2629 (2001).

188. Gupta B, Anjum N, Jain R, Revagade N, Singh H. *J Macromol Sci: Part C: Polym Rev*, C**44**: 275 (2004).

189. Hegazy E-S, Abd El-Rehim H, Khalifa N, Atwa S, Shawky H. *Polym Int*, **43**: 321 (1997).

190. Hegazy EA, Kamal H, Khalifa NA, Mahmoud GA. *J Appl Polym Sci*, **81**: 849 (2001).

191. Abd El-Rehim H, Hegazy E-S, Ali A. *J Appl Polym Sci*, **74**: 806 (1999).

192. Abd El-Rehim HA, Hegazy E-S, El-Hag Aly A. *React Funct Polym*, **43**: 105 (2000).

193. Saito, K, Sugita K, Tamada M, Sugo T. *J Chromatogr A*, **954**: 277 (2002).
194. Choi S-H, Nho YC. *Korea Polym J*, **7**: 38 (1999).
195. Kim M, Saito K. *Rad Phys Chem*, **57**: 167 (2000).
196. Li G-Q, Konishi S, Saito K, Sugo T. *J Membr Sci*, **95**: 63 (1994).
197. Ozawa I, Saito K, Sugita K, Sato K, Akiba M, Sugo T. *J Chromatogr A*, **888**: 43 (2000).
198. Sugi A, Ogawa N, Iinuma Y, Yamamura H. *Talanta* **28**: 551 (1981).
199. Denizli A, Say R, Patir S, Arica Y. *Separat Purificat Technol*, **21**: 181 (2000).
200. Denizli A, Say R, Patir S, Arica Y. *React Funct Polym*, **46**: 157 (2000).
201. Kiyohara S, Sasaki M, Saito K, Sugita K, Sugo T. *React Funct Polym*, **31**: 103 (1996).
202. Kumar V, Bhardwaj YK, Dubey KA, Chaudhari CV, Goel NK, Biswal J, Sabharwal S, Tirumalesh K. *Sep Sci Tech*, **41**: 3123 (2006).
203. Trochimczuk AW, Streat M. *React Funct Polym*, **40**: 205 (1999).
204. Hsien TY, Rorrer GL. *Industr Eng Chem Res*, **36**: 3631 (1997).
205. Rorrer GL, Hsien TY, Way JD. *Indust Eng Chemist Res*, **32**: 2170 (1993).
206. Baba Y, Hirakawa H, Kawano Y. *Chem Lett*, **1**: 117 (1994).
207. Shim S-K, Ryu J-J. *Anal Sci Tech*, **11**: 452 (1998).
208. Tikhonov VE, Radignina LA, Yamskov A. *Carbohydr Res*, **290**: 33 (1996).
209. Qu R, Lui Q. *Huanjing Huaxue*, **15**: 41 (1996).
210. Qu R, Lui Q. *Tianran Chanwu Yu Kaifa*, **8**: 49 (1996).
211. Qu R, Lui Q. *Gaofenzi Cailiao Kexue Yu Gongcheng*, **12**: 140 (1996).
212. Kawamura Y, Mitsuhashi M, Tanibe H, Yoshida H. *Indust Eng Chem Res*, **32**: 386 (1993).
213. Kawamura Y, Yoshida H, Asai S, Tanibe H. *Water Sci Technol*, **35**: 97 (1997).
214. Vieira RS, Beppu MM. *Colloids Surf A*, **279**: 196 (2006).
215. Wan Ngah WS, Ghani SA, Kamari A. *Biores Technol*, **96**: 443 (2005).
216. Kang DW, Choi HR, Kweon DK. *Polymer*, **20**: 989 (1996).
217. Kang DW, Choi HR, Kweon DK. *J Appl Polym Sci*, **73**: 469 (1999).
218. Yazdani PM, Retuert J. *J Appl Poly Sci*, **63**: 1321 (1997).
219. Kim YB, Han S-M, Yun C-S, Kang S-J. *Advan Chitin Sci*, **3**: 391 (1998).
220. Chan WC, Wu JY. *J Appl Polym Sci*, **81**: 2849 (2001).
221. Xu SM, Zhang SF, Lu RW, Yang JZ, Cui CX. *J Appl Polym Sci*, **89**: 263 (2003).
222. Xu SM, Feng S, Yue F, Wang JD. *J Appl Polym Sci*, **92**: 728 (2004).
223. Xu SM, Feng S, Peng G, Wang JD, Yushan A. *Carbohydr Polym*, **60**: 301 (2005).
224. Zhang LM, Chen DQ. *Colloids Surf A: Phys Eng Aspects*, **205**: 231 (2002).
225. Guo L, Zhang S-F, Ju B-Z, Yang J-Z. *Carbohydr Polym*, **63**: 487 (2006).
226. Deng SB, Ting YP. *Langmuir*, **21**: 5940 (2005).
227. Deng SB, Ting YP. *Environ Sci Technol*, **39**: 8490 (2005).
228. Yu J, Tong M, Sun X, Li H. *Biochem Eng J*, **33**: 126 (2007).
229. Yu J, Tong M, Sun X, Li H. *J Hazard Mater*, **143**: 277 (2007).
230. George B, Pillai VNR, Mathew B. *J Appl Polym Sci*, **74**: 3432 (1999).

231. Vinodkumar GS, Mathew B. *Euro Polym J*, **34**: 1185 (1998).
232. George B, Mathew B. *J Macromol Sci: Pure and Appl Chem*, A**38**: 429 (2001).
233. George SC, Ninan KN, Thomas S. *J Membr Sci*, **176**: 131 (2000).
234. Merrifield JD, Davids WG, MacRae, JD, Amirbahman A. *Water Res*, **38**: 3132 (2004).
235. Sanju F, Varshney L. *Rad Phys Chem*, **74**: 310 (2005).
236. Ghoul M, Bacquet M, Morcelle M. *Water Res*, **37**: 729 (2003).
237. Fei B, Wach RA, Mitomo H, Yoshii F, Kume T. *J Appl Polym Sci*, **78**: 278 (2000).
238. Seko N, Tamada M, Yoshii F. *Nucl Instr Methods Phys Res B*, **236**: 21 (2005).
239. Dos Santos AB, Cervantes FJ, Van Lier JB. *Bioresour Technol*, **98**: 2369 (2007).
240. Martel B, Devassine M, Crini G, Weltrowski M, Bourdonneau M, Morcellet M. *J Polym Sci Part A: Polym Chem*, **39**: 169 (2001).
241. McKay G, Blair HS, Gardner JR. *J Appl Polym Sci*, **28**: 1499 (1989).
242. Chiou MS, Li HY. *J Hazard Mater*, B**93**: 233 (2002).
243. Chiou MS, Li HY. *Chemosphere*, **50**: 1095 (2003).
244. Chiou MS, Ho PY, Li HY. *Dyes Pigments*, **60**: 69 (2004).
245. Chao AC, Shyu SS, Lin YC, Mi FL. *Bioresour Technol*, **91**: 157 (2004).
246. Zhang LM, Chen DQ. *Macromol Mater Eng*, **288**: 252 (2003).
247. Abdel-Aal SE, Gad YH, Dessouki AM. *J Hazard Mater*, **129**: 204 (2006).
248. Crini G, Morin N, Rouland JC, Janus L, Morcellet M, Bertini S. *Eur Polym J*, **38**: 1095 (2002).
249. Hooper EW, Kavanagh PR, Wells AF, in *Advances in Ion Exchange for Industry and Research*, PA Williams and A Dyer, eds. (Cambridge: Royal Society of Chemistry, 1999), p. 57.
250. Hegazy E-S, Abd El-Rehim HA, Ali AMI, Nowier HG, Ali HF. *Nucl Instrum Meth Phys Res B*, **151**: 393 (1999).
251. Choi S-H, Nho YC. *Rad Phys Chem*, **57**: 187 (2000).
252. Yamagishi H, Saito K, Furusaki S, Sugo T, Ishigaki I. *Ind Eng Chem Res*, **30**: 2234 (1991).
253. Maziad NA, Sayed MS, Hegazy EA. *Polym Int*, **51**: 150 (2002).
254. Suzuki T, Saito K, Sugo T, Ogura H, Oguma K. *Analytical Sci*, **16**: 429 (2000).
255. Pekel N, Salih B, Güven O, *Macromol Symp*, **169**: 329 (2001).
256. Choi S-H, Nho YC. *Rad Phys Chem*, **58**: 157 (2000).
257. Sekiguchi K, Serizawa K, Konishi S, Saito K, Furusaki S, Sugo T. *React Polym*, **23**: 141 (1994).
258. Muzzarelli RAA, in Proceedings of International COnference On Chitin and Chitosan, RAA Muzzarelli, C Jeuniaux, and GW Gooday, eds. (New York: Plenum, 1986), p. 321.
259. Yazicigil Z. *Desalination*, **212**: 70 (2007).
260. Boucher M, Turcotte N, Guillemette V, Lantagne G, Chapotot A, Pourcelly G, Sandeaux R, Gavach C. *Hydrometallurgy*, **45**: 137 (1997).
261. Jeong J, Kim M-S, Kim B-S, Kim S-K, Kim W-B, Lee JC. *J Hazard Mater*, **124**: 230 (2005).

262. Elmidaoui A, Belcadi S, Houdus Y, Cohen T, Gavach C. *J Appl Polym Sci Part A: Polym Chem*, **30**: 1407 (1992).
263. Elmidaoui A, Cherif AT, Brunea J, Duclert F, Cohen T, Gavach C. *J Membr Sci*, **67**: 263 (1992).
264. Yuee F, Tianyi S. *J Membr Sci*, **39**: 1 (1988).
265. Fang Y-E, Ma C-X, Chen Q, Lu X-B. *J Appl Polym Sci*, **68**: 1745 (1998).
266. Przybycien TM, Pujar NS, Steele LM. *Curr Opin Biotechnol*, **15**: 469 (2004).
267. Demmer W, Nussbaumer D. *J Chromatogr*, **A852**: 73 (1999).
268. Brandt S, Coffe RA, Kessler SB, O'Connor JL, Zale SE. *Bio-Technol*, **6**: 779 (1988).
269. Ghosh R. *J Chromatogr A*, **952**: 13 (2002).
270. Van RR, Zydney A. *Curr Opin Biotechnol*, **12**: 208 (2001).
271. Kawai T, Saito K, Lee W. *J Chromatogr B*, **790**: 131 (2003).
272. Saito K, Tsuneda S, Kim M, Kubota N, Sugita K, Sugo T. *Rad Phys Chem*, **54**: 517 (1999).
273. Koguma I, Sugita K, Saito K, Sugo T. *Biotechnol Prog*, **16**: 456 (2000).
274. Shinano H, Tsuneda S, Saito K, Furusaki S, Sugo T. *Biotechnol Prog*, **9**: 193 (1993).
275. Tsuneda S, Shinano H, Saito K, Furusaki S, Sugo T. *Biotechnol Prog*, **10**: 76 (1994).
276. Hegazy EA, El-Assy N, Rabie A-G, Ishigaki I, Okamoto J. *J Polym Sci, Polym Chem Ed*, **22**: 597 (1984).
277. Hegazy EA, Taher NH, Ebaid AR. *J Appl Polym Sci*, **41**: 2637 (1990).
278. Dessouki A, Taher N, El-Arnaouty M. *Polym Int*, **45**: 67 (1998).
279. Dessouki A, Taher N, El-Boohy H. *Rad Phys Chem*, **36**: 371 (1990).
280. Ringrose B, Kronfli E. *Eur Polym J*, **36**: 591 (2000).
281. Jun L, Min Y, Hongfei H. *Rad Phys Chem*, **46**: 847 (1995).
282. Porath J, Carlsson J, Olsson I, Belfrage G. *Nature*, **258**: 598 (1975).
283. Godjevargova T, Dimov A. *J Appl Polym Sci*, **57**: 487 (1995).
284. Gebauer KH, Thommes J, Kula MR. *Biotechnol Bioengr*, **54**: 181 (1997).
285. Kanani DM, Komkova E, Wong T, Mika A, Childs RF, Ghosh R. *Biochem Eng J*, **35**: 295 (2007).
286. Odian G, Keusch P, Raffo J, Tufano J. *A Study of the Synthesis of Hemodialysis Membranes*. RAI Research Corporation, Progress Report, Pb 182990, December 1968.
287. Aliev R, Starannikova L, Teplyakov V, Burillo G. *Rad Phys Chem*, **53**: 533 (1998).
288. Yang J-S, Hsiue G-H. *J Membr Sci*, **111**: 27 (1996).
289. Nakamura M, Kiyohara S, Saito K, Sugita K, Sugo T. *J. Chromatogr A*, **822**, 53 (1998).

11

IN COATINGS, ADHESIVES, AND LAMINATES

James W. Rawlins and Sharathkumar K. Mendon

INTRODUCTION

Crosslinking is the process of covalent bond formation between independent polymer macromolecules to form matrices in coatings, adhesives, and laminates. The process results in a three-dimensional network structure via physical and/or chemical reactions. Examples of physical crosslinking include trapped entanglements, e.g., polyethylene crosslinked with peroxide; hydrogen bonding, e.g., polyurethanes; nonbond crosslinking, e.g., vinyl monomers polymerization with a large macrocyclic ring [1]. Chemical crosslinking occurs via one or more of the following methods [2]:

(1) Linear polymers crosslinking either directly or by short bridges, e.g., curing of unsaturated polyesters with styrene as the reactive diluent.
(2) Branched polymer molecules crosslinking via relatively few intermolecular chemical bonds, e.g., auto-oxidation of drying oil derivatives.
(3) Two or more low-molecular-weight resins reacting to form a crosslinked polymer networks, e.g., polyols reacting with isocyanates to form polyurethanes.

In graft polymers, adjacent blocks in the main chain, side chains, or both are constitutionally different. General approaches to graft copolymer synthesis are [3]

Polymer Grafting and Crosslinking, Edited by Amit Bhattacharya, James W. Rawlins and Paramita Ray

(1) Chain transfer mechanism
(2) Radiation or photochemical activation of polymer (used to create active grafting sites)
(3) Use of polymers with labile functional groups or chemical modification of polymers to create active grafting sites

COATINGS

Coatings are protective and/or decorative compositions that dry to a cohesive, adherent, tack-free film upon application to a substrate. Technological advances have transformed coatings manufacture from an arts-and-crafts industry to a highly scientific branch of the chemical industry. Global coatings market sales were an estimated $86 billion in 2005 [4]. Of the approximately 4500 coatings companies worldwide, just one-tenth produce 85% of the total coatings volume [5]. The fastest growing coatings markets are in the rapidly industrializing countries of the Asia-Pacific region.

Coatings are broadly classified as architectural, original equipment manufacturer (OEM), and special-purpose coatings. Architectural coatings are applied to interiors and exteriors of stationary structures such as homes, buildings, and fences. OEM coatings include automotive finishes, furniture coatings, container coatings, paper coatings, insulating finishes, and powder coatings. The catch-all category of special purpose coatings includes marine coatings, traffic paints, automotive refinishes, aerosols, and maintenance coatings [6].

Coatings are blends of polymers (resin/medium/binder), pigments (including extenders/fillers), additives, and solvents. Polymers serve as the continuous phase while pigments constitute the discontinuous phase. Additives include wetting and dispersing agents, defoamers, driers, light stabilizers, and biocides. Increasing cognizance of the environmental impact of volatile organic compounds (VOCs) has promoted the development of environmentally responsible systems such as powder coatings, waterborne coatings, and ultraviolet (UV) curable coatings.

A wide variety of natural and synthetic polymers are employed in coatings. Natural resins such as shellac, copal, and asphalt are used without modification, while rosin, cellulose, and rubber need to be derivatized before use. Synthetic polymers such as polyesters, acrylics, epoxies, silicones, and polyurethanes have now replaced natural resins in most coatings applications.

Polyesters are characterized by ester groups in their backbones and are synthesized via condensation polymerization of polyfunctional alcohols with polybasic acids (or anhydrides). Common monomers used in polyester synthesis include maleic anhydride, adipic acid, phthalic anhydride, isophthalic acid, trimellitic anhydride, pentaerythritol, neopentyl glycol, and trimethylol propane. The relative proportion of the various monomers decides the polyester molecular weight and residual functionality (carboxyl and/or hydroxyl). Hydroxyl functional polyesters are typically crosslinked with amino resins or

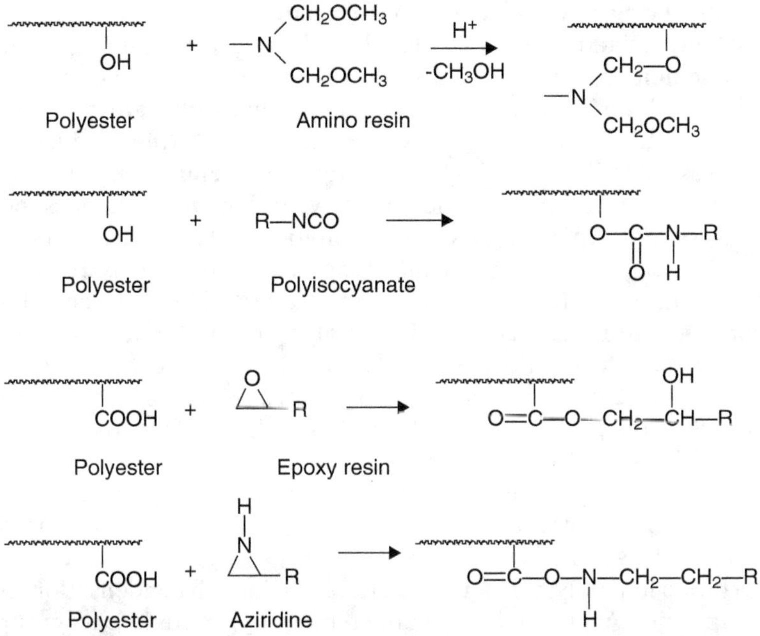

Figure 11.1. Crosslinking reactions of polyester resins.

polyisocyanates, while carboxyl functional polyesters can be crosslinked with epoxy resins or aziridines (Figure 11.1).

Polyesters modified with vegetable oils or fatty acids are termed oil-modified polyesters (alkyds). Vegetable oils are distinguished by their drying ability (related to their iodine value, IV) as nondrying oils (IV < 120, e.g., coconut, cottonseed), semidrying oils (IV = 120 to 150, e.g., safflower, sunflower, soybean), and drying oils (IV > 150, e.g., linseed, tung). Based on oil content, alkyds are classified as short oil alkyds (<40% oil), medium oil alkyds (40%–60% oil), long oil alkyds (60%–70% oil), and very long oil alkyds (>70% oil). Alkyds cure via a two-stage process: a physical drying stage in which the solvent evaporates, followed by auto-oxidative cure. Auto-oxidation proceeds through six stages: induction, initiation, hydroperoxide formation, decomposition, crosslinking, and degradation [7]. Since auto-oxidation is a slow process, driers, which are typically carboxylate (naphthenic acid/2-ethylhexanoic acid/isononanoic acid) salts of metals such as cobalt, zirconium, calcium, and manganese, are employed to enhance the drying rate.

Although cobalt salts are the most popular driers in alkyd coatings, their use is diminishing due to their suspected carcinogenicity [8]. Drawing upon the similarity between the oxidative drying of alkyds and lipid oxidation in biological systems, Micciché adopted a biomimetic approach to develop environmentally responsible alternatives for cobalt driers [9]. A combination of ascorbic acid palmitate and iron 2-ethyl hexanoate in an alkyd system exhibited shorter drying times and comparable hardness to a commercial cobalt drier. Further, the addition of two ligands, *N,N*-bis(2-ethyl-5-methyl-imidazol-4-ylmethyl)aminopropane and 2-ethyl-4-methylimidazole accelerated the drying time and film hardness development of an alkyd coating.

The slow auto-oxidative crosslinking of alkyds is further reduced at low temperatures. Blending with high glass transition temperature (T_g) acrylics hastens the physical drying, but the high solvent demand of acrylics increases the VOC content. To overcome this limitation, Klaasen developed alkyd systems that included thiol-ene chemistry as an additional crosslinking mechanism [10]. Thiol-ene polymerizations are characterized by their lack of oxygen inhibition. In free-radical curing of acrylates, the addition of oxygen to a propagating chain forms a peroxy radical that lacks the reactivity to add to a new unsaturated compound, and results in termination. In the presence of thiols, however, the peroxy radical abstracts a hydrogen from the thiol to yield a thiyl radical (Figure 11.2) that propagates the polymerization [11]. Storage-stable, fast-drying alkyd coatings based on polyfunctional thiols, catalysts such as vanadyl bipyridyl or vanadyl acetylacetonate, and visible light photoinitiators dried in <2 h at 10 °C.

Hyperbranched polymers are synthesized via polycondensation of AB_x ($x > 2$)-type monomers, and are characterized by multiple end groups and

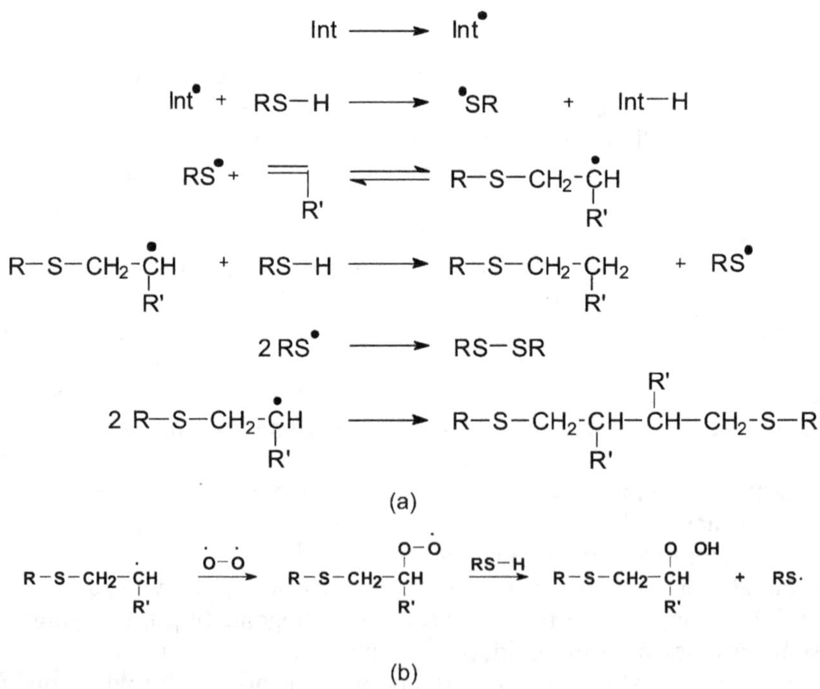

Figure 11.2. (a) Thiol-ene reaction scheme, (b) reaction of an alkylperoxy radical with thiol.

highly branched architecture. Due to their approximately globular morphology, hyperbranched polymers have relatively low molecular volume (and lower viscosity) at a given molecular weight than linear polymers. Hyperbranched polyesters with methacrylate end groups are employed in UV curing applications where they offer high reactivity, fast cure, low shrinkage, toughness, and good chemical abrasion and scratch resistance properties. Dendritic methacrylate polyesters based on 1,2,4-benzenetricarboxylic acid anhydride cured five times faster than traditional, linear epoxy methacrylate oligomers [12]. UV-curable hyperbranched resins have also been synthesized by the Diels-Alder reaction of polyfunctional sorbic esters with acrylates [13]. Hyperbranched aliphatic polyesters grafted with crystalline, linear aliphatic polyester chains, and end-capped with methacrylate moieties exhibit a rheological behavior suitable for low-temperature-curing powder coatings; i.e., films are formed at <80 °C and UV cured [14].

Powder-coating polyesters are generally based upon neopentyl glycol, isophthalic acid, and terephthalic acid. Powder coatings based on super durable polyesters deliver excellent outdoor weathering and are based on isophthalic acid, the most weatherable of aromatic diacid monomers [15]. Super durable coatings meet the American Architectural Manufacturers Association standard 2604–98, which is the highest standard for architectural coatings [16].

Phosphorus-containing lactone modified polyesters employed in two-component (2K) polyurethane flame-retardant coatings are synthesized by polycondensation of dimethyl phenyl phosphonate with polycaprolactone, adipic acid, and trimethylolpropane [17]. Coatings with 20 wt. % dimethyl phenyl phosphonate content were self-extinguishing. Unlike halogen-based coatings that liberate toxic compounds during a fire, phosphorous-based coatings are nontoxic and exhibit two to four times higher flame retardancy than those containing halogen compounds [18].

In systems containing a double bond, optimum grafting efficiency is achieved by direct addition of the grafting agent to the double bond. Due to their lack of steric hindrance, acrylates typically result in high levels of grafting. Sterically hindered grafting agents such as methacrylates are predisposed to allylic hydrogen abstraction as the hydrogens allylic to double bonds are good leaving groups. This creates a relatively stable and unreactive radical on the resin that terminates the abstracting methacrylate chain and results in a loss of grafting efficiency and rate of polymerization [19]. Grafting 30%–40% by weight of methyl methacrylate (MMA) onto a dehydrated castor oil alkyd resulted in faster drying, better mechanical properties, and improved outdoor weathering properties [20]. Besides the unsaturation sites on the fatty acids, graft copolymerization can also be initiated on the active methylene group via hydrogen abstraction in the presence of an initiator like di-*tert*-butyl peroxide. Methacrylate grafting resulted in increased molecular weight and lower viscosity as the branched polymer is less viscous than the linear polymer. The improved tensile strength was ascribed to dense intermolecular packing of grafted MMA branches. However, beyond 40% MMA, high amounts of an MMA homopolymer caused phase separation and resulted in lower tensile strength.

Styrene was successfully grafted onto unsaturated polyesters via alternating copolymerization with maleic anhydride. Mechanical properties of the grafted polyesters varied with the composition of the grafted side chain. Grafting of an aromatic unsaturated monomer onto aromatic polyesters improved the elongation-to-break property of the grafted polymer. Grafting of an aliphatic unsaturated monomer onto aliphatic polyesters improved the elongation-to-break of the grafting polymer [21].

Polyesters formulated for waterborne systems employ surfactants or hydrophilic groups to achieve water dispersibility. However, such coatings often suffer from poor water resistance and adhesion properties as the hydrophilicity is retained in the final film. Alternatively, water dispersibility can be achieved via the grafting of hydrophilic acrylics onto the polyester. The solvency of the reaction medium, concentration of unsaturated bonds, and the reactivity ratios of unsaturated polyesters with acrylic monomers strongly influence the particle size [22].

Acrylic resins are copolymers of methacrylate esters such as butyl acrylate (BA), ethyl acrylate, and MMA, and may include small amounts of other comonomers such as styrene and α-methyl styrene. Acrylic resins are characterized by their water-white color, excellent clarity, and good exterior durability. The use of comonomers such as hydroxy ethyl methacrylate, glycidyl methacrylate (GMA), and methacrylic acid impart the relevant functionality to acrylic resins. Self-crosslinking acrylic resins contain hydroxyl and carboxyl functionalities, and an amide monomer such as methacrylamide, *N,N*-bisbutoxymethyl methacrylamide, or *N*-butoxyacrylamide.

Acrylic polyols crosslinked with polyisocyanates are the dominant technology in OEM refinishes, as they provide attractive appearance, rapid ambient cure, and excellent mechanical and chemical resistance properties. Due to their high solvent requirements and high molecular weight (10,000–20,000 g/mol), acrylic polyols yield coatings with <50% solids at application viscosities. Low-molecular-weight acrylics (<2500 g/mol) suited for higher solids coatings have to be formulated at higher average functionality to achieve the desired performance characteristics. However, this results in shorter pot lives and longer drying times relative to higher molecular weight acrylic resins. The dry times may be reduced by adding catalysts but this reduces the pot life even further. Low-molecular-weight polyols have to be processed to very high levels of conversion since any nonfunctional molecules will act as plasticizers while monofunctional species will contribute a dangling chain to the network. The design and synthesis of low-molecular-weight acrylic polyols is thus a challenging task.

Powder coatings based on glycidyl functional acrylic resins crosslinked with dodecanedioic acid are employed in automotive clear coats and wheel trims. Bifunctional polyesters designed to cure with glycidyl functional acrylics and blocked isocyanates yield super durable matte and semigloss powder coatings

[23]. Gloss reduction occurs due to the incompatibility between the acrylic and polyester resins, and the desired gloss can be achieved by varying the relative resin proportions.

The ready reactivity of carbonyl groups with hydrazides at ambient temperature is utilized in developing one-component (1K) self-crosslinkable acrylic emulsions by copolymerizing methacrylate monomers with diacetone acrylamide, and post-adding adipic dihydrazide as the crosslinker. The emulsions exhibit sandability, high mar resistance, and good hardness and chemical resistance properties [24].

As mentioned earlier, a common limitation of waterborne coatings is that the hydrophilic groups stabilizing the resins in water adversely affect film properties such as weathering and water resistance. Harui et al. proposed an ambient temperature cured, 2K waterborne coating system in which the hydrophilic functional groups were consumed via crosslinking during film formation [25]. The system consists of an acid-functional acrylic resin (dispersed in water by neutralization with triethylamine) that is crosslinked with γ-glycidoxypropyl trimethoxysilane. Curing occurs via three reactions: (1) between epoxy groups and *tert*-amino groups of the acrylic resin, (2) between epoxy groups and carboxylic acid groups of the acrylic resin, and (3) hydrolysis and condensation of silyl groups.

UV-curable urethane coatings based on acrylic polyols offer better outdoor durability than similar systems based on polyester or polyether polyols. Judicious choice of reaction conditions and monomers is essential for synthesizing low viscosity, storage-stable, acrylate functional, aliphatic urethane prepolymers [26]. UV-curable systems based on tetrahydrofurfuryl acrylate, oligoetherglycol oligourethane methacrylate, isobornyl acrylate, and liquid diene rubber oligourethane methacrylate have been developed for coating high-density digital versatile discs (HD-DVD) [27].

Matsuno et al. grafted polymethyl methacrylate (PMMA) onto acrylic polyolss for 2K polyurethane coatings [28]. The acrylic resins were synthesized by reacting carboxyl-terminated homopolymers of various acrylate monomers (prepared via solution polymerization) with GMA. The graft copolymers displayed smaller radii of gyration (and therefore lower viscosity at the same solids), improved film toughness, better rheology, and greater reactivity with polyisocyanates over conventional acrylics.

Amino resins have been employed for over 60 years as crosslinkers for polyesters, acrylics, and epoxies as they offer good adhesion and excellent chemical and solvent resistance properties. Primarily, four amines are used in the synthesis of amino resins, i.e., urea, melamine, benzoguanamine, and glycoluril. The exceptional chemical stability of the triazine ring has made melamine resins the preferred amino resin crosslinker for exterior surface coatings. Urea-based resins cure faster, but they lack the exterior durability of melamine resins and are employed in interior finishes and primers. Benzoguanamine resins are employed in coatings requiring high detergent and chemical resistance. Glycoluril resins offer the advantages of lower formaldehyde release during curing and improved corrosion protection, but require a strong acid catalyst and will not cure with systems catalyzed only by the copolymer acidity.

Amino resins are synthesized by reacting the appropriate amine with formaldehyde and subsequent alkoxylation with an alcohol and are characterized by the presence of three functional groups: i.e., imino (–NH), methylol ($-CH_2OH$), and alkoxy ($-CH_2OR$). Early amino resins were oligomeric reaction products of urea or melamine with formaldehyde and butanol and were supplied at 50%–80% solids in butanol or butanol/xylene. These resins had high self-condensation tendencies and yielded tightly crosslinked films with highly functional polymers but imparted limited flexibility. Subsequently developed resins such as hexamethoxymethylmelamine (HMMM) are highly etherified, highly reactive, low-molecular-weight condensation products of melamine with formaldehyde and methanol that are supplied at solids close to 100% and have less tendency to self-condense [29].

Amino resins cure by reacting with the copolymer hydroxyl or ether groups via etherification or transetherification. Curing with amide groups and carboxyl groups on the copolymer occur via formation of methylene bridges and esters, respectively. Amino resins also participate in self-condensation reactions. Together, these reactions provide a polymer network well beyond the gel point in properly cured films [30]. The curing of amino resins is promoted by acid catalysts or by acid groups on the copolymer. Fully alkylated resins require strong acid catalysts such as sulfonic acids, e.g., *p*-toluene sulfonic acid, dodecylbenzene sulfonic acid, and dinonylnaphthalene disulfonic acid. Amino resins containing methylol groups and imino groups are preferably catalyzed by weak acids such as alkyl acid phosphates or organic acids such as citric and maleic acids.

Polyurethanes, formed by reacting polyols with polyisocyanates, are extremely versatile polymers and are widely used in coatings, foams, adhesives, and elastomers. Polyols employed in polyurethanes are based on polyesters, acrylics, polycaprolactones, or castor oil, and also include hydroxyl-terminated polybutadiene (HTPB) and polyether diols, e.g., poly(propylene oxide) glycol, copolymers of (propylene/ethylene oxides) glycols, and polytetramethylene ether glycol (PTMEG). The volatility and high reactivity of isocyanates are moderated by converting them to polyisocyanates such as biurets and isocyanurates. However, the conversion is achieved at the expense of half the initial isocyanate moieties. The isocyanurate ring enhances the thermal stability, chemical and flame resistance, and film-forming characteristics [31].

Polyurethanes exhibit a wide range of properties due to the variety of morphologies resulting from their chemical and physical crosslinks. Physical crosslinking arises from hydrogen bonding and domain formation. Hydrogen bonds are formed primarily among N–H groups and carbonyl groups, and to a lesser extent with polyether oxygen atoms (weak bonds). Domain formation results from incompatibility between the polyol soft segments and the isocyanate hard segments. Crystalline regions may also form in the hard and soft domains [32].

Polyurethane coatings are classified into various categories based on the polyisocyanate:

(1) Blocked isocyanates are adducts of isocyanates with compounds containing dissociable protons such as diethyl malonate, ε-caprolactam, methylethylketoxime (MEKO), and dimethylpyrazole, and are employed in 1K systems. The various blocked isocyanates differ from each other in their deblocking temperature and curing chemistries during subsequent heat cure [33, 34]. Dimerization of isocyanates to form a uretdione is referred to as self-blocking or internally blocked isocyanates. While powder coatings formulated with ε-caprolactam blocked isocyanates exhibit weight losses of 4%–5%, uretdione-based powder coatings do not liberate any VOCs. As the sole crosslinker, uretdiones yield high gloss powder coatings while diamine-modified uretdiones yield highly matt coatings (8 units at 60°) with excellent flow properties. The matt appearance stems from the uniform surface microstructure generated due to incompatibility between the urea and urethane moieties [35].

(2) Oil-modified urethanes are formed by reacting isocyanates [typically toluene diisocyanate (TDI)] with hydroxy-functional drying oil derivatives and curing via auto-oxidation [36]. These systems are used extensively in the wood floor finish market due to their ease of use, fast drying times, mar and scuff resistance, high abrasion resistance, and good stain and chemical resistance.

(3) Moisture-cured polyurethane systems are based on isocyanate-capped low-molecular-weight adducts with polyester or polyether glycols [37]. Upon application, the isocyanates react with atmospheric moisture to form carbamic acid, which quickly dissociates to carbon dioxide and a primary amine. The amine then reacts with isocyanate moieties to form ureas. Biurets and allophanates are also likely to form as a result of successive reactions. Temperature and humidity significantly affect the curing rate of moisture-cured polyurethanes.

(4) Thermoplastic polyurethanes contain fully reacted isocyanates and dry by solvent evaporation.

(5) Waterborne polyurethane dispersions (PUDs) offer very good performance properties at low VOC content. PUDs are synthesized either by polymerizing the polyurethane in solvent, and then dispersing in water, or an isocyanate-terminated prepolymer is prepared in a melt or an aprotic solvent and then chain extended with a diamine in the water phase in the presence of a neutralizing tertiary amine [38]. Epoxy-acrylic graft copolymers modified with ethylene diamine have been employed as chain extending agents to synthesize PUDs with improved scratch hardness and impact resistance [39].

(6) Acrylic or polyester polyols cured with polyisocyanates are 2K systems curable at ambient or low temperatures (80°C).

(7) Water-dispersible coatings involving emulsifiable polyisocyanates are 2K systems in which the polyisocyanate is rendered hydrophilic enough to make it water dispersible. The synthesis of hydrophilic polyisocyanates is the subject of several patents [40].

Isocyanates have also been employed as crosslinkers for polyaspartic esters. The moderate reactivity of the polyaspartic esters leads to coatings with acceptable pot life, high hardness, good gloss, and excellent weathering and environmental etch resistance [41].

Polyurethanes with flexible polyol backbones such as HTPB and PTMEG retain their properties at low temperatures making them suitable candidates for use in adhesives and coatings employed in cryogenic conditions. The low-temperature mechanical performance of such polyurethanes can be augmented by grafting polymers onto the backbone. For instance, a macrodiol synthesized by grafting BA to a 1-mercapto-2,3-propane diol was formulated into polyurethanes by reacting with polyethylene glycol, trimethylol propane, and TDI [42]. Increasing the graft concentration for a fixed chain length resulted in improved mechanical properties and an initial decrease in T_g. On the other hand, increasing the graft chain length for a fixed concentration decreased both tensile strength and T_g. The trends suggested that the cohesive force induced by the polar urethane moieties determined the properties of the grafted polymer rather than the free volume or entanglement effects of the graft.

Yukawa et al. synthesized self-crosslinkable graft copolymers with complementary reactive groups, i.e., a hydroxyl group in the second segment and a pendant blocked isocyanate in the first segment [43]. The polymerizable isocyanate, *m*-isopropenyl-α,α-dimethyl benzyl isocyanate (TMI), was copolymerized with BA by free-radical solution polymerization. Subsequently, some of the isocyanate groups were reacted with 2-hydroxyethyl acrylate, and the remaining isocyanate groups were blocked with MEKO. The prepolymer was then copolymerized with methacrylates (including hydroxyl ethyl methacrylate). In another process, the TMI/BA copolymer was reacted with MEKO to block approximately 70% of isocyanate groups and then reacted with a polyester polyol. The incorporation of neutralizable functionalities such as carboxyl or tertiary amino groups into the second segment of the graft copolymer enabled the resins to be water dispersible. The coatings were curable at 100 °C–120 °C.

Polyureas are formulated from isocyanate prepolymers and amine-functional resins and are characterized by rapid cure speed (requires plural spraying equipment), high-build film formation in a single coat, zero VOC, and low-temperature flexibility. The crosslinking reaction is autocatalytic and independent of humidity and temperature. Consequently, polyureas cure well at low temperatures where polyurethanes and epoxy systems will not cure. The extremely rapid reaction rates enable the coated products to be used within a few hours after application.

The most commonly used isocyanate in polyureas is methylene diphenyl isocyanate (MDI), while aliphatic isocyanates are used in applications requiring UV stability. The amine-functional polymers are generally a mixture of amine-terminated ethylene oxide and/or propylene oxide polyethers. Diethyltoluenediamine (DETDA) is the standard chain extender used in aromatic polyurea spray coatings. DETDA contributes to the hard segment and improves the heat resistance of the cured film. Chain extenders like dimethylthiotoluenediamine, *N,N′*-di(*sec*-butyl)aminobiphenyl methane, or 4,4′-methylenebis(3-chloro,2,6-diethyl)aniline slow down the reaction rate significantly [44].

Polyureas based on hindered amine adducts (HAA) are recommended for use in direct-to-metal coatings. HAAs are synthesized by reacting sterically hindered primary amines with acrylate oligomers and/or epoxy resins, e.g., by reacting trimethylol propane triacrylate with tertiary $C_{12–14}$ primary amine. The proximity of the amine to the carbonyl group reduces the amine reactivity due to inductive effects and results in longer gel times and good flow and leveling. These coatings exhibit excellent adhesion, corrosion protection, cure down to −18 °C, and good UV and chemical resistance. Although not as elastomeric as the aromatics, their flexibility is comparable to conventional polyurethanes and better than epoxies [45].

CARBAMATES

Polyol/melamine formaldehyde systems in automotive clearcoats yield excellent appearance. However, environmental pollutants such as acid rain cause permanent spotting and pitting known as "environmental etch" due to the acid catalyzed hydrolysis of the ether linkages [46]. Since urethane groups are more resistant to acid hydrolysis than ether linkages, polyurethanes offer better environmental etch resistance than polyol/melamine formaldehyde systems. The use of carbamate functional moieties crosslinked with melamine formaldehyde resins offer a nonisocyanate avenue for 1K finishes with good etch resistance and favorable performance properties.

Carbamates are carbamic acid esters (RNH_2COOH), and they structurally resemble urethanes. Oligomeric dimethylol propionic acid carbamate and citric acid carbamate formulated into clearcoats with melamine formaldehyde resin displayed higher solids, lower VOCs, comparable etch resistance, and better scratch and mar resistance than 2K polyurethane clearcoats [47]. Coatings with excellent performance characteristics have been reported by crosslinking oligomeric glycidyl carbamate with multifunctional amines [48]. Carbamates dispersed in amino resins are used in powder slurry compositions [49]. The use of carbamates as crosslinkers in coatings is the subject of several patents [50].

ACETOACETATES

Acetoacetates are characterized by the presence of a reactive methylene group between two carbonyl groups and react readily with amines at ambient temperature. To ensure adequate pot life, coatings formulated with acetoacetates employ aldimines or ketimines as the blocked crosslinker (deblocking occurs upon reaction with atmospheric moisture after application). Since aliphatic aldimines impart low gloss and poor solvent resistance, benzaldimines (benzaldehyde-aliphatic primary amine adducts) are preferred as crosslinkers, yielding ambient temperature curable, 1K coatings that compare well in performance tests against 2K polyurethane coatings [51]. The chemistry is also applicable in waterborne coatings since the acetoacetate aromatic aldimine curing system is stable in water [52]. Waterborne acetoacetate-functionalized alkyd resins (synthesized by transesterification of an alkyd resin with alkyl acetoacetates) exhibited superior tack-free time and through-dry properties relative to the alkyd coating [53]. A ketimine-acetoacetate system with good pot life/reactivity balance was formulated to a 2K high solids primer for aerospace alloys with excellent corrosion resistance properties [54].

Acetoacetate-functional acrylic polymers have been synthesized via the transacetylation of a hydroxyl-functional acrylic monomer with an acetoacetate and subsequent copolymerization with acrylic comonomers, or via the transesterification of a hydroxyl-functional polymer with an alkyl acetoacetate. Acetoacetylation of polyester polyols yields acetoacetate-functional polyesters [55]. Copolymerization of acetoacetoxyethyl methacrylate with acrylic comonomers yields stable latexes that can be formulated into 2K coatings using trimethylolpropane tri(polyoxypropylene amine) as the crosslinker [56].

Aziridines, characterized by the presence of two carbon atoms and one nitrogen atom in a three-membered ring, are the simplest heterocyclic systems and are extremely susceptible to cleavage due to ring strain [57]. Acid-functional latexes crosslinked with aziridines have been employed as 2K, clear, interior wood coatings. Polyfunctional aziridines such as trimethylolpropane tris-[1-(2-methyl)aziridine] propionate, trimethylolpropane tris-3-(1-aziridinopropionate), and pentaerythritol tris-3-(1-aziridinopropionate) are favored due to their low vapor pressure (reduced toxicity) and low levels required in coatings [58].

Carbodiimides (N=C=N) are employed as ambient crosslinkers for acid-functional aqueous polymers such as polyurethanes, acrylics, and latexes (Figure 11.3). Commercial polycarbodiimides are water-dispersible, zero-VOC systems that offer longer pot lives and comparable or even better performance properties than isocyanates.

The reaction between carbodiimides and carboxylic acids proceeds via the formation of an *O*-acyl urea, which either rapidly rearranges to form an *N*-acyl urea or reacts with a second carboxyl group [59]. At high temperatures or in the presence of a strong carboxylic acid, the *N*-acyl urea reversibly dissociates into an isocyanate and an amide. In a crosslinked system, the newly formed isocyanate and amide remain within each other's vicinity, and the restricted mobility is likely to promote the formation of *N*-acyl urea. Reaction of the *O*-acyl urea with a carboxyl group results in the formation of an anhydride and a urea group, but mobility restrictions are likely to hamper this reaction.

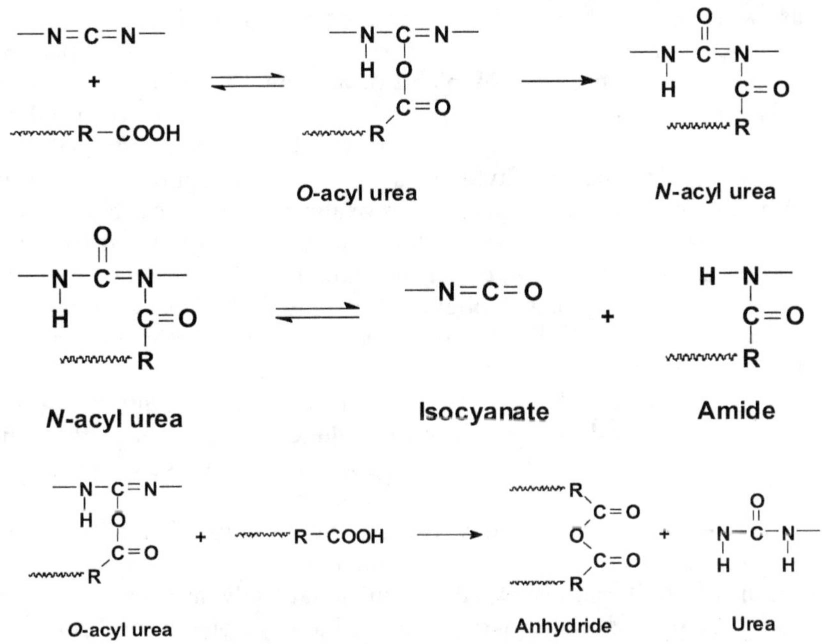

Figure 11.3. Crosslinking reactions of carbodiimides with acid-functional polymers.

Conversion of the polar –COOH group to the less polar *N*-acyl urea group reduces the number of residual acid groups in the final film and improves its water resistance.

Pham and Winnik synthesized cyclohexylcarbodiimidoethyl methacrylate and *tert*-butylcarbodiimidoethyl methacrylate, and formulated them into latexes for crosslinking with acid functional latexes [60]. Polymer diffusion was studied by labeling the acid functional latex with a donor dye and the carbodiimide latex with an acceptor dye. High gel content in the final film was promoted by factors favoring polymer diffusion over crosslinking, i.e., high annealing temperature, low functional group content, and less reactive carbodiimides.

Oxazolidines are five-membered heterocyclic ring compounds employed in polyurethane and polyurea coatings as moisture scavengers and reactive diluents. Bisoxazolidines contain oxazolidine rings connected by a carbonate bridge and afford low viscosity due to restricted intermolecular hydrogen bonding. Upon application, oxazolidines react with atmospheric moisture to yield amine and hydroxyl groups that react with isocyanates to form urea and urethane groups, respectively [61]. Oxazolidines are also employed in epoxy coatings as a reactive diluent [62].

Phenolic resins were one of the first synthetic resins to be developed and are synthesized via condensation polymerization of phenol (or substituted phenols) with aldehydes (or aldehyde precursors). Phenolic resins are classified as resoles and novolacs, depending on the pH and the ratio of phenol to formaldehyde during synthesis. A blend of aqueous formaldehyde and pure phenol in equal proportions has a pH of 3–3.1. In the presence of alkaline catalysts (pH 7–11) and a molar excess of formaldehyde over phenol, the reaction product contains free methylol groups and is termed a resole. The use of excess phenol and acidic catalysts such as oxalic acid yields novolacs that are essentially thermoplastic as they lack the reactive methylol functionality.

Resoles are self-crosslinking under the influence of heat. At temperatures below 130 °C, the methylolated phenols form ethers by liberating water. At higher temperatures, the ether bond breaks to form a methylene bridge by liberating formaldehyde. Resoles can also be cured by the addition of an acid catalyst just prior to use. The methylol groups condense readily with phenol rings under acidic conditions to form methylene bridges even without heat activation. Any methylene ethers that are formed decompose with the evolution of formaldehyde.

Resoles can crosslink with epoxy resins in the presence of alkaline catalysts such as benzylamine or morpholine. Phenolic-epoxy coatings are characterized by excellent flexibility, adhesion, and chemical and solvent resistance properties, and are employed as lacquers on the inside of tinplated food containers. U.S. patent 5,548,015 describes an aqueous epoxy-phenolic dispersion that employs a protective colloid such as polyvinyl alcohol for stabilizing the epoxy and phenolic resins [63]. Resoles cured with polyvinyl butyral are employed as "wash primers" to protect sheet metals before painting. Resoles cured with polyvinyl formal are employed as insulating varnishes in wire coatings.

Novolacs crosslink upon heating with compounds that can form methylene bridges, e.g., hexamethylene tetramine (HMTA), and yield networks with high crosslink densities. Reactive diluents such as furfuryl alcohol or furfuraldehyde enhance wetting and toughness. Epoxy novolac systems offer excellent thermal and chemical resistance. Crosslinking via the reaction of the epoxy groups with the phenolic hydroxyl requires extended periods (>5 h) at temperatures above 200 °C. Triphenylphosphine is typically employed as the catalyst and aids in the reduction of the shrinkage associated with curing. U.S. patent application 2007/0100116 describes an epoxy novolac coating that cures at lower temperatures and shorter curing schedules with low shrinkage, improved processability, and good thermal and chemical resistance [64].

Aralkylated phenolic resins, synthesized by reacting phenols with styrene derivatives and aryl diolefins [65, 66], exhibit reduced discoloration upon exposure to light. Alkoxylated phenolic resins are synthesized by reacting phenolic resins with alkylene carbonates and have been used to develop moisture-curable polyurethane coatings with very high solids content and much-improved chemical resistance [67]. Alkoxylation converts the phenolic hydroxyls to aliphatic hydroxyls and facilitates their reactivity with isocyanates.

Alkoxylated phenolic resins have also been employed to augment the performance of polyurea systems [68]. In general, polyureas lack resistance to concentrated acids, alkalis, and strong solvents. Incorporation of the alkoxylated phenolic resin with the amine component of polyurea systems results in improved chemical resistance, better adhesion, and lower moisture vapor transmission properties.

Not all phenolic resins involve the use of formaldehyde derivatives. Phenolic alcohols such as 2-[3,5-bis(*tert*-butyl)-4-hydroxybenzyl]-1,3-propanediol have been proposed for synthesizing thermo-oxidatively stable polyesters [69]. A phenolic epoxidized polyurethane coating based on linseed oil has been proposed for coating the insides of petroleum tanks [70].

Epoxy resins are characterized by the presence of at least two oxirane moieties per molecule, and react readily due to their high oxirane bond angle tension and polarity. By judicious selection of a crosslinking agent and curing conditions, epoxy resins can yield coatings with excellent toughness and adhesion, high chemical and electrical resistance, good heat resistance, and mechanical properties ranging from extreme flexibility to high strength and hardness. Epoxy resins are employed in a wide range of protective coatings, adhesives, caulking compounds, flooring, molding resins, and composites. The most important class of epoxy resins are condensation products of epichlorohydrin and bisphenol-A with the number of repeat units (n) ranging from 0.1 to 26. The threshold between liquid and solid resins is approximately n = 1.1 [71].

Depending on the curing agent, epoxy resins react via the epoxy group and/or the hydroxyl group. Epoxy resins can be crosslinked with amines, anhydrides, carboxylic resins, phenolic resins, and isocyanates as well as by cationic UV radiation (Figure 11.4). Crosslinkers containing primary and/or secondary amine groups are supplied as 2K systems due to their short reaction times. Primary amines generate films with high crosslink density and good chemical resistance but the high volatility and unpleasant odor of amines are an inconvenience. Moreover, primary amines are also prone to "surface blush" as amines tend to react with atmospheric carbon dioxide and form carbamates. The surface defect can be alleviated by allowing an induction period of ~30 min between blending and application. Polyamidoamines, i.e., condensation products of dimer fatty acids and an excess of diamines, offer good flexibility with reasonable cure times, although the performance properties are compromised to some extent. Blocked amines such as ketimines cure after deblocking by atmospheric moisture and significantly enhance the pot life of epoxy coatings.

Cycloaliphatic, araliphatic, and aromatic amines react relatively slowly with epoxy resins. Cycloaliphatic diamines such as isophorone diamine and bis-*p*-aminocyclohexylmethane offer light color and good chemical resistance in ambient and thermally cured epoxy systems. Without sacrificing color or chemical resistance, 1,3-bisaminocyclohexane provides fast curing. Imidazoles offer long pot life, rapid cure at elevated temperatures, and good high-temperature resistance. Dicyandiamide (DICY) is used as a crosslinker for epoxy powder coatings, and yields tightly crosslinked films with good chemical and solvent resistance properties.

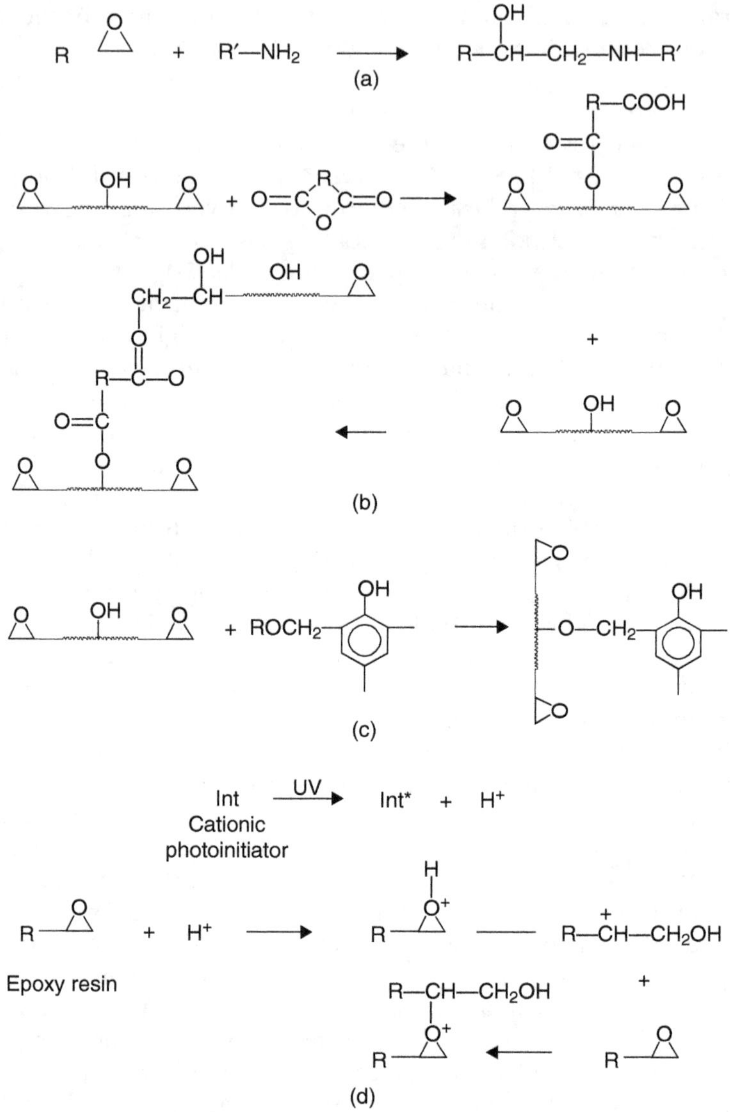

Figure 11.4. Crosslinking reactions of epoxy resins with (a) amines, (b) anhydrides, (c) phenolic resins, and (d) cationic cure.

Polycarboxylic anhydrides react with epoxy resins at elevated temperatures to yield films with good performance properties. Epoxy resins crosslinked with aromatic isocyanates via their hydroxyl groups are used in low-temperature-cure coatings. Cathodic electrodeposition coatings are formulated with water-dispersible, amine-modified epoxy resins crosslinked with blocked isocyanates.

Epoxy esters, i.e., fatty acid esters of epoxy resins, are classified as short-oil, medium-oil, and long-oil esters based on their fatty acid content (similar to alkyds). The long-oil esters are used in air-drying coatings while short-oil esters are employed in heat-cured coatings with amino resins as crosslinkers. Epoxidized oils are used in conjunction with high acid value polyesters (~100 mg KOH/g) for air-dry/heat-curable 2K coatings in applications that demand good appearance as well as toughness and wear resistance (e.g., floor finishes, tool finishes, marine finishes, and transportation finishes). Carboxylic acid functional polyesters are extensively used as crosslinkers for epoxy resins in powder coatings.

Cycloaliphatic epoxy resins are widely used in cationic UV curable coatings due to their excellent mechanical properties, good adhesion to a variety of substrates, lack of oxygen inhibition, low volume shrinkage after cure (<5%), and good heat and chemical resistance [72]. Polymerization is initiated via photoexcitation of diaryliodonium or triarylsulfonium salts to form an excited singlet state that cleaves to yield a Bronsted acid. The acid protonates the epoxide monomer and propagates the polymerization by reacting with other epoxides. An alternate mechanism has been proposed for cationic polymerization of epoxides in the presence of alcohols [73].

Waterborne epoxy coatings evolved into two fundamentally different technologies commonly identified as Type I and Type II systems. In Type I systems, the amine curing agent is employed as the emulsifier for the liquid epoxy resin. These systems have short pot lives but slow drying rates and yield hard films that are more suited for cementitious substrates than metals. Type II systems utilize solid epoxy resins predispersed in water and cosolvents, and have longer pot lives and fast drying rates but incomplete coalescence that often results in heterogeneous films with epoxy-rich and amine-rich domains [74]. The next generation of waterborne epoxy systems employs self-emulsifying epoxy resins crosslinked with amines modified with polyaromatic segments and nonionic segments for enhanced compatibility and coalescence with the epoxy resin. Low-VOC coatings formulated with these systems display fast drying, high hardness, and good corrosion protection [75]. Waterborne 2K acrylic-epoxy coatings are used in industrial (moderate duty) and high-performance architectural applications due to their ease of handling, good application and performance properties, and low-VOC content.

In coatings, silicon is employed in an inorganic matrix form (zinc-rich coatings) as well as organic-inorganic hybrid systems. In zinc-rich coatings, hydrolysis of silicon-based precursors such as tetraalkoxy silanes results in the formation of silanol moieties that undergo rapid polycondensation to form a polymeric matrix characterized by Si-O-Si linkages. As the Si–O bond is intrinsically stronger (452 kJ/mol) than the C–C bond (350 kJ/mol), silicon polymers offer high heat and UV resistance. The inorganic coating is highly durable and offers excellent corrosion resistance via galvanic protection.

Organic-inorganic hybrid systems combine the processability, toughness, and durability of organic polymers with the abrasion resistance, hardness, chemical resistance, weatherability, and UV resistance of silicon polymers. Organic polymers employed for this purpose include hydrogenated epoxies, acrylated urethanes, and modified acrylics (with epoxy, hydroxyl, amine, and alkoxysilyl functionalities). Epoxy side chains provide flexibility, processability, good resistance to chemicals and water, and controlled reactivity to balance flexibility and toughness [76]. Epoxy polysiloxane coatings are supplied at 90% volume solids (VOC 120 g/L) and display excellent weatherability (90% gloss retention after exposure for 12 months at 45° South Florida). A two-coat system consisting of either a zinc silicate or a zinc epoxy primer (75 μ) followed by a high solids epoxy polysiloxane hybrid coating (125 μ), applied on SA 22 dry abrasive blast-cleaned steel passed the Norsok test specification of 6000 h salt spray, 6000 h condensation chamber, and 4200 h cyclic testing by intermittent exposure to salt spray and UV light. This test program is one of the most severe performance tests for protective coatings systems in the coatings industry [77].

The high electronegativity of fluorine and its small atomic size afford fluoropolymers unique properties such as hydrophobicity, chemical stability, low coefficient of friction, water impermeability, weathering resistance, and low surface energy. Hydroxy functional fluoroethylene–alkyl vinyl ether alternating copolymers are commercially available in solvent, emulsion, and powder grades for polyurethane coatings [78]. Hydroxy-terminated polyfluoropolyether (PFPE) resins can be crosslinked with aliphatic and cycloaliphatic polyisocyanates to formulate easily cleanable coatings with very high solids (>80%) that form films rapidly even at temperatures <20 °C [79]. Fluoroepoxide monomers synthesized by allylating fluorinated diols and subsequent epoxidation are photopolymerizable in the presence of triphenylsulfonium hexafluoroantimonate [80].

Orthoesters contain three alkoxy groups attached to the same carbon atom and are easily transesterified with alcohols [81]. Bicyclic orthoesters (BOEs), two rings sharing a pair of bridgehead carbon atoms, and spiroorthoesters (SOEs, orthoesters sharing a common atom), are useful as latent polyols as they yield two –OH groups upon acid-catalyzed hydrolysis. BOEs are synthesized via acid-catalyzed transesterification of polyfunctional alcohols and acyclic orthoesters while SOEs are synthesized by reacting an epoxy-functional compound such as butyl glycidyl ether with lactones such as caprolactone or butyrolactone [82]. Due to reduced hydrogen bonding, orthoesters possess low viscosity and are attractive candidates for formulating high solids coatings without the problems of high functionality, short pot life, and high catalyst concentration that plague systems formulated with low-molecular-weight polyols [83]. Both melamines and polyisocyanates can be employed as crosslinkers with orthoesters. The silicon analog, i.e., spiroorthosilicates, synthesized by reacting tetraethyl orthosilicate with 2-butyl-2-ethyl-1,3-propanediol or 2-ethyl-1,3-hexanediol, has been employed as latent reactive diluents in fast curing, 2K polyurethane coatings with long pot lives [84].

ADHESIVES

The global adhesives and sealants market was worth more than $20 billion in 2004. The terms "adhesives" and "sealants" are often used together; however, they are designed to perform different functions. While adhesives keep at least two surfaces together, sealants are designed to fill the space between two surfaces to provide a barrier or protective coating [85]. Commodity and general purpose adhesives, which are sold in relatively high volumes, include formaldehyde resins, starch and dextrin adhesives, styrene-butadiene latex binders, natural rubber latex, hot melts, solvent-based adhesives, pressure-sensitive adhesives, and inorganic adhesives. Specialty adhesives, which are formulated for specific applications, employ polyurethanes, epoxies, cyanoacrylates, anaerobics, reactive acrylics, and radiation-cured adhesives [86].

In simple terms, adhesion is the phenomenon of causing two materials to be bound together, while an adhesive is the material utilized in achieving this phenomenon [87]. The materials bonded by the adhesive are referred to as "adherends" or "substrates." While coatings dry to a tack-free (top surface) adherent film on the substrates to which they are applied, adhesives maintain intimate contact with substrates at all their interfaces. The strength of bonded surfaces depends not only on the degree of adhesion to the adherends, but also on the cohesive strength of the adhesive.

Natural materials employed as adhesives include mud, clay, animal glue (collagen), plant derivatives (starches and dextrins), milk protein (casein), gums, resins, bitumen, and rubber. The development of synthetic polymers quickly displaced natural-product-based adhesives from industrial applications. For instance, vinyl and acrylic emulsion polymers replaced collagen glue used in furniture and cabinet industries. Phenol-formaldehyde and urea-formaldehyde resins gained control of the structural and decorative plywood industries. Resorcinol-formaldehyde resins were preferred in lumber laminating and other construction specialties over casein glue. Although synthetic adhesives are generally considered superior to natural adhesives, certain biological adhesives such as those noted in barnacles and mussels display exemplary adhesion. The trend towards environmentally responsible products has revitalized interest in adhesives derived from natural materials.

Synthetic adhesives are primarily classified as thermoplastic adhesives or thermosetting adhesives. Modern adhesives are complex formulations containing multiple ingredients that perform specialty functions. The choice of ingredients and their proportion depends on the application, properties desired, application and processing conditions, and cost. Apart from the resin itself, an adhesive formulation may contain crosslinkers, accelerators, inhibitors (retarders), solvents, diluents, pigments, extenders, plasticizers, tackifiers, and additives such as thickeners, antioxidants, antifungal agents, and surfactants.

Apart from the adhesive itself, other factors such as surface preparation, joint design, and protection from hostile conditions in service influence the performance of an adhesive joint. Adhesives are preferred over other bonding methods such as welding and fastening in automotive and aerospace applications to reduce weight and increase fuel efficiency. Additionally, adhesive bonding offers the following advantages [88]:

(1) Improves joint stiffness resulting in uniform stress distribution over a larger area
(2) Reduces noise and vibration
(3) Seals the joint against moisture and debris ingress
(4) Reduces stress concentrations at the joint edges resulting in good fatigue resistance
(5) Joins incompatible materials or dissimilar metals (which may otherwise lead to galvanic corrosion)

Low-molecular-weight polymers allow facile processing but possess limited cohesive strength. The curing step subsequent to application increases the molecular weight and increases the adhesive's cohesive strength. Curing in adhesives can be classified as follows [89]:

(1) 1K, heat-cured: The resin and crosslinker are supplied as single component adhesives and react at elevated temperatures. These adhesives are commonly used for structural applications and surface component mounting in electronics. Adhesive chemistries include epoxies, acrylics, silicones, phenolics, and polyimides.
(2) 1K, moisture-cured: Polyurethanes and silicones can be formulated to cure in the presence of moisture. Curing is influenced by the environmental conditions that occur during curing such as relative humidity, temperature, and joint design, and may result in defects such as bubbling, uneven cure, and fast skin generation. These systems typically require a bond line depth of less than 10–15 mm to enable moisture to diffuse and react with the material in the joint center. Generally, moisture-cured adhesives cure slower and have less cohesive strength than their 2K counterparts.
(3) 2K, ambient cure: The resin and crosslinker are supplied separately and mixed immediately before application. Adhesive chemistries include epoxies, acrylics, silicones, and polyurethanes.
(4) Anaerobic: The adhesive is cured in the absence of oxygen in enclosed areas through the catalytic action of moisture.
(5) Cyanoacrylates: These "instant" adhesives cure very rapidly when exposed to air.
(6) UV curing: Curing is initiated by UV light and proceeds very rapidly to completion.
(7) Solvent-based: The adhesive is supplied in an organic solvent or water that must be evaporated before bonding can occur. Polymers

employed in solvent-based adhesives include natural rubber, urea-formaldehyde resins, polyvinyl acetate, polyurethane, polychloroprene, and acrylics.

(8) Pressure-sensitive: Pressure-sensitive adhesives are based on low T_g polymers that maintain permanent tack due to their viscoelastic characteristics. The adhesives are typically based on elastomeric materials such as styrene-butadiene rubber, butyl rubber, silicone rubber, nitrile rubber, and acrylic rubber or block copolymers, along with small quantities of tackifiers, plasticizers, waxes, and oils.

(9) Hot melt: These adhesives are solids at ambient temperatures and melt on heating to yield low-viscosity fluids that set rapidly upon cooling. Hot-melt adhesives are primarily employed in bookbinding, product assembly, and box and carton heat-sealing applications. Polymers employed in hot-melt adhesives include paraffins, ethylene vinyl acetate copolymers, styrene-isoprene-styrene copolymers, polyimides, polyurethanes, and ethylene ethyl acrylate copolymers.

Adhesives may also be grouped as structural or nonstructural. Structural adhesive joints are expected to provide a useful service life equivalent to that of the product containing the joint. Nonstructural applications require low strength or temporary bonding, e.g., pressure-sensitive tapes, hot melts, and packaging adhesives.

Hot-melt adhesives (HMAs) are applied at elevated temperatures as a liquid mel and gradually cure to their long-term performance via moisture diffusing from the environment [90, 91], crystallization [92], vitrification, and phase-separation processes. HMAs are well suited for high-speed production as they are solvent-free and form bonds very quickly after application. HMAs are typically isocyanate-terminated polymers based on polyesters and polyether diols, and aromatic isocyanates such as MDI. Hydrogenated MDI is used in HMAs designed for external applications.

Polyester crystallites contribute to good green strength due to their high viscosity at elevated temperatures. Upon cooling, polyester crystallites may also serve as reinforcement, enhancing the adhesive performance properties, particularly at elevated temperatures [93, 94]. Polyethers contribute to adhesive toughness via their flexibility. Acrylic copolymers are often added to improve compatibility between the polyether and polyester [95, 96], and enhance the green strength owing to their high T_g and viscosity [97, 98]. Hsu et al. studied the miscibility behavior of HMAs containing polyester, polyether, and acrylics, and concluded that polyethers played a crucial role in expanding the region of miscibility whereas polyesters did not affect the blend miscibility [99].

Aqueous polymer isocyanate (API) adhesives are synthesized by blending vinyl latexes or carboxylated styrene–butadiene rubber latexes with polymeric isocyanates based on MDI or TDI [100]. Hydroxyl functionality is provided via incorporation of ~10 wt. % polyvinyl alcohol to the latexes. The polyisocyanates are either dissolved in a mutually compatible organic solvent or modified with nonionic surfactants to blend with the latex. API adhesives have replaced urea-formaldehyde resins in wood laminates and finger-jointing applications, and offer high bonding strength with rapid cure at elevated temperatures.

Pressure-sensitive adhesives (PSAs) are used extensively in tapes, labels, and protective films due to their peel-and-press simplicity and are one of the biggest and fastest growing sectors in adhesives. A PSA is a soft, viscoelastic solid composed of a lightly crosslinked high molecular weight polymer that remains in a permanently tacky state and is characterized by instantaneous adhesion upon application of light pressure. The degree of crosslinking dictates the shear and cohesive strength of the adhesive.

PSAs are formulated from natural rubber, styrene-butadiene block copolymers, styrene-isoprene block copolymers, or acrylic resins. Natural rubber–based PSAs were the first to be developed and are still in use for general purpose applications. Styrene-based PSAs are solvent-free systems applied as hot melts above 160 °C and facilitate high production rates. Both these PSAs have poor environmental stability and tend to degrade under thermo-oxidative or UV exposure. Acrylic PSAs typically contain 50%–90% methacrylate esters of higher alcohols (C_4–C_{17}) and 2%–20% of a monomer with the desired functional groups, e.g., acrylic acid [101]. PSAs are typically applied as 20–100 micron thick films in latex, solution, or melt forms. The first two are primarily used for acrylic adhesives while the last form is employed with block copolymer-based adhesives. Latex PSAs offer low viscosities and are free of organic solvents, but the presence of residual surfactants and additives may affect adhesive performance. Although characterized by high-VOC content, solution PSAs are still widely used as the synthesis process facilitates better control over molecular structure and contains fewer impurities.

Photopolymerizable PSAs offer faster processing times, require less space and energy, and contain lower VOCs. Commercial UV-curable acrylic hot-melt adhesives are based on acrylic polymers that contain a copolymerized benzophenone derivative [102]. Since the extent of crosslinking depends only on the UV dosage, production rates can be increased by using successive irradiation units. The adhesive chromophore filters UV light, resulting in reduced conversion at higher thicknesses.

Hot-melt PSAs reduce or eliminate the use of organic solvents during synthesis and application and typically comprise low-molecular-weight rubber or acrylic block copolymers. The inherent physical crosslinking of block copolymers results in a good balance of cohesive strength and processability. Rubber block copolymers combine polystyrene hard blocks with soft blocks such as polyisoprene, polybutadiene, or ethylene–butylene copolymers. Well-defined structures of block copolymers require more stringent reaction conditions and greater knowledge of reactive pathways than homopolymerization [103]. In recent years, controlled or "living" radical polymerizations such as atom transfer radical polymerization (ATRP), nitroxide-mediated polymerization, and reversible addition-fragmentation chain transfer (RAFT) polymerization have enhanced the opportunities for block copolymer synthesis [104]. Ionic, electron-beam, and UV-initiated crosslinking have been employed in hot-melt PSAs for various applications [105].

One way to enhance the cohesive strength of an adhesive is via ionic crosslinking by introducing metal ions into an acid-functional adhesive or by generating acid-base interaction between polymer chains. Ionic crosslinking offers the following benefits [106]:

(1) Does not require special curing equipment
(2) No interference from pigments or fillers
(3) Adhesive thickness does not limit the curing efficiency or process speed
(4) Crosslinking may be thermally reversible or may be designed to form a thermosetting system

Everaerts et al. employed acid-base interactions in hot-melt acrylate PSAs using a copolymer of isooctylacrylate (IOA) and acrylic acid (AA), and copolymers with *N,N*-dimethylaminoethyl methacrylate (DMAEMA), e.g., 40/60 IOA/DMAEMA and 40/58/2 IOA/DMAEMA/AA as the crosslinker [31]. Other crosslinkers evaluated in the study were polyetherimines (PEI), DICY, and a polyfunctional aziridine. The authors noted that in contrast to block copolymers, the onset of flow was not controlled by the polymer T_g but by the strength of the acid-base interactions. Rheological data indicated that the strength of the acid-base interactions and the structure of the base had a significant effect on the adhesive properties.

Crosslinked epoxy resins offer many useful properties for structural engineering applications such as high modulus and failure strength, low creep, and good performance at elevated temperatures. Although adhesion is controlled primarily by surface energies, other performance attributes such as cohesive strength, hardness, and durability are dramatically impacted by the choice of curing agent used to crosslink the epoxy [107].

Unmodified DGEBA epoxy adhesives are toughened either by adding flexible elements to the matrix or by including a secondary phase. Matrix flexibility can be provided inherently via the epoxy resin and/or the curing agent, or by employing a reactive modifier. Flexible epoxy resins include epoxy-dimer acid adduct, epoxidized vegetable oils, elastomer-modified epoxy resins, diglycidyl ethers of polyalkylene glycols, aliphatic polyols, castor oil, or ethoxylated resorcinol. Crosslinkers such as polyamidoamines, polyamides, oligomeric polyamines, silyl-terminated polyethers [108], and amine-terminated silicone resins influence performance properties including toughness, tensile strength, and thermal and chemical resistance.

One of the means of adding a secondary phase is via the incorporation of a reactive liquid rubber such as carboxyl or amine-terminated nitrile rubber (NBR, copolymer of acrylonitrile and 1,2- or 1,3-butadiene). During curing, the rise in molecular weight lowers the compatibility between the epoxy resin and rubber phases. The *in situ* phase separation results in a matrix that contains finely dispersed rubber-rich domains (0.1–5 μ) within the cured epoxy resin. The rubbery particles enhance the adhesive toughness via interactions of the stress field ahead of the crack tip and enhance plastic deformation of the epoxy matrix. The enhancement in fracture energy depends upon the particle size, volume fraction, and size distribution of the dispersed phase and on the chemical structure and T_g of the epoxy resin and curing agent employed. Cure conditions strongly influence the morphology of the dispersed phase in the cured resin, and consequently, the performance of the epoxy adhesive [109]. Hydroxy methyl bisphenol-A has been shown to be an effective reactive interfacial agent in an NBR-epoxy adhesive by imparting higher adhesive joint strength, toughness, thermal stability, T_g, and lower elongation-at-break [110].

Another approach to improving the mechanical performance of epoxy adhesives is via the use of nanoparticles or fibers having average particles sizes between 5–50 nm. Kinloch et al. incorporated nanosilica into an NBR-epoxy adhesive cured with a blend of *N*,*N*-dimethyl-1,3-diaminopropane and a polyamide [111]. Fracture energy (G_e) increased from 1200 J/m^2 for the control rubber-toughened epoxy to 2300 J/m^2 at 4.1 wt. % of nanosilica. Improvements in peel energy, lap shear strength, and toughness were also noted. Reinforcement of epoxy adhesives with nanoparticles grafted with alkoxysilanes improved viscoelastic properties compared with the neat epoxy adhesive. Increased stiffness and a T_g shift of about 20K were also observed [112].

Epoxy-phenolic adhesives are used for bonding metal, glass, ceramic, and phenolic composites and possess very good resistance to weathering, oil, solvents, and moisture. However, the high aromatic content of these adhesives results in relatively low peel and impact strength, and limited thermal shock resistance.

Epoxy-nylon adhesives were one of the first structural adhesives designed specifically to possess high shear strength and extremely high peel strength. Epoxy-nylon adhesives also offer excellent tensile lap shear strength, and good fatigue and impact resistance [113]. However, the inherent hydrophilicity of nylons results in poor moisture resistance.

Epoxy-sulfide adhesives offer excellent flexibility, chemical resistance, and good low-temperature performance, but lack shear strength. A tertiary amine is employed as the accelerator (Figure 11.5). A key limitation in using sulfides is their distinct unpleasant odor, but the use of polysulfides with minimal low-molecular-weight impurities helps mitigate odor issues. Epoxy-sulfide adhesives are typically used for bonding rubber to metal.

Waterborne epoxy dispersions offer an environmentally responsible alternative to solvent-based epoxy adhesives, and are either based on bisphenol-A derivatives of varying molecular weights, polyfunctional epoxies, or epoxy resins modified with urethanes or elastomers. Any curing agent used with solvent-based epoxy systems and that can be stabilized in water can be employed for curing waterborne epoxy dispersions, e.g., polyamides, polyami-

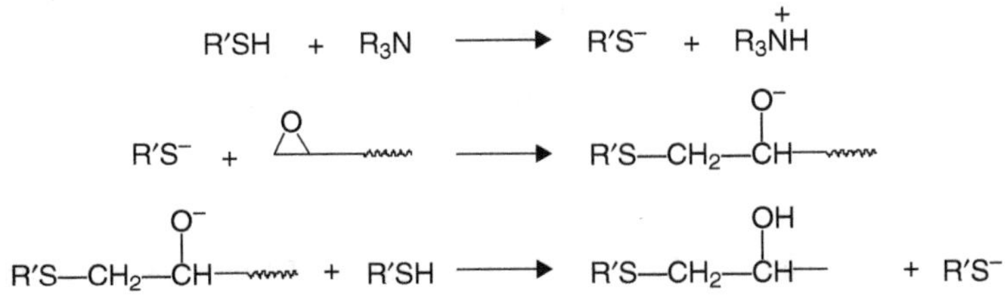

Figure 11.5. Epoxy sulfide crosslinking.

doamines, DICY (often with 2-methyl imidazole as an accelerator), phenolics, and amino resins. Addition of 10–20 wt. % of waterborne epoxy dispersions to carboxyl and amine functional latex adhesives enhances their tensile strength and water resistance. It is important to ensure that the pH of the epoxy dispersion is close to the latex pH to avoid latex coagulation. Such blends are employed as laminating adhesives to various plastic and metal foils and films.

Liquid crystalline epoxy resins contain a rigid backbone and are characterized by high toughness. Liquid crystalline epoxy resins containing a mesogenic group and crosslinked with a curing agent with neighboring active hydrogens possess a very weak glass transition, and thus a very high rubbery modulus [114]. Ochi et al. investigated the bonding properties of an adhesive system comprising diglycidyl ether of terephthalylidene-bis-(4-amino-3-methylphenol) crosslinked with 4,4′-diaminodiphenylmethane, and related it to their phase structure (Figure 11.6) [115]. The lap shear strength of the nematic system was lower than that of the isotropic system, although the nematic adhesive layer showed larger deformability than the isotropic one. The high bonding strength of the isotropic system was attributed to the localization of hydroxyl groups (formed during cure) at the adhesive interface. The

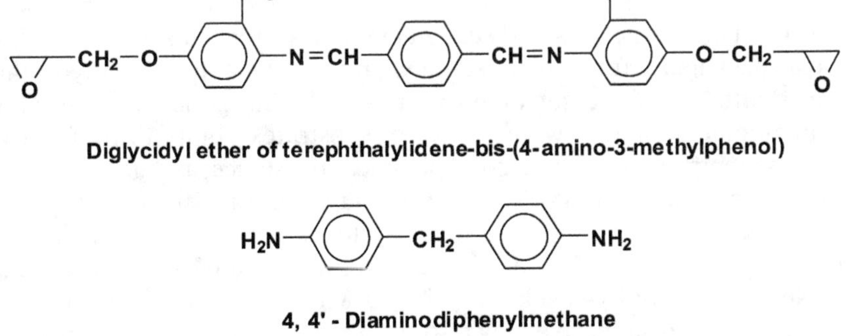

Figure 11.6. Liquid crystalline epoxy adhesive system.

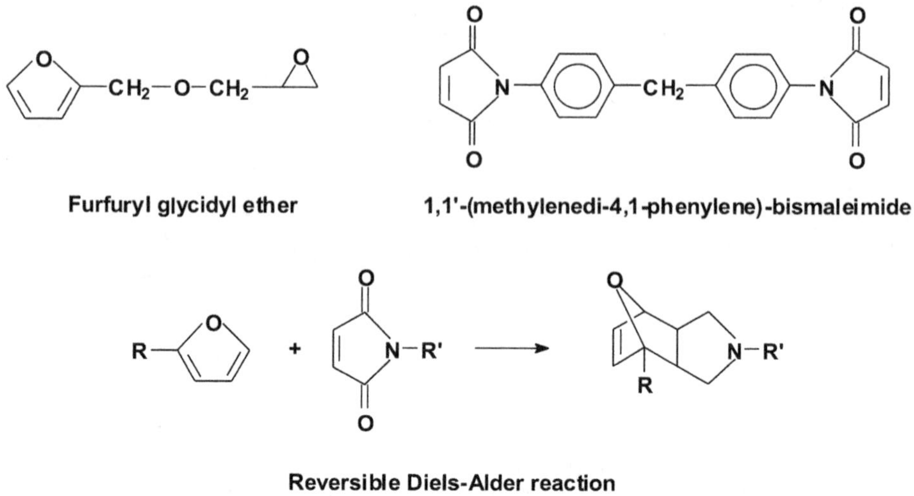

Figure 11.7. Epoxy resin of a reversible adhesive.

restricted molecular motion in the ordered domains of the nematic system resulted in low concentration of hydroxyl groups at the adhesive interface and low bonding strength.

While adhesives are normally expected to maintain their performance indefinitely, recycling and/or repairing requires that the adhesion be reversed. Thermally reversible adhesives have been synthesized by reacting an aliphatic diamine with a diepoxy compound containing two Diels-Alder adducts [116]. The Diels-Alder adduct was formed by reacting a furfuryl glycidyl ether (diene) with 1,1′-(methylenedi-4,1-phenylene)-bismaleimide (dienophile) (Figure 11.7). The adhesive displayed a T_g of 74 °C and a constant shear modulus up to approximately 90 °C. Above 90 °C, a retro Diels-Alder reaction led to a significant loss in modulus. The loss of modulus was reversible with temperature.

Polyurethanes have established a prominent position in the formulation of high-performance adhesives and sealants in applications in such industries as footwear, packaging, furniture assembly, and plastic bonding. Polyurethane adhesives offer a unique set of properties including good bond durability, substantial adherence to a wide variety of substrates, ability to perform well over a wide temperature range, good chemical resistance, and good balancing of lap shear and peel strength as well as good solvent resistance [117].

One-component (1K) polyurethane systems are based on isocyanate-terminated prepolymers and cure by reacting with ambient moisture (Figure 11.8). Relative humidity levels of 40%–60% are most desirable to ensure optimum curing. The adhesive may also react with hydroxyl groups present on substrates such as wood and paper. Excessive moisture on the substrate will

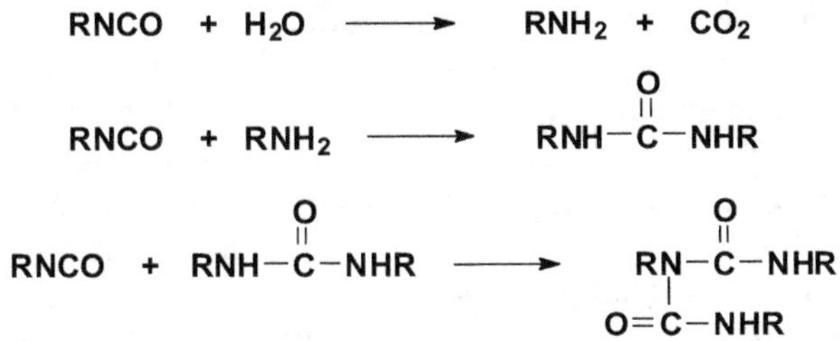

Figure 11.8. Curing reactions of 1K polyurethane adhesives.

lead to pinholes or foaming from the CO_2 released during cure. Use of latent isocyanates such as blocked isocyanates and ketimines helps avoid the problems associated with reactivity of the free isocyanate.

Anaerobic adhesives are employed for augmenting the seal or holding force of a mechanically joined appliance assembly, e.g., locking threaded fasteners; sealing pipe fittings; retaining bearings, pulleys, and gears to shafts; sealing flanged surfaces; and replacing cut gaskets in automotive, construction, and farming equipment. Anaerobic compositions are stable and remain in a liquid state in the presence of oxygen but polymerize to a solid in the absence of oxygen. Curing occurs via ambient temperature bulk polymerization of monomeric and telechelic methacrylates initiated via decomposition of hydroperoxides by amines and is catalyzed by transition metals, particularly copper and iron. Therefore, anaerobic adhesives cure rapidly on iron- and copper-rich surfaces, but on cobalt- or zinc-plated metals they may cure slowly or not at all. Metals and alloys that contain very low amounts of iron and copper are generally insensitive to anaerobic cure.

U.S. patent 5,411,998 discloses the use of anaerobic compositions in 1K and 2K systems for sealing surfaces having thick bondlines [118]. The 1K systems are based on a combination of saccharin and onium salts (phosphonium or quaternary ammonium salts other than iodides). In the 2K system, one part contains cumene hydroperoxide, saccharin, and an onium salt, while the other includes saccharin and acetylphenyl hydrazine.

Structural acrylic adhesives [119] are typically formulated with methacrylates, a rubber hardener such as chlorosulfonated polyethylene, and a redox initiator. Some free radicals are also generated by the initiator's reaction with the hardener and graft the acrylic polymer to the rubber particles. The adhesive is supplied as a 2K package with the initiator components separated. Each component is applied to an adhered end, and when joined, curing is completed in ~10 min. The components may also be premixed before application.

Silanols, i.e., polydimethylsiloxanes terminated with hydroxyl groups (molecular weights 15,000–150,000), are susceptible to condensation prior to final application under both mildly acidic and basic conditions and are employed as intermediates for most room temperature vulcanizable (RTV)

1K and 2K adhesives. RTV silicones are widely used as sealants for window panes and ceramics as they offer outstanding weather resistance. Their excellent moisture and chemical resistance after crosslinking makes them ideal for application in moist environments such as bathrooms, kitchens, and marine applications. Silicones maintain their adhesion across temperature extremes from −60 °F to 300 °F, and serve as useful sealants in automotive engine applications, ovens, and industrial processes.

One-component RTV systems employ multifunctional silanes in a two-stage reaction. In the first stage, the silanol is reacted with an excess of moisture-sensitive multifunctional silane, resulting in a silicone resin with the multifunctional silane group at each end. The RTV is supplied in this form; upon exposure, the end groups rapidly react with moisture. The most common chemistries employed for moisture-cure RTVs are acetoxy, oxime, alkoxy, amine, aminoxy, and benzamide (Figure 11.9) [120]. The acetoxy RTVs are characterized by their distinct vinegary odor during cure due to the acetic acid liberated upon crosslinking. This may be a concern for corrosion if the sealant is used around metal parts. Amine- and aminoxy-cured systems have an ammoniacal odor and are typically used in automotive and industrial applications.

Two-component RTV silicones are platinum-catalyzed systems, with one component containing vinyl-ether-terminated silicone resin and the other containing a hydride-functional silicone polymer. Two-component RTVs cure faster than 1K RTV adhesives and also offer deep section cure as they do not liberate volatiles during cure. However, these systems require mixing, degassing, and dispensing apparatus for application. Moreover, a primer is often necessary to ensure good adhesion.

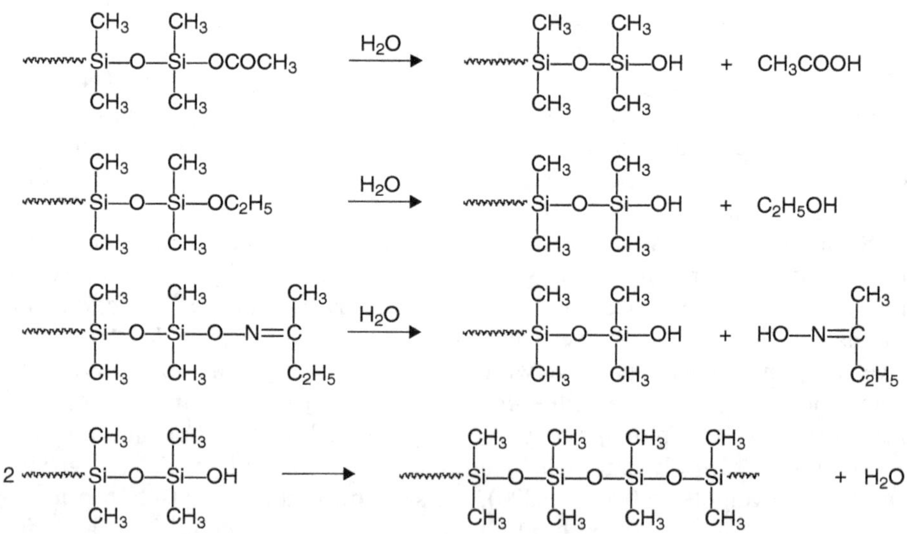

Figure 11.9. Curing reactions of 1K RTV silicone adhesives.

Cyanoacrylates are known for their extremely rapid cure and are employed with rubber and plastic parts in printed circuit board wires and components, disposable plastic medical devices, toys, appliances, automotives, and cosmetic packaging. The "instant adhesives" are also a quick fix for repairs in the home. Cyanoacrylates are characterized by the presence of two strongly electron-withdrawing groups (–CN and –COO) that facilitate anionic polymerization. Polymerization is initiated by surface moisture but is inhibited by oxygen and does not proceed to completion unless the adhesive joint is closed. Methyl cyanoacrylates work best on metal and rubber parts while ethyl cyanoacrylates are recommended for work on plastic parts. Octyl cyanoacrylates are favored for medical applications as they exhibit excellent biocompatibility and flexibility.

Cyanoacrylates are employed in surgery as a tissue adhesive to replace traditional suturing techniques. Potential benefits include better cosmetic results, more rapid wound closure, less pain, and lower risk of transmitting infectious diseases. On the other hand, concerns with cyanoacrylates include difficulty in precise delivery due to low product viscosity, poor shear strength (particularly in moist environments), high stiffness of cured adhesives on soft tissues leading to mechanical incompatibility and adhesive failure, and heat generation during cure. U.S. patent 5,350,798 disclosed the use of methoxypropyl cyanoacrylate as the adhesive precursor and a triethylene glycol oxalate ester as a modifier; however, incompatibility between the polycyanoacrylate and the modifier resulted in phase separation [121]. U.S. patent 6,699,940 discloses polymer chains end capped with alkyl or alkoxyalkyl cyanoacrylate esters to overcome the limitations of the earlier patent [122]. While cyanoacrylates form extremely strong bonds between tissues, they act as barriers to biological repair. To overcome this limitation, U.S. patent application 2006/0147479 proposes a tissue adhesive comprising a cyanoacrylate in combination with an angiogenic factor such as butyric acid and its derivatives to promote the growth of new blood vessels from preexisting vessels [123].

Extensive harvesting of old growth forests and current conservation policies have resulted in a significant reduction in the size of logs used in construction materials. Nonetheless, wood remains the classic renewable resource. Wood composites are manufactured by combining wood furnish with suitable polymeric adhesives under the influence of pressure and heat. The wood furnish used in composites is obtained from relatively small trees and includes tree residues such as bark and shavings, thereby utilizing a significant portion of the tree. Consequently, composites offer an excellent means to maximize useful production from limited resources. For instance, several markets such as molded products, inorganic-bonded products, and panel products (plywood, oriented strand board, fiberboard, and particleboard) have emerged to meet the strong demand for wood composites.

Presently, the vast majority of wood adhesives are based on formaldehyde derivatives such as urea-formaldehyde (UF) or phenol-formaldehyde (PF) resins. UF resins are characterized by low cost, easy application, low cure temperatures, water solubility, resistance to microorganisms, hardness, lack of color in the cured resin, and excellent thermal and mechanical properties when cured. A major limitation of UF resins is their low water resistance, especially in combination with heat, as the methylene and methylene-ether bridges are easily hydrolyzed and liberate formaldehyde. Formaldehyde is extremely toxic and is classified as a carcinogen [124]. The level of formaldehyde emissions from formaldehyde-based products is regulated by law, especially with respect to indoor use.

Several attempts have been made to reduce formaldehyde emissions from composite products. Mami ski et al. showed that incorporation of glutaraldehyde in a UF resin with low water resistance markedly improved the glue's resistance to boiling water [125]. The improved resistance was attributed to 5-carbon bridges formed during the cure of glutaraldehyde with UF resin. U.S. patent 5,593,625 describes methods of preparing a legume-based thermosetting resin to produce composites by blending soy flour with methyl diphenyl isocyanate [126]. U.S. patent 6,306,997 refers to a soybean-based adhesive that contains 70% soybean flour and 30% phenol-formaldehyde resin [127]. Wang and Sun described particleboards derived from wheat straw blended with MDI, corn stalk pith, and sodium hydroxide-modified soy protein isolate [128]. However, isocyanates present toxicity issues of their own during handling and processing.

Lignin, second only to cellulose as a source of fixed carbon in the biosphere, is highly resistant to rapid biological degradation. Chemically, lignin is a complex polymer of phenylpropane units that are crosslinked to each other via a variety of different chemical bonds. The abundance of lignin as a waste product in pulp mills presents an attractive proposition for incorporation in adhesives, primarily as a substitute for UF or PF resins. However, the inherent low reactivity of lignin towards formaldehyde resins results in long pressing times. Methylolated lignin (synthesized by reacting formaldehyde with lignin) reacts faster with PF resins, and the blends have been employed in plywood manufacture as the rate of pressing does not have a significant effect on the economics of the manufacturing process. However, other wood composites such as particleboard, oriented strand board, and medium density fiberboard do not favor the use of lignin as the rate of pressing is critical in determining the cost and profitability of both process and product [129]. To obviate the reactivity limitations, Pizzi et al. developed a system involving polymeric MDI, methylolated lignin, and PF resin [130, 131]. Crosslinking occurs via formation of lignin-lignin, lignin-phenol, and phenol-phenol methylene bridges as well as via urethane bridges between the lignin methylol groups and PF resin with the isocyanate, and yielded panels at industrially significant pressing times.

In an effort to develop more environmentally responsible materials, lignin derivatives have been synthesized by replacing formaldehyde with glyoxal, a nontoxic aldehyde [132]. Particleboards manufactured using adhesive blends of glyoxylated lignin with polymeric MDI exhibited internal bond strength values exceeding the specifications for exterior-grade panels. Moreover, the reactivity was sufficient to yield panels in press times comparable to those of formaldehyde-based commercial adhesives.

Liu and Li investigated the use of demethylated kraft lignin (DKL), a by-product in the production of dimethyl sulfoxide from kraft lignin, as a wood adhesive using PEI as the crosslinker [133]. The authors proposed that the phenolic hydroxyl groups of DKL oxidize to form quinones that react with PEI to form Schiff bases and Michael addition products, and yield a water-insoluble, highly crosslinked polymeric network. Lap shear tests on maple veneer specimens indicated that 1:1 DKL:PEI weight ratio was optimum in terms of enhancing the strength and water resistance. PEI molecular weight had little effect on the adhesive performance.

Geng and Li proposed lignin adhesives synthesized by treating lignin and/or lignocellulosic materials such as decayed wood and solubilized decayed wood with PEI and adhesion promoters such as sodium borohydride and boric acid [134]. The boron compound is believed to form esters by reacting with the lignin catechol moieties and also possibly with the hydroxyl groups of carbohydrates naturally occurring in lignocellulosic materials. Solubilized decayed wood adhesives formulated with an adhesion promoter exhibited the highest shear strength of all the compositions tested.

Thames et al. developed formaldehyde-free, soybean protein-based adhesives for particleboards. Unlike other approaches to making formaldehyde-free adhesives, the adhesives of Thomas et al. are totally based on agricultural derivatives, primarily soybean protein. Soybean seeds contain about 40% protein. After oil extraction, the defatted soy flakes are processed to remove the nonprotein components to yield soybean protein isolate (SPI). SPI is the most highly refined soybean protein product sold commercially and contains >90% protein on a moisture-free basis. An alternate source of soybean protein is defatted soybean flour (DSF) that contains ~53% protein and 35% carbohydrates.

Thames et al., U.S. patent 6,790,271, discloses a powder adhesive composed of soybean protein, a polyol plasticizer such as glycerine, and a vegetable oil derivative such as maleinized methyl ester of tung oil [135]. The maleinized ester served as a crosslinker and enhanced the water resistance of the particleboard. In a second patent, U.S. 7,081,159, Thames et al. describes a liquid adhesive composition based on soybean protein in conjunction with zinc sulfate heptahydrate, calcium oxide, sodium benzoate, pine oil, wax emulsion, and nonsulfonated kraft lignin [136]. The particleboards were evaluated as per American National Standards Institute (ANSI) specifications. DSF-based particleboards met M-1 and M-S grade particleboard requirements while SPI-based particleboards exceeded the requirements of M-1, M-2, M-3, and M-S grades. Particleboard grades M-S, M-2, and M-3 are used in kitchen and vanity cabinets. Industrial particleboard grades used in office and residential furniture, kitchen cabinets, case goods, and other applications are typically M-S, M-2, or M-3 grades. Countertops and shelving applications also require particleboards that pass the M-2 grade. Soil biodegradability (AATCC TM 30–199,

1:1:1 soil:compost:sand mixture maintained at 37 °C) studies indicated that the soybean protein-based particleboards lost an average of 27.9% of their total weight while UF-based particleboards lost an average of 11.6%, indicating that soybean protein-based particleboards biodegraded faster than commercial particleboards.

Recent improvements in adhesive formulation have resulted in formaldehyde-free particleboards with improved water resistance properties that perform competitively against UF resin-based commercial particleboards [137]. A key distinction between the soybean protein liquid adhesive and UF resins are the rheological characteristics and specifically higher viscosity. UF resin is supplied at waterlike consistency at 65% solids, while the soybean protein-based liquid adhesive has a pastelike consistency even at 32% solids. However, the soybean protein adhesive blends efficiently with the wood furnish, and its performance has been validated at plant trials.

Marine environments severely challenge the performance of synthetic adhesives. However, biological adhesives synthesized by barnacle shells, kelp adhesives, mussel glues, and oyster cements form in a few seconds and function exceedingly well over wide temperature ranges, varying salinities, humidity levels, and tidal currents of oceans. Biofouling, i.e., the settlement and accumulation of living organisms on man-made physical structures in an aquatic or marine environment, costs the U.S. Navy alone more than $1 billion a year due to reduced fuel efficiency and costs involved in removing the biofouling material [138].

Mussel adhesive proteins (MAP) contain tandem repeat sequences of approximately 5–15 amino acids, of which one or more residues are 3,4-dihydroxyphenylalanine (DOPA) [139]. DOPA is characterized by two key functional groups—an amino group and a catechol group. Various reactions between the amino group and the catechol group solidify and crosslink the MAP, thus converting the MAP into a very strong and very water-resistant adhesive [140]. Interestingly, the transition metal (e.g., copper, iron, manganese, and zinc) content of mussel adhesives is up to 100,000 times that of open ocean water [141]. Wilker studied the effects of chloride and nitrate salts of various metals on MAP extract and concluded that Fe^{+3} induced the greatest degree of crosslinking on the adhesive precursor [142].

Various researchers have employed DOPA in adhesives or to develop biomimetic strategies for surface modification. Liu and Li disclosed adhesives for lignocellulosic composites that were synthesized by reacting phenolic hydroxyl-augmented PSI with DOPA [143]. Deming and Yu developed water-soluble copolypeptides containing DOPA and L-lysine by ring-opening polymerization of α-amino acid *N*-carboxyanhydride monomers [144]. Aqueous solutions of these copolymers, when mixed with a suitable oxidizing agent (e.g., O_2, mushroom tyrosinase, Fe^{3+}, H_2O_2, or IO_4^-), formed crosslinked, moisture-resistant adhesive bonds with aluminum, steel, glass, and plastics. Successful adhesive formation was dependent on oxidation conditions; chemical oxidants gave the best results.

U.S. patent 6,867,188 describes a nonirritating, nonallergenic, and nontoxic bioadhesive composition suitable for covering and attaching structures to eye tissues such as the cornea. The composition is provided as a 2K kit comprised MAP and polysaccharides, preferably negatively charged, and optionally, an oxidizing agent such as hydrogen peroxide nitroprusside ions or periodate ions. Bioinspired polymers based on MAP for antifouling and medical applications are detailed in the works of Messersmith [145].

LAMINATES

In many applications, a single material may not possess all the required properties. The process of lamination, i.e., joining two or more layers of materials (films, foils, paper, etc.) via adhesive bonding, enables the properties of different materials to be combined advantageously into a single product with multifunctional properties, e.g., high tensile strength and high gas permeability.

Lamination processes are generally classified as dry bond laminating and wet bond laminating. In dry bond laminating, a liquid adhesive is coated on a substrate, dried with heat and airflow, and then laminated to a second substrate via a heated compression nip. In wet bond laminating, a liquid adhesive is applied to a substrate and immediately laminated to a second substrate. The laminate is passed through a heated dryer before proceeding to a lamination nip/cooling section to achieve the desired laminate bond strength. In wet bonding, drying occurs only after the substrates are joined, and at least one substrate must be porous to allow evaporation of water or solvent. Multilayer plastic films can also be formed via extrusion lamination and coextrusion lamination techniques [146]. In extrusion lamination, a thin layer of plastic film is extruded between two other film layers (plastic, foil, or paper) and is particularly useful for bonding printed layers of film together. Coextrusion lamination involves simultaneous extrusion of different polymer melts through a die and is generally used for very high volume laminate production.

Waterborne laminating adhesives are based on naturally occurring materials such as dextrin, starches, and natural rubber, or synthetic emulsions based on acrylic-vinyl esters or polyurethanes. The former are more commonly employed for labeling and other packaging applications, while the latter are mostly used for either wet or dry laminating. Surfactants are also incorporated in the adhesive formulation to achieve proper adhesion to low surface energy substrates such as polyolefins or silicone-coated release liners. Depending on the tolerance, alcohols may also be used as adhesion enhancers in lieu of surfactants. In general, however, surfactants lower the wet bond strength and water resistance of the laminate. If one of the substrates is a printed surface, surfactants may result in discoloration or blurring of the printed ink. U.S. patent 4,617,343 discloses latex compositions with high alcohol tolerance that do not have free surfactants present after film formation [147]. Alcohol tolerance was imparted by the use of at least 12% by weight of vinyl ester monomers such as vinyl acetate, vinyl propionate, or vinyl butyrate. One to two percent of methacrylic acid or its hydroxyalkyl ester such as hydroxyethyl methacrylate was also incorporated as a comonomer to achieve the required alcohol tolerance. Latex synthesis was accomplished using polymerizable surfactants, typically polyoxyethylene alkylphenyl ether derivatives of unsaturated acids such as maleic, fumaric, or methacrylic acids. Maleic surfactants are particularly attractive because of their simple synthesis and lack of homopolymerization [148].

Printed circuit boards (PCBs) are typically manufactured using copper-clad laminates with a reinforced polymeric material serving as the dielectric between the metal. For electronic products that require only a single side of an electrical connection, the dielectric material is made of inexpensive, paper-reinforced phenolic laminates. Equipment requiring multilayer interconnect structures employ laminates based on reinforced glass fiber, partially brominated, bisphenol-A-based epoxy resin (commonly termed as Flame Resistant-4, or simply FR-4) due to its good copper peel strength, excellent insulation, and processability. DICY is generally used as the curing agent.

The trend towards lighter, shorter, smaller, and more reliable electronic products has necessitated the development of PCBs with improved dielectric properties including a lower dielectric constant (D_k) and loss factor (D_f), superior thermal properties such as high T_g and decomposition temperature (T_d), and good processability. With T_gs between 110 °C and –140 °C, FR-4 has an excessive coefficient of thermal expansion in the z-axis direction and poor thermal resistance resulting in resin smearing during hole drilling. To improve the T_g and thermal resistance, multifunctional epoxy resins such as epoxidized novolacs and tetraphenylethane tetraglycidyl ether are added at varying levels. However, their proportion needs to be controlled since very high crosslink density leads to brittleness, poor adhesion, and low copper peel strength [149].

The move toward lead-free soldering has resulted in higher soldering temperatures, and consequently, laminates are now required to handle soldering temperatures of 260 °C and higher. Laminates with higher T_g values are based on polyimides, cyanate esters, allylated polyphenylene ethers, bismaleimide triazine, and tetrafunctional epoxy systems [150]. Cyanate esters cure through the cyclotrimerization of reactive cyanate functional groups into densely cross-linked triazine resins, which have low dielectric constants and high T_gs. Cyanate esters are less expensive than polyimides and are better suited for high-frequency applications and thick multilayer structures than FR-4 because of their lower thermal coefficient of expansion, lower dielectric constant, and low loss factors. However, cyanate ester PCB laminates are prone to form micro-cracks during board fabrication [151].

Allylated polyphenylene ethers exhibit T_gs > 200 °C with good electrical properties and are used in high-speed digital applications with triallyl cyanurate or triallyl isocyanurate as crosslinkers. Allylated polyphenylene ethers are typically synthesized by dehydrogenating polyphenylene ethers with *n*-butyl lithium followed by reaction with allyl halide.

Polyimides are noted for their flexibility and outstanding mechanical, thermal, and electrical properties, and are widely used in flexible PCBs, ribbon cables, reinforcing materials, and multilayer PCBs. Flexible PCBs utilize the thinnest dielectric substrate available for electronic transmission and significantly reduce the weight of electronic packages. Polyimide-based flexible PCBs are in demand for miniature devices such as mobile phones, automobile electronics, and aerospace electronics. To ensure adequate safety, the copper conductors must adhere firmly to the polyimide to withstand soldering as well as stresses from vibrations and shocks, and temperature and humidity fluctuations.

The polyimide layer is bonded to the metal substrate by an adhesive, usually based on epoxy resins or acrylates. However, the poorer thermal and dielectric properties of the adhesive relative to the polyimide compromise the PCB performance characteristics. Moreover, the adhesive layers add to the thickness of the PCBs. Adhesive-free laminates offer advantages over adhesive-based laminates in terms of thinness, thermal resistance, and dimensional stability. Several attempts have been made to enhance the adhesion between the polyimide resin and metal, e.g., incorporating bismaleimide modified with barbituric acid or its derivatives [152], employing a polyimide upper layer and a polyimide-bismaleimide lower layer [153], or a polyimide prepared by the polymerization reaction involving imidazole- or benzimidazole-containing amine monomers [154]; however, all of them have their own limitations.

U.S. patent 6,133,408 discloses a polyimide laminate produced by casting a polyamic acid precursor onto the metal foil followed by imidization via heating above 250°C [155]. The polyamic acid precursor was synthesized by reacting aromatic tetracarboxylic acid anhydrides such as 3,3′,4,4′-biphenyltetracarboxylic acid dianhydride, pyromellitic dianhydride, and/or 3,3′,4,4′-diphenylsulfonetetracarboxylic acid dianhydride, with aromatic diamines such as 4,4′-oxybisbenzenamine, *m*-phenylenediamine, and/or *N*,*N*′-diphenylmethylenediamine in a polar aprotic solvent.

Lee et al. evaluated the influence of pyridine and triazole derivatives on the adhesion between polyimides and sputter-deposited copper. Thus, 4,4′-oxydianiline (ODA) was separately mixed with 2,6-diaminopyridine (DAP) and 3,5-diamino-1,2,4-triazole (DATA) to form codiamines. Polyimides were synthesized in two steps by reacting the codiamines with 3,3′,4,4′-pyromellitic dianhydride. The adhesion strength of sputter-deposited copper to the polyimide films was proportional to the functional group content. At a molar ratio of DAP to ODA of 1:6, the 90° peel strength of the copper/polyimide laminates reached a maximum of 990 J/m^2. At a similar molar ratio of DATA to ODA, the 90° peel strength reached a maximum of 696 J/m^2. X-ray photoelectron spectroscopy studies indicated that laminates with low to moderate adhesion failed mostly at the interface near the polyimide, while the laminates with high adhesion failed mostly in the polyimide [156].

Multilayer laminates have been proposed wherein the polyimide is bonded directly to metal without an adhesive. Such laminates employ formable polyimides based on diphenyl tetracarboxylic acid and are inferior in thermal stability to fully aromatic polyimides that are no longer formable. Moreover, formable polyimides may flow during laminating and promote undesirable direct contact between the metal layers.

Bismaleimide resins are used in glass or carbon fiber reinforced laminates due to their excellent thermal and mechanical properties and low moisture sensitivity. Maleimides possess two carbonyl groups conjugated to the double bond and present a highly electron poor double bond that reacts with nucleophiles to yield linear, high molecular weight polyimides [157, 158].

In general, bismaleimides are soluble in very few solvents such as *N*-methylpyrrolidone. Various attempts have been made to improve the solubility of maleimides and improve their processability. For instance, U.S. patent 4,460,783 discloses that certain aromatic ether bismaleimide compounds, e.g., bis(malimidophenoxyphenyl) propane, are highly soluble in common solvents such as acetone, toluene, and methyl ethyl ketone [159]. The solubility of these bismaleimides are attributed to the presence of significant amounts of acid group-containing precursors, i.e., terminal amic acid groups that have not been converted to imide groups. The bismaleimide solutions tend to separate over time when the resin concentration exceeds 50%, but this can be avoided by adding equimolar amounts of polyamines such as diethylene triamine, triethylene tetramine, or PEI [160].

Liaw et al. synthesized several novel polyimides based on new diamines containing flexible units such as sulfonyl [161], cyclododecyl [162], tricyclodecyl [163], norbornyl and *t*-butylcyclohexyl [164], and adamantyl and diphenylmethylene groups [165]. The structural modifications lowered the polymer chain interactions by reducing the stiffness of the polymer chain, but maintained the high thermal resistance properties. The Michael addition reaction of bismaleimides with diamines results in the formation of polyaspartimides that are easier to process than linear, high-molecular-weight condensation polyimides [166, 167]. A series of amorphous polyaspartimides were synthesized by reacting bis (3-ethyl-5-methyl-4-maleimidophenyl) methane with various diamines. The polyaspartimides were soluble in *N*-methylpyrrolidone, dimethylformamide, dimethylsulfoxide, and tetrahydrofuran, possessed high T_gs (242 °C–308 °C), and good thermal stability (10% weight loss at 408 °C–459 °C in air and 396 °C–421 °C in nitrogen) [168].

Bismaleimide resins have been blended with epoxy resins to improve their processing characteristics. However, the use of aromatic amines as curing agents for the epoxy component results in reduced shelf stability. The use of phenolic resins as curing agents (with imidazoles as accelerators) overcomes this limitation [169]. The composition is recommended for electrical laminates and structural composites applications.

PCB laminates are required to meet V0 level of fire retardancy according to standard UL-94 of the American Underwriters Laboratories. To meet this requirement, fire-retardant chemicals such as aluminum hydroxide, antimony oxide, talc, and clay are added to the polymer system. Alternatively, bromine or phosphorous are incorporated into the polymer backbone, e.g., FR-4.

However, the products of combustion are often toxic and corrosive. Polymers containing phosphorous yield phosphoric acid upon combustion while brominated polymers liberate hydrogen bromide. Environmental legislation requires the removal of bromine-containing compounds from PCBs. Fluoropolymers and phosphorous-modified epoxy resins similar to FR-4 are available as bromine-free alternatives to FR-4.

Sethumadhavan et al. proposed polymeric systems comprising covalently bound polyhedral oligomeric silsesquioxane (POSS), with additional dispersed POSS if necessary [170]. The use of covalently bound POSS imparted flame retardancy with acceptable dielectric constants and dissipation factors. POSS is characterized by a chemical structure intermediate between silica (SiO_2) and silicone (R_2SiO) and is represented by the generic formula $RSiO_{1.5}$ wherein R is an organic moiety. POSS molecules can be considered as the smallest particles of silica possible. However, unlike silica or modified clays, POSS contains covalently bonded reactive functionalities suitable for polymerization or grafting POSS monomers to polymer chains, and nonreactive organic functionalities to enhance its solubility and compatibility characteristics [171].

Metal laminates comprising metal sheets joined together by an adhesive offer weight reduction without losing stiffness when replacing metal sheets. Metal laminates with polypropylene adhesives lack adequate thermal resistance since polypropylene melts at ~160 °C. U.S. patent 4,690,856 discloses a high-temperature-resistant adhesive based on a polyamide, a polyolefin grafted with an unsaturated polar moiety such as maleic anhydride, and an aminosilane such as *N*-β-(amino ethyl)-γ-aminopropyltrimethoxysilane [172]. However, the block copolymer nature of such adhesives results in high viscosity and low tensile modulus, especially at elevated temperatures. The grafted groups also tend to react with each other, resulting in poorer interaction with the substrate.

The limitations of the above adhesive can be overcome by replacing the aminosilane with a reactive copolymer such as styrene-maleic anhydride copolymer, which enhances bonding by reaction with the substrate [173]. Moreover, the copolymer reacts with the amine end groups and limits crosslinking between the grafted polyethylene and amine end groups. The addition of small amounts of epoxy resin is also recommended to improve adhesion to the metal sheets.

Conductive polymers such as polyacetylene, polyaniline, and polypyrrole (PPY) are characterized by the presence of π-conjugated systems formed by the overlap of carbon *p* orbitals and alternating C—C bonds [174]. On the other hand, polytetrafluoroethylene (PTFE) is a highly insulating polymer with very low dielectric loss and good thermal resistance properties, but its extremely low surface energy prevents it from adhering to any other substrate. Kang et al. developed a process of surface graft copolymerization of a plasma-pretreated PTFE film with concurrent lamination to a PPY film [175]. PTFE films were subject to 60 s of argon plasma treatment. Argon plasma caused defluorination of the PTFE film and upon exposure to the atmosphere, resulted in the form of peroxides and hydroperoxides on the film's surface [176]. The plasma-pretreated PTFE films were then placed in a Pyrex® tube containing 20% by volume of GMA in 1,4-dioxane. The reaction mixture was purged with dry argon gas for 30 min, sealed, and subjected to UV irradiation (>290 nm) for 2 h in a photochemical reactor. The GMA graft-modified PTFE film was extracted with acetone to remove the unreacted monomer, lapped with a PPY film in the presence of GMA containing 20% by volume of hexamethylene diamines, and heated at 170 °C for 3 h. T-peel adhesion tests on the PPY/PTFE laminate exhibited cohesive failure in the PTFE substrate indicating excellent adhesion between the two polymers. Such multilayer structures consisting of laminated conductive and dielectric layers are of value in sensor and microelectronic applications [177].

Laminated glass is employed as a safety feature in automobile windshields, bulletproof glass, and aircraft windows. Laminated glass is produced by bonding two or more layers of glass with a polyvinyl butyral interlayer. When subjected to sudden impact, the interlayer holds the glass layers together and prevents the formation of large pieces. For applications that require the laminated glass to withstand severe impact, e.g., architectural applications in coastal areas, the interlayer is based on an ionomer resin composed of ~80% ethylene and ~20% methacrylic acid (partly neutralized with sodium ions) [178]. Such laminated glass may also be used as side windows for automobiles.

In general, automobile windows (other than windshield windows) employ unlaminated, tempered glass, which when shattered, produce a shower of glass splinters termed as "spall." Even laminated glass windows generate sharp pieces that can cause lacerations to the vehicle occupants. An optically clear laminate that reduces spall and lacerative consequences on impact fracture is described in U.S. patent 7,238,401 [179]. The laminate is composed of two or more layers of biaxially oriented polyester films joined together by a PSA. One face of the laminate is protected by a scratch-resistant layer while the other face of the laminate is bonded to the glass by means of the adhesive. The adhesive is based on two acrylate resins synthesized via solution polymerization and cured with 1,1′-isophthaloyl bis(2-methylaziridine). One of the acrylate resins is a copolymer of isooctyl acrylate, methyl acrylate, and acrylic acid while the other is a copolymer of butyl acrylate and acrylic acid. The scratch-resistant coating for the laminate face was synthesized by polymerizing 3-(trimethoxysilyl)propyl methacrylate, pentaerythritol acrylate and *N,N*-dimethyl acrylamide in the presence of an aqueous dispersion of colloidal silica particles.

Decorative laminates are widely employed as counters and table tops, bathroom and kitchen work surfaces, furniture, cabinets, wall paneling, partitions and doors. Decorative laminates are typically manufactured by combining phenolic-resole-impregnated kraft paper layers and melamine-formaldehyde-resin-impregnated decorative papers for top layers under pressure (up to 1300 psi) and temperature (up to 300 °F). The pressed stock is then sized to standard market dimensions and sanded on the reverse to enhance laminate adhesion to a substrate. Typically, seven or eight core sheets are consolidated with only a single print and single overlay sheet to form a conventional 1/16-inch decorative laminate. U.S. patent 4,126,725 discloses the use of 0.002 inch thick perforated steel foils to enhance the dimensional stability of decorative laminates. Such reinforced laminates resist shrinkage of the melamine layer and exhibit enhanced impact resistance [180].

CONCLUSIONS

Coatings, adhesives, and laminates are complex materials that make extensive use of the principles of multiple disciplines such as surface chemistry, polymer chemistry, organic chemistry, inorganic chemistry, physical chemistry, and physics. This chapter only briefly describes the main crosslinking and grafting chemistries used in these applications. Many polymeric systems now routinely employ more than one crosslinking chemistry in the same system to obviate the limitations of individual crosslinkers. Challenges such as environmental compliance, cost, aesthetics, and weight savings without compromising material performance will continue to push the technological level of these polymeric applications, and usher in improved and exciting products.

REFERENCES

1. Zada A, Avny Y, Zilkha A. *Eur. Polym J*, **35**: 1159 (1999).
2. Brock T, Groteklaes M, Mischke P. *European Coatings Handbook*, U Zorll, ed. (Hanover: Vincentz Verlag, 2000), p. 24.
3. Lenz R. *Organic Chemistry of Synthetic High Polymers* (New York: Interscience Publishers, 1967), p. 711.
4. Reisch M. *Chem. Eng. News*, **84(42)**: 13 (2006).
5. Wright T. *Coatings World*, **12(8)**: 6 (2007).
6. U.S. Census Bureau, www.census.gov/mcd, accessed March 2008.
7. Erich SJF, van der Ven LGJ, Huinink HP, Pel L, Kopinga K. *J Phys Chem B*, **110**: 8166 (2006).
8. European Ecolabel for Indoor Paints and Varnishes, Draft report, November 2001. Available at http://ec.europa.eu/environment/ecolabel/pdf/paints_varnishes/interimreport_nov2001.pdf.
9. Miccichè F. "The Combination of Ascorbic Acid Derivatives/Iron Salts as Catalyst for the Oxidative Drying of Alkyd-based Paints: A Biomimetic Approach," Ph.D. dissertation, Eindhoven University of Technology, The Netherlands, 2005.
10. Klaasen RP, van der Leeuw RPC. *Progr Org Coat*, **55(2)**: 149 (2006).
11. Hoyle CE, Lee TY, Roper T. *J Polym Sci Part A: Polym Chem*, **42(21)**: 5301 (2004).
12. Huang H, Zhang JZ, Shi WF. *J Appl Polym Sci*, **80**: 499 (2001).

13. Frings RB, Wend M. *DIC Technical Review*, **9**: 43 (2003).
14. Johansson M, Malmstrom E, Jansson A, Hult A. *J Coat Tech*, **72(906)**: 49 (2000).
15. Stevenson TA, Subramanian R, Starcher RV, Smedley L, Dove HH. *Flexible, Super Durable Powder Coatings Composition*, U.S. Patent Application 2006/0079650.
16. Merck Y. *Surf Coat Int Part B: Coatings Transactions*, **84(B3)**: 231 (2001).
17. Park HS, Kwon SY, Seo KJ, Im WB, Wu JP. *J Coat Tech*, **71(899)**: 59 (1999).
18. Nishizawa H. *The Technology and Application of Polymeric Materials: Market and Outlook* (Tokyo: CMC Co. Ltd., 1996), p. 269.
19. Schork FJ, Tsavalas J. *Progr. Colloid Polym. Sci.*, **124**: 126 (2004).
20. Majumdar S, Kumar D, Nirvan YPS. *J Coat Tech*, **879(70)**: 27 (1998).
21. Shimizu T, Higashiura S, Ohguchi M. *J Appl Polym Sci*, **75**: 1149 (1999).
22. Shimizu T, Higashiura S, Ohguchi M. *J Appl Polym Sci*, **72**: 1817 (1999).
23. Dumain E, Toman A, Iwamura G. *Burnish Resistant Powder Coating Composition*, U.S. Patent 6,472,071, October 29, 2002.
24. Lee S, Billiani J, Pfohl WF, Lee KI, in *Proc. 30th International Waterborne, High Solids and Powder Coatings Symposium*, RF Storey, SF Thames, MW Urban, DW Wicks, eds. (The University of Southern Mississippi, 2003), p. 523.
25. Harui N, Hibi T, Tagaito C, Ooka M, in *Proc. 24th International Waterborne, High Solids and Powder Coatings Symposium*, RF Storey, SF Thames, MW Urban, DW Wicks, eds. (The University of Southern Mississippi, 1997) p. 79.
26. Pourreau DB, Smyth S. *J Coat Tech*, **2(15)**: 42 (2005).
27. Moon JH, Han HS, Shul YG, Jang DH, Ro MD, Yun DS. *Progr Org Coat*, **59**: 106 (2007).
28. Matsuno Y, Adachi T, Numa N. *Progr Org Coat*, **35**: 117 (1999).
29. High Solids Amino Crosslinking Agents, Cytec Product Brochure.
30. Santer JO. *Progr Org Coat*, **12**: 309 (1984).
31. Kordomenos PI, Kresta JE. *Macromolecules*, **14**: 1434 (1981).
32. Chiou BS, Schoen PE. *J Appl Polym Sci*, **83**: 212 (2002).
33. Wicks, ZW. *Progr Org Coat*, **3(1)**: 73 (1975).
34. Wicks, DA, Wicks ZW. *Progr Org Coat*, **36**: 148 (1999).
35. Schmitt F, Wenning A, Weiss, JV. *Progr Org Coat*, **34**: 227 (1998).
36. Caldwell RA. *J Coat Tech*, **2(14)**: 30 (2005).
37. Chattopadhyay DK, Prasad PSR, Sreedhar B, Raju KVSN. *Progr Org Coat*, **54**: 296 (2005).
38. Kim BK, Lee JC. *J Polym Sci, Part A Polym Chem*, **34**: 1095 (1996).
39. Parmar R, Patel K, Parmar J. *Polym Int*, **54**: 488 (2005).
40. (a) Huckestein B, Renz H, Kothrade S, Haberle K *Water-Emulsifiable Polyisocyanates*, U.S. Patent 5,780,542, July 14, 1998; (b) Knight AE, Godelieve B. *Uretonimine-Modified Emulsifiable Polyisocyanate Compositions*, U.S. Patent 5,364,897, November 15, 1994; (c) Haberle K. *Water-Emulsifiable Polyisocyanates*, U.S. Patent 5,583,176, December 10, 1996; (d) Erdem B, Jimenez J, Bhattacharjee D, Argyropoulos JN. *Emulsifiable Polyisocyanate*, WO 2006/029140.

41. Yeske PE, Wicks DA, in *Proc. 20th International Waterborne, High Solids and Powder Coatings Symposium*, RF Storey and SF Thames, eds. (The University of Southern Mississippi, 1993), p. 49.
42. Nair PR, Nair CPR, Francis DJ. *Eur Polym J*, **33**(**1**): 89 (1997).
43. Yukawa Y, Yabuta M, Tominaga A. *Progr Org Coat*, **24**(**1**): 359 (1994).
44. Broekeart M. *Polyurea Spray Coatings: The Technology and Latest Developments*, www.huntsman.com, accessed August 2007.
45. Takas TA. *J Coat Tech*, **1**(**5**):40 (2004).
46. Gregorovich BV, Hazan I. *Progr Org Coat*, **24**: 131 (1994).
47. Green ML. *J Coat Tech*, **73**(**918**): 55 (2001).
48. Edwards PA, Striemer G, Webster DC. *JCT Research*, **2**(**7**): 517 (2005).
49. Harris PJ, Ohrbom WH, Rehfuss JW. *Powder Slurry Compositions with Solid Particulate Carbamate Resin Component Dispersed in Liquid Aminoplast Resin Carrier*, U.S. Patent 6,391,969, May 21, 2002.
50. (a) Rehfuss JW, St. Aubin DL *Curable Carbamate-Functional Polymer Composition*, European Patent 0,594,142; (b) Blank WJ. *Certain Hydroxyalkyl Carbamates, Polymers and Uses Thereof*, U.S. Patent 4,820,830, April 11, 1989; (c) Rehfuss JW, St. Aubin DL, *Composite Color-Plus-Clear Coating Utilizing Carbamate-Functional Polymer Composition in the Clearcoat*, U.S. Patent 5,474,811, December 12, 1995; (d) Gupta RB, Wu KJ. *Curable Compositions Containing 1,3,5-Triazine Carbamates and Epoxy Compounds*, U.S. Patent 5,852,133, December 22, 1998; (e) Webster DC, Crain AL. *Carbamate Functional Oligomers and Coating Therefrom*, U.S. Patent 6,465,679, October 15, 2002; (f) Pourreau DB, Whitman PJ, Goldstein SL, Harris SH. *Multifunctional Allyl Carbamates and Coating Therefrom*, U.S. Patent 6,780,951, August 24, 2004; (g) Orhbom WH, Richards TS, December TS, Laugal JA. *Use of Carbamate-Modified Aminoplast Resins to Improve the Appearance and Performance of Powder Coatings*, U.S. Patent 6,992,149, January 31, 2006; (h) Campbell DH, St. Aubin DL, Ohrbom WH. *Carbamate Functional Addition Polymers And A Method For Their Preparation*, U.S. Patent 7,179,861, February 20, 2007.
51. Kim KJ, Williams RC, in Proc. 201 International Waterborne, High Solids and Powder Coatings Symposium, RF Storey and SF Thames, eds. (The University of Southern Mississippi, 1993), p. 57.
52. Mao CL, Chen N, Tien CF. *Waterborne Crosslinkable Compositions*, U.S. Patent 5,536,784, July 16, 1996.
53. Kao T, Grosso PV, Spilman GE, Clark MD. *Waterborne Acetoacetate-Functionalized Alkyd Coating Compositions*, U.S. Patent 6,780,523, August 24, 2004.
54. Zabel KH, Boomgaard RE, Thompson GE, Turgoose S, Braun HA. *Progr Org Coat*, **34**: 236 (1998).
55. Bartman B, Swift G. *Acetoacetate Functionalized Polymers and Monomers Useful for Crosslinking Formulations*, U.S. Patent 4408018, October 4, 1983.
56. Geurink PJA, van Dalen L, Van der Ven LGJ, Lamping RR. *Progr Org Coat*, **27**: 73 (1996).
57. Metzger JO, Fürmeier S. *Eur. J Org Chem*, **3**: 661 (1999).
58. Roesler RR, Danielmeier K. *Progr Org Coat*, **50**: 1 (2004).

59. Posthumus W, Derksen AJ, van den Goorbergh JAM, Hesselmans LCJ. *Prog. Org. Coat.*, **58**: 231 (2007).
60. Pham HH, Winnik WA. *Macromolecules*, **39**: 1425 (2006).
61. Carter NG. *Surf Coat Int*, **82(10)**: 497 (1999).
62. Howarth GA. *Surf Coat Int*, **86(B2)**: 91 (2003).
63. Bourlier KA, Mulrenin PS. *Aqueous Phenolic Resin Dispersions*, U.S. Patent 5,548,015, August 20, 1996.
64. Zaldivar RJ. *Low Temperature Processed Resin for Thermal and Chemical Protective Coatings*, U.S. patent application 2007/0100116.
65. Hutchings DA, Mills JL, Bourlier K. *Phenolic Polymers Made by Aralkylation Reactions*, U.S. Patent 5,889,137, March 30, 1999.
66. Hutchings DA, Hariharan R, Lucas E, Elahi SA, Randall AK, Bourlier K. *Alkyd and Aralkyd Derivatives of Phenolic Polymers*, U.S. Patent 5,770,750, June 23, 1998.
67. Rodriguez CL, Goldstein SL. *Urethane Compositions*, U.S. Patent 6,245,877, June 12, 2001.
68. Primeaux DJ, Davis TE. *Phenolic/Polyurea Coating Copolymer Compositions and Process*, U.S. Patent 6,797,789, September 28, 2004.
69. Keck MH, Kline RH. *Thermo-Oxidative Stable Copolyesters*, U.S. Patent 3,951,915, April 20, 1976.
70. Shakeb AMF. *Surf Coat Int Part B: Coatings Transactions*, **88(B2)**: 139 (2005).
71. Brock T, Groteklaes M, Mischke P. *European Coatings Handbook*, U Zorrl, ed. (Hanover: Vincentz Verlag, 2000), p. 74.
72. Nash HA, Docktor HJ, Webster DC. *JCT Research*, **1(3)**: 153 (2004).
73. Sangermano M, Bongiovanni R, Malucelli G, Roppolo I, Priola A. *Progr Org Coat*, **57**: 44 (2006).
74. Dubowik DA, Ross GC. *Surf Coat Int*, **84(A3)**: 126 (2001).
75. Elmore JD, Kincaid DS, Komar PC, Nielsen JE. *J Coat Tech*, **74(931)**: 63 (2002).
76. Mowrer NR, Foscante RE, Rojas JL. *Epoxy-Polysiloxane Polymer Composition*, U.S. Patent 5,804,616, September 8, 1998.
77. Keijman JM. Paper presented at the conference 2nd Jornadas Da Revista Corrosão E Protecção de Materiais, Lisboa, November 2000.
78. www.lumiflon.com, accessed September 2007.
79. Temtchenko T, Turri S, Novelli S, Delucchi M. *Progr Org Coat*, **43**: 75 (2001).
80. Montefusco F, Bongiovanni R, Sangermano M, Priola A, Harden A, Rehnberg N. *Polymer*, **45**: 4663 (2004).
81. Isaka H, Yonehara Y. *J Coat Tech*, **75(937)**: 59 (2003).
82. van den Berg KJ, Hobel K, Klinkenberg H, Noomen A, Van Oorschot JC. *Coating Composition Comprising a Bicyclo- Or Spiro-Orthoester Functional Compound*, U.S. Patent 6,297,329, October 2, 2001.
83. Noomen A. *Progr Org Coat*, **45**: 225 (2002).
84. van den Berg KJ, Van Oorschot R, Van der Ven LGJ. *Surf Coat Int Part B*, **89(B4)**: 285 (2006).
85. Petrie EM. *Handbook of Adhesives and Sealants* (New York: McGraw-Hill, 2000), p. 3.

86. Dunn DJ. Adhesives and Sealants—Technology, Applications and Markets, Rapra Market Report, 2003.
87. Patrick RC. *Treatise on Adhesion and Adhesives, Vol. 2*, (New York: Marcel Dekker, 1969), p. 5.
88. Barnes, TA, Pashby IR. *J Mater Proc Technol*, **99**: 72 (2000).
89. Duncan B, Abbott S, Court R, Roberts R, Leatherdale D. *A Review of Adhesive Bonding Assembly Processes and Measurement Methods*, National Physical Laboratory Report MATC(A)135, UK, 2003.
90. Comyn J, Brady F, Dust RA, Graham M, Haward A. *Int J Adhesion Adhesives*, **18**: 51 (1998).
91. Lepene BS, Long TE, Meyer A, Kranbuehl DE. *J Adhesion*, **78**: 297 (2002).
92. Longhenry JL, Love BJ, Murthy NS. *J Mater Sci*, **32**: 2283 (1997).
93. Cui YJ, Chen DH, Wang XL, Tang XZ. *Int J Adhesion Adhesives*, **22**: 317 (2002).
94. Sanchez-Adsuar MS. *Int J Adhesion Adhesives*, **20**: 291 (2000).
95. Datta S, Lohse DJ. *Polymeric Compatibilizers: Uses and Benefits in Polymer Blends* (Munich: Hanser, 1996), p. 197.
96. Utracki LA. *Commercial Polymer Blends*, 1st ed. (London: Chapman & Hall, 1998), p. 85.
97. De Genova R, Harper MD, Clay WA, Cranley PE, Hunter MK. *TAPPI J.*, **79**: 196 (1996).
98. Hung JM. *Adhesives Age*, **40**: 52 (1997).
99. Duffy DJ, Heintz AM, Stidham HD, Hsu SL, Suen W, Chu W, Paul CW. *J Adhesion*, **79(11)**: 1091 (2003).
100. Hu H, Liu H, Zhao J, Li J. *J Adhesion*, **82(1)**: 93 (2006).
101. Satas, D, ed. *Handbook of Pressure Sensitive Adhesive Technology* (New York: Van Nostrand Reinhold, 1989).
102. Tauber A, Scherzer T, Weiss I, Mehnert R. *J Coat Tech*, **74(927)**: 41 (2002).
103. Jerome R, Henrioulle-Granville M, Robin JJ. *Prog Polym Sci*, **16**: 837 (1991).
104. Matyjaszewski K, Shipp DA, Mcmurtry GP, Gaynor SG, Pakula T. *J Polym Sci: Part A: Polym Chem*, **38**: 2023 (2000).
105. (a) Rehmer G, Boettcher A, Portugall M *Benzophenone Derivatives and Their Preparation*, U.S. Patent 5,264,533, November 23, 1993; (b) Su S, Ozari Y, Vargas RR, Lum H. *Low Viscosity Acrylic Hot Melt Adhesives*, U.S. Patent 5,252,662, October 12, 1993; (c) Ramharack R, Chandran R. *Radiation Cured Hot Melt Pressure Sensitive Adhesives From Photoinitiator-Functionalized, Crosslinkable Copolymers*, U.S. Patent 5,536,759, July 16, 1996.
106. Everaerts E, Zieminski K, Nguyen L, Malmer J. *J Adhesion*, **82(4)**: 375 (2006).
107. Petrie EM. *Handbook of Adhesives and Sealants* (New York: McGraw-Hill, 2000), p. 361.
108. Devroey DRE, Homma M. *Int. Journal Adhesion Adhesives*, **21**: 275 (2001).
109. Gillham JK, in *Structural Adhesives—Developments in Resins and Primers*, AJ Kinloch, eds. (London: Elsevier Applied Science Publishers, 1986), p. 13.
110. Achary PS, Ramaswamy R. *J Adhesion*, **78(8)**: 695 (2002).
111. Kinloch AJ, Lee JH, Taylor AC, Sprenger S, Eger C, Egan D. *J Adhesion*, **79(8/9)**: 867 (2003).

112. Bauer F, Decker U, Ernst H, Findeisen M, Langguth H, Mehnert R, Sauerland V, Hinterwaldner R. *Int J Adhesion Adhesives*, **26**: 567 (2006).
113. Petrie EM. *Handbook of Adhesives and Sealants* (New York: McGraw-Hill, 2000), p. 372.
114. Ochi M, Tsuyuno N, Nakanishi Y, Murata Y. *J Appl Polym Sci*, **56**: 1161 (1995).
115. Harada M, Watanabe Y, Ochi M. *J Adhesion*, **81(6)**: 587 (2005).
116. Aubert JH. *J Adhesion*, **79(6)**: 609 (2003).
117. Forschner TC, Gwyn DE, Xiao HX, Suthar B, Sun LQ, Frisch KC. Paper available at www.shellchemicals.com/chemicals/pdf/pdo/PolyHotMelt.pdf?section=our_products. 2001.
118. McArdle CB, Burke J. *Curing Anaerobic Compositions Through Thick Bondlines*, U.S. Patent 5,411,998, May 2, 1995.
119. Comyn J. *Adhesion Science* (Cambridge: RSC Paperbacks, 1997), p. 46.
120. Flackett D, in *Silicon Compounds: Silanes and Silicones*, A Arkles and G Larson, eds. (Morrisville, PA: Gelest, Inc., 2004), p. 435.
121. Linden CL, Shalaby SW. *Absorbable Tissue Adhesives*, U.S. Patent 5,350,798, September 27, 1994.
122. Shalaby SW. *Cyanoacrylate-Capped Heterochain Polymers and Tissue Adhesives and Sealants Therefrom*. U.S. Patent 6,699,940, March 2, 2004.
123. Atkin NJ, Ansell C, Jackson AP. *Angiogenic Medical Cyanoacrylate Adhesive*, U.S. patent application 2006/0147479.
124. International Agency for Research on Cancer, Press Release #153, 2004.
125. Mami ski ML, Pawlicki J, Parzuchowski P. *J Adhesion*, **82(6)**: 629 (2006).
126. Riebel MJ, Torgusen PL, Roos KD, anderson DE, Gruber C. *Biocomposite Material and Method of Making*, U.S. Patent, 5,593,625, January 14, 1997.
127. Kuo M, Myers DJ, Heemstra H, Curry D, Adams DO, Stokke DD. *Soybean-Based Adhesive Resins and Composite Products Utilizing Such Adhesives*, U.S. Patent 6,306,997, October 23, 2001.
128. Wang D, Sun XS. *Industrial Crops and Products*, **15(1)**: 43 (2002).
129. Stephanou A, Pizzi A. *Holzforschung*, **47**: 439 (1993).
130. Stephanou A, Pizzi A. *Holzforschung*, **47**: 501 (1993).
131. Batubenga DB, Pizzi A, Stephanou A, Cheesman P, Krause R. *Holzforschung*, **49**: 84 (1995).
132. El Mansouri N, Pizzi A, Salvadó J. *Holz Roh Werkst*, **65**: 65 (2007).
133. Liu Y, Li K. *J Adhesion*, **82(6)**: 593 (2006).
134. Geng X, Li K. *Adhesive Composition and Methods of Using and Making the Same*, U.S. patent application 2004/0220368.
135. Thames SF, Cook RC, Mendon SK. *Soy Protein Based Adhesive Containing A Vegetable Oil Derivative*, U.S. Patent 6,790,271, September 14, 2004.
136. Thames SF, Sankovich BG, Shera JN, Mendon SK, Evans JM. *Soy protein based adhesive and particleboard*, U.S. Patent 7,081,159, July 25, 2006.
137. Thames-Rawlins Research Group, unpublished results, patent pending.
138. http://www.marinebiotech.org/antifoul.html, accessed July 2007.
139. (a) Waite JH *J Biol Chem* 1983, **258**: 2911; (b) Waite JH, Housley TJ, Tanzer ML. *Biochemistry* 1985, **24**: 5010; (c) Laursen RA, in *Structure, Cellular Synthesis and*

Assembly of Biopolymers, ST Case, ed. (Berlin: Springer-Verlag, 1992), p. 55; (d) Yamamoto H. *Biotechnol Genet Eng Rev*, **13**: 133 (1996).

140. Rzepecki LM, Waite JH, in *Bioorganic Marine Chemistry*, PJ Scheuer, ed. (Berlin: Springer-Verlag, 1991), p. 119.
141. (a) Coombs TL, Keller PJ *Aquat. Toxicol.* 1981, **1**: 291; (b) George SG, Pirie BJS, Coombs TL. *J Exp Mar Biol Ecol* 1976, **23**: 71; (c) Donat JR, Bruland KW, in *Trace Elements in Natural Waters*, B Salbu and E Steinnes, eds. (Ann Arbor, MI: CRC Press, 1995), p. 247.
142. Monahan J, Wilker JJ. *Chem Commun*, **2003(14)**: 1672 (2003).
143. Liu Y, Li K. *Modified Protein Adhesives and Lignocellulosic Composites Made from the Adhesive*, U.S. Patent 6,060,798, May 9, 2000.
144. Yu M, Deming TJ. *Macromolecules*, **31**: 4739 (1998).
145. (a) Dalsin JL, Messersmith PB. *Materials Today*, **8(9)**: 38 (2005); (b) Dalsin JL, Hu BH, Lee BP, Messersmith PB. *J Am Chem Soc*, **125**: 4253 (2003); (c) Statz AR, Meagher RJ, Barron AE, Messersmith PB. *J Am Chem Soc*, **127**: 7972 (2005); (d) Lee BP, Dalsin JL, Messersmith PB, in *Biological Adhesives*, AM Smith and JA Callow, eds. (Berlin: Springer, 2006),p. 257; (d) Patz T, Cristescu R, Narayan R, Menegazzo N, Mizaikoff B, Messersmith PB, Stamatin I, Mihailescu IN, Chrisey DB. *Appl Surf Sci*, **248(1–4)**: 416 (2005); (e) Catron ND, Lee H, Messersmith PB. *Biointerphases*, **1(4)**: 134 (2006).
146. Ayton F, in *Handbook of Beverage Packaging*, GA Giles, eds. (Boca Raton, FL: CRC Press, 1999), p. 131.
147. Walker JL, Mitry MT. *Laminating Adhesives Containing Polymerized Surfactant*, U.S. Patent 4,617,343, October 14, 1986.
148. Abele S, Zicmanis A, Graillat C, Monnet C, Guyot A. *Langmuir*, **1**: 1033 (1999).
149. Kehong F. *Circuit World*, **30(4)**: 16 (2004).
150. Goosey M, Poole M. *An Introduction to High-Performance Laminates and the Importance of Using Chemical Processes in PCB Fabrication*, Rohm and Haas Electronic Materials publication.
151. Haddick GT. *Process For Producing Defect Free Multilayer Printed Circuit Boards*, U.S. Patent 5,283,108, February 1, 1994.
152. Chen KM, Pan JP, Shiau GY, Wang TH, Chang WC. *Polyimide Adhesive Composition Including Barbituric Acid Modifier*, U.S. Patent 5,200,474, April 6, 1993.
153. Yu TA, Pan JP, Chen HL, King JS, Wang TH. *Flexible Copper/Polyimide and Barbituric Acid Modified Bismaleimide Blend/Polyimide Laminate*, U.S. Patent 5,372,891, December 13, 1994.
154. Chen HL, Ho SM, Wang TH, Pan JP. *Polyimide Composition for Polyimide/ Copper Foil Laminate*, U.S. Patent 5,290,909, March 1, 1994.
155. Chiu CH, Sun DJ, Hsu YH, Shiang FT, Chen CH, Wu PSC. *Polyimide Resin For Cast on Copper Laminate and Laminate Produced Therefrom*, U.S. Patent 6,133,408, October 17, 2000.
156. Chun-Kang KU, Chang-Hong HO, Lee YD. *J Adhes Sci Technol*, **19(11)**: 909 (2005).
157. Mather BD, Viswanathan K, Miller KM, Long TE. *Prog Polym Sci*, **31**: 487 (2006).

158. Chandra R, Rajabi L. *J Macromol Sci, Rev Macromol Chem*, **C37**: 61 (1997).
159. Nishikawa A, Shitara M, Era S, Fukushima T, Suzuki H, Kohkame H. *Ether Imides and Process for Producing The Same*, U.S. Patent 4,460,783, July 17, 1984.
160. Dahms RH. *Polyimide Laminates*, U.S. Patent 4,861,650, August 29, 1989.
161. Liaw DJ, Liaw BY. *Eur Polym J*, **33**: 1423 (1997).
162. Liaw DJ, Liaw BY. *Polymer*, **40**: 3183 (1999).
163. Liaw DJ, Liaw BY, Chung CY. *J Polym Sci, Part A: Polym Chem*, **37**: 2815 (1999).
164. Liaw BY. Liaw DJ, *Polym J*, **31**: 1270 (1999).
165. Liaw DJ, Liaw BY. *Macromol Chem Phys*, **200**: 1326 (1999).
166. Crivello JV. *J Polym Sci, Polym Chem Ed*, **11**: 1185 (1973).
167. Bell VL, Young PR. *J Polym Sci, Part A: Polym Chem*, **24**: 2647 (1986).
168. Liaw DJ, Liaw BY, Chen JJ. *Polymer*, **42**: 867 (2001).
169. Young GC, *Bismaleimide Resin Composition Containing Epoxy Resin and a Phenolic Curing Agent Therefore*, U.S. Patent 4,816,531, March 28, 1989.
170. Sethumadhavan M, Landi VR, Borges LD. *Circuit Materials, Circuits, Multi-Layer Circuits, and Methods of Manufacture Thereof*, U.S. Patent Application 2005/0173780.
171. www.hybridplastics.com, accessed July 2007.
172. Ito B, Fukuda K, Hara R. *Metal Polyamide Adhesive Laminates*, U.S. Patent 4,690,856, September 1, 1987.
173. Gauriat A, Calvez C, Schmit F, Nazikkol C, Filthaut C, Smith RH, Pac J. *Metal-Polyamide/Polyethylene-Metal Laminate*, U.S. patent application 2006/0269701.
174. Heeger AJ, Kivelson SA, Schrieffer JR, Su WP. *Rev. Mod. Phys*, **60**: 781 (1998).
175. Lim VWL, Kang ET, Neoh KG, Tan KL. *J Appl Polym Sci*, **80**: 716 (2001).
176. Kang ET, Neoh KG, Tan KL, Morikawa M, Uyama Y, Ikada Y. *Macromolecules*, **25**: 1959 (1992).
177. Sacher E. *Prog Surf Sci*, **47**: 273 (1994).
178. Smith CA. *Glass Laminates Having Improved Structural Integrity Against Severe Impacts*, U.S. Patent 6,737,151, May 18, 2004.
179. Dietz PT. *Glazing Element and Laminate for Use in the Same*, U.S. Patent 7,238,401, July 3, 2007.
180. Shiflet FL. *High Pressure Laminates*, U.S. Patent 4,126,725, November 21, 1978.

12

IN COMMODITY PLASTICS

JAMES W. RAWLINS AND JAMES WHITTERMORE IV

INTRODUCTION

Plastics that are in high demand, produced at high volumes, and relatively inexpensive (typically less than $1 per pound) are considered commodity plastics. The four primary commodity plastics (based on volume production) are polyethylene, polyvinyl chloride, polypropylene, and polystyrene, all of which are used for mass production consumer items [1]. Often, however, the intrinsic properties of commodity plastics are not perfectly compatible with the properties expected of the end application. Developing and using entirely new polymeric materials to meet application parameters is financially and technologically undesirable, especially when available materials can simply be modified for a specific application. Grafting and/or crosslinking are two commonly used modifications to manipulate commodity plastic properties. For example, surface adhesion and increased breadth of blending compatibility is possible by grafting into the bulk or onto the surface of materials. Varying degrees of crosslinking are used to enhance structural stability at higher temperatures, improve impact resistance, and inhibit stress crack formation.

Modifications to commodity plastics are typically effected via reactive processing, irradiation techniques, and chemical techniques, or by a combination of the three in order to achieve the desired characteristics. Reactive extrusion, in conjunction with free-radical initiators and irradiation, is the most commonly used technique to modify commodity plastics. Free-radical initiators are used more often than irradiation to initiate the reaction of the monomer onto

Polymer Grafting and Crosslinking, Edited by Amit Bhattacharya, James W. Rawlins and Paramita Ray

the polymeric backbone. When adding grafting agents, the goal is to expose as much surface area of the commodity plastic as possible to grafting agents prior to initiation of the reaction. Excess grafting agents are then removed by vacuum prior to and during the extrusion process [2].

Efficient blending of the polymer and grafting agent is necessary to minimize homopolymerization of the grafting agent (although homopolymerization is desirable in some applications). The graft chain length from the polymer backbone depends on a number of variables such as grafting agent concentration, initiator concentration, polymer concentration, processing temperature, and processing time. For some applications, the graft chain length is kept so short that the mechanical properties are almost unchanged while the chemical properties of the plastics are altered [2].

Chemical grafting techniques involve the addition of free-radical initiators into the bulk and surface of the polymer. However, this process is also likely to result in the inclusion of chemical impurities that can significantly impact the properties of the final product. Irradiation techniques employ radiation sources such as ultraviolet (UV), gamma ray, beta ray, or plasma.

POLYETHYLENE

Polyethylene (PE) is the highest volume commodity plastic and is commonly distributed in one of three forms: high-density polyethylene (HDPE), low-density polyethylene (LDPE), and linear low-density polyethylene (LLDPE). LDPE and LLDPE are most commonly used in film applications, while HDPE is used in blow molding, injection molding, and piping applications [3].

PE crosslinking is conducted to improve impact resistance, reduce stress crack formation, and improve deformation properties at higher temperatures. Crosslinking is achieved via free-radical initiators, silane coupling, or through irradiation [4]. Electron-beam (e-beam) radiation is commonly employed to crosslink PE below its melting temperature for such applications as heat-shrink packaging and electrical connector manufacturing. Although crosslinking by e-beam radiation is limited by penetration depth, the inherent lack of by-products offers benefits such as the absence of dielectric breakdown of materials, improved toughness, and chemical resistance, along with extended temperature utility [5].

Grafting is used to introduce polar functionality on the PE backbone for applications that require enhanced adhesion to polar polymers or fillers. Song et al. grafted a reactive, nonionic surfactant onto LLDPE via beta-ray irradiation in air with a twin-screw extruder and confirmed the grafting via infrared spectroscopy studies. Differential scanning calorimetry (DSC) studies indicated that grafting increased the crystallization temperature by 3 °C with concurrent reduction in melting enthalpy. Upon grafting, the nonionic surfactant is thought to act as a nucleating agent. Grafting did not affect the tensile properties of LLDPE [6].

Mahmoud developed an ion exchange membrane using gamma-radiation-induced grafting of styrene and acrylic acid onto LDPE. The study explored the optimum conditions for obtaining a high grafting yield on LDPE. The grafted membranes were hydrophilic and had high chemical and thermal stability. Adsorption of metal ions to the membrane showed dependence on the adsorption time, the pH of the medium, and the initial feed concentration. The membrane had a higher affinity for Zn^{2+} ions over Cd^{2+} and Pd^{2+}, although Pd^{2+} showed the greatest affinity in mixtures of the three ions. Regeneration of the ion-exchange membrane was easily performed with a 1 M HNO_3 treatment. The ion exchange membranes were amenable for removing Cd^{2+}, Pd^{2+}, and Zn^{2+} ions from wastewater [7].

Lin et al. sought to improve the haemocompatibility of LDPE by ozone grafting a sulfoammonium zwitterionic polymer, *N,N′*-dimethyl(methacryloyl ethyl)ammonium propane sulfonate (DMAPS). X-ray photoelectron spectroscopy and attenuated total reflection Fourier transform infrared spectroscopy (ATR-FTIR) showed that sulfobetaine structures had been grafted to the LDPE surface. As the DMAPS concentration was increased, more DMAPS was grafted to LDPE. Grafting also increased the hydrophilicity of the films. A platelet-rich plasma adhesion study showed that no platelet adhesion occurred on the grafted LDPE films when the films were maintained at 37 °C for 180 min. The grafting technique and protocol shows promise for use in biomedical applications [8].

The adhesive properties of LLDPE powder were significantly enhanced by grafting with acrylic acid. The degree of grafting increased with increase in irradiation time. The acrylic acid-grafted LLDPE exhibited improved hydrophilicity and peel strength when used as a hot-melt adhesive between HDPE and a steel joint [9].

Riande et al. studied the gas permeability of an LLDPE film with acrylic acid grafted onto its surface via UV radiation. Grafting reduced the permeability coefficient of the grafted film to transport oxygen, nitrogen, carbon monoxide, carbon dioxide, methane, ethane, ethylene, propane, and argon by a factor of approximately 1/6. The reduced gas permeability was attributed to surface crosslinking and/or the formation of ordered structures on the surface [10].

Nazockdast et al. investigated free-radical melt grafting of maleic anhydride (MA) on HDPE in the presence of 2,5-dimethyl-2,5-di(t-butylperoxy)hexane peroxide (DTBPH). Increasing the rotor speed enhanced the degree of grafting while lowering the extent of competitive side reactions. The use of styrene as a comonomer decreased crosslinking side reactions and increased the degree of grafting [11].

POLYVINYL CHLORIDE

Polyvinyl chloride (PVC) is the second largest volume commodity plastic. PVC is the largest consumer of chlorine and is widely used in electronics, construction, packaging, and transportation. As PVC is difficult to process, the industry is highly dependent upon formulation technologies. Although grafting of PVC is feasible, it is not employed commercially because PVC can be easily compounded with less expensive plasticizers and/or elastomers to enhance its property range [5].

Guillermina et al. grafted poly(ethylene glycol) methacrylate (PEGMA) onto PVC tubes by first illuminating them with UV light then submerging them in an aqueous solution of PEGMA and heating at 50 °C to 60 °C for 24 h [12]. Malygin et al. used atomic-force microscopy (AFM) to investigate corona-discharge-induced grafting of acrylic acid from a PVC surface. Confirmation of monomeric grafting from the PVC surface was obtained via physicochemical study of the grafted products. The modified PVC surface also exhibited improved adhesion properties toward inks used in offset printing [13].

Choudhary and Maiti accomplished melt-graft polymerization of *n*-butyl methacrylate on PVC using a free-radical initiator. Initiator and monomer concentrations, shear rate, residence time, and temperature were optimized to maximize the grafting efficiency at 14%. Processability, thermal, and mechanical properties all showed improvement subsequent to grafting [14]. Ghaffari and Taghizadeh investigated the kinetics and mechanisms of graft polymerization of methyl methacrylate onto PVC using benzoyl peroxide as the free-radical initiator. It was noted that grafting efficiency was related to initiator, monomer and polymer concentrations, reaction time, and processing temperature [15].

Lee et al. investigated UV graft polymerization of *N*-vinyl-2-pyrrolidinone (NVP) onto a porous surface of PVC as a means to enhance its surface wettability and to decrease adsorptive fouling. The graft yield was maximized upon 5 min of UV irradiation. Increased NVP grafting from the PVC surface resulted in decreased cake layer adsorption. It is believed that the higher hydrophilicity of the NVP-grafted PVC contributed significantly to lower sludge adhesion [16].

Sreenivasan grafted 2-hydroxyl ethyl methacrylate (HEMA) onto PVC by employing Williamson's reaction between the HEMA hydroxyl and metallic sodium. Scanning electron microscopy (SEM) studies showed a change in surface morphology subsequent to grafting. A key advantage of this process was that it retained the acrylic double bond for further reaction [17].

Grafting reactions have been used to modify the inner surfaces of commercial PVC tubes to increase their biocompatibility. Tetrafluoroethylene was grafted onto PVC tubes using atmospheric pressure glow plasma and yielded surfaces that were similar to polytetrafluoroethylene. Contact angle studies indicated that grafting reduced the surface roughness. Optimum flow rates for the formation of a uniform and thick surface coating were established in this investigation [18].

The PVC grade used in the manufacture of blood bags employs di(2-ethylhexyl) phthalate (DEHP) as a plasticizer to achieve the desired flexibility and low-temperature properties. The leaching of DEHP from blood bags into blood, plasma, or serum is a major concern as DEHP is classified as a hypolipidemic hepatocarcinogen and is toxic to humans. Grafting polyethylene glycol (PEG), a blood compatible polymer, on the surface of PVC can increase the blood compatibility and migration resistance of flexible PVC. Lakshmi and Jayakrishnan used Williamson's ether synthesis to graft PEG onto the surface of plasticized PVC. Migration of the plasticizer from PEG was decreased significantly post grafting, even into petroleum ether, an organic extractant. The grafted PVC exhibited improved blood compatibility with minimal adhesion to platelets [19].

Modification of PVC can also be accomplished by crosslinking using high-energy irradiation such as e-beam; however, some material degradation is likely to occur during the process. Ratnam et al. investigated e-beam irradiation crosslinking of unplasticized PVC (UPVC) formulations in the presence of two different stabilizers, trimethylolpropane triacrylate (TMPTA) and Irganox 1010. Stabilized UPVC was found to be crosslinkable by e-beam irradiation with minimal degradation. TMPTA increased the rate of crosslinking while Irganox 1010 was seen to inhibit crosslinking. Interestingly, the glass transition temperature increased when UPVC was irradiated in the presence of TMPTA [20]. E-beam irradiation of PVC-ethylene-vinyl acetate (EVA) blends yielded higher gel contents and a reduction in chlorine loss with increasing levels of EVA incorporation. It was also noted that a higher number of vinyl acetate groups in EVA resulted in better mechanical properties after irradiation [21].

POLYPROPYLENE

Polypropylene (PP) is the third highest volume commodity plastic after PE and PVC. PP is intrinsically rugged and resistant to solvents, bases, and acids, making it well suited for a variety of consumer applications. Grafting and crosslinking techniques have been explored for PP to further enhance its applicability. For instance, increasing the polarity of PP improves its compatibility with polar polymers, glass, and metals [22].

Some recent studies that involve the grafting of materials off a PP backbone use grafting monomers such as maleic anhydride (MA) [23, 24], phospholipids [25], and glycidyl methacrylate [26]. Tan and Xu successfully grafted PP by irradiating a PP suspension in the presence of MA without adding an initiator. X-ray diffraction and DSC studies showed that grafted PP exhibited lower crystallinity than pure PP. The percentage of grafting onto PP reached a maximum at 4.82% [23].

Ishihara et al. studied the graft polymerization of 2-methacryloyloxyethyl phosphorylcholine (MPC) onto the surface of PP to increase its biocompatibility and lubricity. The graft polymerization was induced by irradiation. It was observed that MPC grafting of PP increased its surface hydrophilicity and mobility, and imparted surface lubricity similar to that of human joints [25].

Ahmad et al. grafted glycidyl methacrylate onto the surface of a biaxially oriented PP film without the use of an initiator. Active species were formed in air by swift-silver ion irradiation. Interestingly, as the grafting degree decreased, so did the contact angle. Ion irradiation induced grafting only in small regions of the polymer and the overlapping of these regions reflected a reduction in grafting yield [26].

The crosslinking of PP is complicated by beta scission and disproportionation along the PP chain [27]. For instance, gamma-irradiation crosslinking in PP has been shown to increase thermal degradation [28]. The thermal degradation was dependent on the irradiation dose and initiator concentration. Irradiation has also been shown to cause chain scission, resulting in lower molecular weight and reduced tensile strength [29]. Up to a certain dosage, irradiation does not significantly affect the crystallinity of crosslinked PP. The use of coagents such as tetramethylolmethane tetraacrylate and trimethylolpropane triacrylate has been successful in inhibiting the fragmenting of PP and increasing crosslinking efficiency [30, 31].

POLYSTYRENE

Polystyrene (PS) was introduced by Dow Chemical [32] and I.G. Farben [33] in 1930 and is presently the fourth largest plastic sold by volume. PS is characterized by its low cost and decent mechanical properties. However, PS does suffer from a few limitations such as poor impact strength, brittleness, low heat distortion, and sensitivity to some common organic solvents.

Grafting reactions onto PS are low-cost techniques that increase the functionality of PS without involving copolymerization. A facile means of grafting is via the use of PS as a chain transfer agent during the polymerization of methyl methacrylate [34, 35]. A similar technique has been used to graft brominated styrene onto PS to enhance flame resistance.

Using aromatic electrophilic substitution, PS has been chloromethylated and then reacted with a tertiary amine to yield PS with ion-exchange capabilities [36, 37]. Reacting PS with sulfonic acid yields a water-soluble PS sulfonate [38]. Anionic grafting of acrylonitrile onto PS via lithiation yielded a graft polymer with much greater char resistance then nongrafted PS; however, the grafted samples also ignited earlier than the nongrafted PS. Ultimately, it was concluded that the grafting of acrylonitrile onto PS was not an effective technique to produce flame-retardant polystyrene [39].

Stimuli responsive surfaces have been formed on the surface of PS by grafting poly(*N*-isopropylacrylamide) via a free-radical technique using helium atmosphere plasma treatment. Contact angle studies with water showed that the grafted PS exhibited a swift temperature response by changing from hydrophilic to hydrophobic beyond the lower critical solution temperature. The use of Mohr's salt increased the graft yield, decreased homopolymerization, and increased the temperature sensitivity of the grafted polymer [40].

CONCLUSION

Commodity polymers are widely used every day in a myriad of applications including transportation, electronics, food packaging, and toys. However, these commodity polymers individually lack the necessary performance characteristics to meet all the application requirements, e.g., poor adhesion and compatibility with other materials, inferior tensile properties, and sensitivity to chemicals or solvents, breadth of thermal stability, and creep resistance. Without appropriate modification, commodity polymers would not be as prevalent as they are today. Grafting and crosslinking are two of the common modifications that enable commodity polymers to have such widespread applications.

REFERENCES

1. Kaplan WA. *Modern Plastics Encyclopedia* (Highstown, NJ: McGraw-Hill, Inc., 1996), pp. A12–A18.
2. Xanthos M. *Reactive Extrusion: Principles and Practice* (Munich: Hanser Publishers 1992), pp. 101–102.
3. Whiteley KS. *Ullmann's Encyclopedia of Industrial Chemistry*, 5th ed., B Elvers, S Hawkins, and G Schultz, eds. (Weinheim, Germany: VCH Publishers, Inc., 1992), pp. 488–518.
4. Singh A, Silverman J, in *Radiation Processing of Polymers*, A Singh and J Silverman, eds. (New York: Hanser, 1992), p. 15.
5. Meister JJ. *Polymer Modification: Principles, Techniques, and Applications* (New York: Marcel Dekker, Inc., 2000), p. 262.
6. Yao Z, Yin J, Song Y, Jiang G, Song Y. *Polymer Bulletin*, **59(1)**: 135 (2007).
7. Mahmoud GA. *J Appl Polym Sci*, **104(5)**: 2769 (2007).
8. Shan B, Yan H, Shen J, Lin S. *J Appl Polym Sci*, **101(6)**: 3697 (2006).
9. Lei J, Gao J, Jiang L. *J Appl Polym Sci*, **100(3)**: 2549 (2006).
10. Costamagna V, Strumia M, Lopez-Gonzalez M, Riande E. *J Polym Sci, Part B: Polymer Physics*, **44(19)**: 2828 (2006).
11. Sheshkali HRZ, Assempour H, Nazockdast H. *J Appl Polym Sci*, **105(4)**: 1869 (2007).
12. Arenas E, Bucio E, Burillo G. *Polym Preprints* (American Chemical Society, Division of Polymer Chemistry), **48(1)**: 545 (2007).
13. Kucherenko OB, Kohlert C, Sosnov EA, Malygin AA. *Russian Journal of Applied Chemistry*, **79(8)**: 1316 (2006).
14. Maiti AK, Choudhary MS. *J Appl Polym Sci*, **92(4)**: 2442 (2004).
15. Taghizadeh MT, Ghaffari S. *Iran Intl J Sci*, **4(1)**: 23 (2003).
16. Kim DS, Kang JS, Kim KY, Lee YM. *Desalination*, **146(1–3)**: 301 (2002).
17. Sreenivasan K. *J Appl Polym Sci*, **74(1)**: 113 (1999).
18. Babukutty Y, Prat R, Endo K, Kogoma M, Okazaki S, Kodama M. *Langmuir*, **15(20)**: 7055 (1999).

19. Lakshmi S, Jayakrishnan A. *Artificial Organs*, **22**(**3**): 222 (1998).
20. Ratnam CT, Nasir M, Baharin A. *Polymer Testing*, **20**(**5**): 485 (2001).
21. Wang S, Zhang Y, Zhang Y, Zhang C, Li E. *J Appl Polym Sci*, **91**(**3**): 1571 (2004).
22. Borsig E. *J Macromol Sci, Part A*, **A36**(**11**): 1699(1999).
23. Tan X, Xu Y. *Radiat Eff Defects Solids*, **163**(**2**): 107 (2008).
24. Tong G-S, Liu T, Hu G-H, Hoppe S, Zhao L, Yuan W-K. *Chem Eng Sci*, **62**(**18–20**): 5290 (2007).
25. Kitano K, Matsuno R, Konno T, Takai M, Ishihara K. *Trans Mater Res Soc Japan*, **32**(**2**): 579 (2007).
26. Chawla S, Ghosh AK, Avasthi DK, Kulriya P, Ahmad S. *J Appl Polym Sci*, **105**(**6**): 3578 (2007).
27. Lazar M, Rado R, Rychly J. *Adv Polym Sci*, **95**: 149 (1990).
28. Minkova L, Lefterova E, Koleva T, Nedkov E, Nikolova M. *Colloid Polym Sci*, **266**: 898 (1988).
29. Busfield WK, O'Donnell JH. *Eur Polym J*, **15**: 379 (1979).
30. Borsig E, Malcherova E, Lazar M. *Polym Int*, **30**: 367 (1993).
31. Sawasaki T, Nojiri A. *Radiat Phys Chem.*, **31**: 877 (1988).
32. Boundy RH, Boyer RF. *Styrene: Its Polymers, Copolymers, and Derivatives* (New York: Reinhold, 1952), p. 11.
33. Bock W, Tschunkur E. *Rubberlike Mass Containing Styrene Polymerizates*, U.S. Patent 1,938,731, December 12, 1933.
34. Brydon A, Burnett GM, Cameron GG. *J Polym Sci, Polym Chem Ed*, **11**: 3255 (1973).
35. Ceresa RJ. *Block and Graft Copolymers* (London: Butterworths, 1962), p. 15.
36. Brydson JA. *Plastics Materials, 4th ed.* (London: Butterworths Scientific, 1982), p. 386.
37. Elias HG. *Macromolecules*. (New York: Plenum, 1977), p. 799.
38. Chauvin Y, Commereuc D, Dawans F. *Prog Polym Sci*, **5**: 95 (1977).
39. Xue TJ, Wilkie CA. *Recent Advances in Flame Retardancy of Polymeric Materials*, **7**: 105 (1996).
40. Wang X, McCord MG. *J Appl Polym Sci*, **104**(**6**): 3614 (2007).

FUTURE DIRECTIONS

AMIT BHATTACHARYA

"Grafting" and "crosslinking" are surely breakthroughs that involve the manufacturing of polymers that will be useful in different applications. In commercial endeavors, however, they have not been explored thus far, but they have a great future.

The living-radical and radiation techniques have been used more readily in grafting and crosslinking. Radiation-induced methods are, however, cleaner in this respect. The processes rapidly expanding now through e-beam curing processes can be achieved in a fraction of second and, without further purification, yield products at any stage. They open up various uses in biomedical applications. The reaction processes that use supercritical CO_2 are finding a foothold in this field.

As is shown in biomedical applications, hydrogels, nanogels, dendrimers, and cyclodextrins are promising. Reaction processes are already being used for in vivo class materials such as sutures, surgical adhesives, catheters, heart valves, and for in vitro class materials in the ophthalmic, dental, and transdermal fields. In the textile industries, the processes are capable of optimizing natural and synthetic fibers (as soil adhesion, durability, moisture regain, dyeing). The processes can produce synthetic-polymer garments that have the comfort of natural-polymer garments. In cable applications, the processes have a significant role, and in the future, e-beam and ultrasonic curing will be used more often.

To get the requisite properties, different plasticizers, antioxidants, and fillers will be the choice. Current research activities are concentrated on these

Polymer Grafting and Crosslinking, Edited by Amit Bhattacharya, James W. Rawlins and Paramita Ray

materials. In separation applications, the subject of the research is to find the proper functionalities to achieve better selectivity. Finished polymer-based coating systems that are mostly based on crosslinking chemistry attract more attention in terms of aesthetics and environmental protection. Much of the research takes steps toward the "biodegradability of polymers" and may solve some of the problems of environmental pollution. The future of the research in these prospective fields is highly promising.

INDEX

Polymer Grafting and Crosslinking, Edited by Amit Bhattacharya, James W. Rawlins and Paramita Ray